Application of Nanotechnology in Food Science, Processing and Packaging

Chukwuebuka Egbuna
Jaison Jeevanandam
Kingsley C. Patrick-Iwuanyanwu
Eugene N. Onyeike
Editors

Application of Nanotechnology in Food Science, Processing and Packaging

Springer

Editors
Chukwuebuka Egbuna
Faculty of Natural Science Department of Biochemistry
Chukwuemeka Odumegwu Ojukwu University
Uli, Nigeria

Kingsley C. Patrick-Iwuanyanwu
University of Port Harcourt
Rivers State, Nigeria

Jaison Jeevanandam
Universidade da Madeira
Funchal, Portugal

Eugene N. Onyeike
University of Port Harcourt
Rivers State, Nigeria

ISBN 978-3-030-98819-7 ISBN 978-3-030-98820-3 (eBook)
https://doi.org/10.1007/978-3-030-98820-3

This Springer imprint is published by the registered company Springer Nature Switzerland AG
The registered company address is: Gewerbestrasse 11, 6330 Cham, Switzerland

Preface

The application of nanotechnology in the food and processing industry has found wide adoption and application in the present decade for value addition of food products, enhanced food quality, shelf-life improvement, safety assurance, cost reduction, and improved nutritional benefits. The application of nanotechnology has also helped to improve the bioavailability of certain bioactive food ingredients and their target delivery by nanoencapsulation and other techniques. This book entitled *Application of Nanotechnology in Food Science, Processing and Packaging* presents complete coverage of the application of nanotechnology in the food and processing industry.

The chapters are structured sequentially to aid flow and continuity. The contributors of this book are senior researchers from both academia and industry around the globe. The primary audience includes food chemists, nutritional scientists, food nanotechnologists, public and private health practitioners, students, etc. We are hopeful that the users of this book will find it extremely useful. Our special thanks go to everyone who contributed to the success of this book, especially the chapter contributors. We are also indebted to our families for their support. To the Springer team in charge of this book, we will remain thankful for their guidance and support. To the reader, we hope you will find this book helpful and informative.

Port Harcourt, Rivers State, Nigeria — Chukwuebuka Egbuna
Funchal, Portugal — Jaison Jeevanandam
Port Harcourt, Rivers State, Nigeria — Kingsley C. Patrick-Iwuanyanwu
Port Harcourt, Rivers State, Nigeria — Eugene N. Onyeike

Contents

About the Editors

Chukwuebuka Egbuna is a chartered chemist, a chemical analyst, and academic researcher. He is a member of the Institute of Chartered Chemists of Nigeria (ICCON), the Nigerian Society of Biochemistry and Molecular Biology (NSBMB), the Society of Quality Assurance (SQA) (USA), and the Royal Society of Chemistry (RSC) (United Kingdom). Egbuna is also an editor of new Elsevier series on Drug Discovery Update. The series includes books, monographs, and edited collections from all areas of drug discovery, including emerging therapeutic claims for the treatment of diseases. He has also edited books with Springer Nature on functional foods and nutraceuticals and John Wiley on poisonous plants and phytochemicals. He has been engaged in a number of roles at New Divine Favor Pharmaceutical Industry Limited, Akuzor Nkpor, Anambra State, Nigeria, and Chukwuemeka Odumegwu Ojukwu University (COOU) in Nigeria. He has collaboratively worked and published quite a number of research articles in the area of phytochemistry, nutrition, and toxicology and related areas. He is a reviewer and editorial board member of various journals, including serving as a website administrator for the *Tropical Journal of Applied Natural Sciences* (TJANS), a journal of the faculty of Natural Sciences, COOU. He is the publishing director of IPS Intelligentsia Publishing Services. His primary research interests are in phytochemistry, nutrition and toxicology, food and medicinal chemistry, and analytical biochemistry. Africa Centre of Excellence in Public Health and Toxicological Research (ACE-PUTOR), University of Port-Harcourt, Choba, Rivers State, Nigeria

Jaison Jeevanandam obtained his Ph.D. at the Department of Chemical Engineering, Faculty of Engineering and Science from Curtin University, Malaysia. His Ph.D. project is on 'Novel nanomedicine for reversing insulin resistance in Type 2 diabetes' which involved cross-disciplinary research between physics, chemistry, biology, medicine, and engineering. He has experience in nanoparticle synthesis, especially in green synthesis using plant extracts, characterization, cytotoxic analysis of nanoparticles, and also in vitro diabetic models. His current research focuses on the application of bionanotechnology in the fabrication of nanoformulation for efficient delivery of food bioactives and drugs for the treatment of several diseases, especially diabetes. Dr. Jaison has authored over 80 publications including original research articles, books, and book chapters. He has edited a book entitled *Research Advances in Dynamic Light Scattering* for Nova Science Publishers. Dr. Jaison also serves as an editor for few Elsevier books as well as two monographs. He is a recipient of many awards and has presented papers in both local and international conferences. Currently, he is a senior researcher at CQM—Centro de Química da Madeira, Universidade da Madeira, Campus da Penteada, Funchal, Portugal. CQM – Centro de Química da Madeira, Universidade da Madeira, Funchal, Portugal

Kingsley C. Patrick-Iwuanyanwu is an Associate Professor in the Department of Biochemistry. Dr. Patrick-Iwuanyanwu is a recipient of The World Academy of Science–International Centre for Chemical and Biological Sciences (TWAS-ICCBS) Postgraduate fellowship award in 2009 to the University of Karachi, Pakistan. He was among the 100 young scientists from around the world selected to attend The World Life Sciences Forum by BioVision (Lyon, France). He is a recipient of Society of Toxicology (SOT)/AstraZeneca IUTOX fellowship award in 2009. He served as a pioneer African representative and one of the founding Executive Members of Student Advisory Council (SAC) of Society of Environmental Toxicology and Chemistry (SETAC, EUROPE) from 2009 to 2011. Dr. Patrick-Iwuanyanwu is a member of several professional bodies including Nigerian Society of Biochemistry and Molecular Biology (NSBMB); Africa Education Initiative (NEF, CT, USA); West African Society of Toxicology (WASOT); Society of Environmental Toxicology and Chemistry (SETAC, Europe); Society of Toxicology (SOT, Reston, USA); and Society for Experimental Biology (SEB, UK). Africa Centre of Excellence in Public Health and Toxicological Research (ACE-PUTOR), University of Port-Harcourt, Choba, Rivers State, Nigeria

Eugene N. Onyeike is working as a Professor at the Department of Biochemistry, Faculty of Chemical Sciences, College of Natural and Applied Sciences, University of Port Harcourt, Nigeria. His international experience includes various programmes, contributions, and participation in different countries for diverse fields of study. Professor Onyeike has supervised many M.Sc. and Ph.D. students among whom are now professors in various universities. His research interests reflect in his wide range of publications in various national and international journals. He is an editorial board member of many journals and serves as a member of various associations, apart from being an author for many books. Africa Centre of Excellence in Public Health and Toxicological Research (ACE-PUTOR), University of Port-Harcourt, Choba, Rivers State, Nigeria

Chapter 1
Application of Nanotechnology in the Food Industry

Bishnu Kumar Pandey, Sonam Pandey, Ravindra Dhar, and Kanti Bhooshan Pandey

1.1 Introduction

The food industry is a global network engaged in the supply of food products all over the world. This industry covers a series of industrial activities including production, processing, conversion, distribution, preservation, and packaging of foods. Likewise, storage and safe distribution of processed foodstuffs are also important since most food items get wasted in these processes. Use of chemical-based preservatives and traditional methods of packaging in the food industry, not only compromise the nutritive values of the foods and health issues but also results in significant economic loss to the industry. Nanotechnology (NT) has opened new avenues and benefits in the food industry for the development of new flavors and sensations, new textures in food components, encapsulation of food additives, and enrichment in the nutritional values of the foods [1]. Nano-foods are those foods that are generated by using NT in production, processing, packaging, and security with enhanced taste and bioavailability. In the present chapter, applications of NT in different industrial activities related to the food industry have been discussed.

B. K. Pandey
Department of Physics, SPM College, University of Allahabad, Prayagraj, India

S. Pandey · R. Dhar
Centre of Materials Science, University of Allahabad, Prayagraj, India

K. B. Pandey (✉)
CSIR-Central Salt and Marine Chemicals Research Institute, Bhavnagar, Gujarat, India

C. Egbuna et al. (eds.), *Application of Nanotechnology in Food Science, Processing and Packaging*, https://doi.org/10.1007/978-3-030-98820-3_1

1.2 Nanotechnology in Food Processing

Initially, most of the agriculture products are in general not used as direct food items, they are processed qualitatively before utilized and the method is called food processing. Among various advantages, food processing is used to eliminate microorganisms and to enhance the shelf-life of the foods to transform them into edible, safe, and sound food products [2, 3]. Mainly there are three vital steps involved in food processing (FP); primary, secondary, and tertiary methods. In primary food processing, agricultural products are tuned into edible items. Secondary FP is a day-to-day process of manufacturing foods from ingredients and tertiary FP is the commercial production of edible items which are commonly called processed foods.

Nanotechnology application in the food industry (Fig. 1.1) has shown great potential in food processing; it provides several opportunities in FP including enhancement in taste, physiological availability, shelf-life, and consistency [4]. Application of NT in FP is also useful in concealing the bitter taste, unpleasant order, and size by modifying the particle size and their distribution in the food items [5]. Nano-encapsulation based on the use of nanocapsules is a promising technique that has gained enormous attention in nano-food technology. Nano-encapsulation can deliver food ingredients at a targeted place and maintain active compounds at that place for a longer time. Recent advances in nano-encapsulation have resulted in the development of nanoparticles with improved properties to be used as nanocapsules [6]. Nano-capsules have many properties that can be used in the food industry for betterment; impotently they can mask odors and biological degradations [7]. Many bioactive compounds of food such as polyphenols, vitamins, and other secondary metabolites are very sensitive in an acidic environment and enzymatic activities in the stomach and duodenum. Encapsulation of these ingredients makes sure to allow them to assimilate effectively as well as readily in the digestive tract. Moreover, nano-encapsulation of the bioactive compound increases the shelf-life of food products. Edible nano-coatings on the food items not only provide resistance

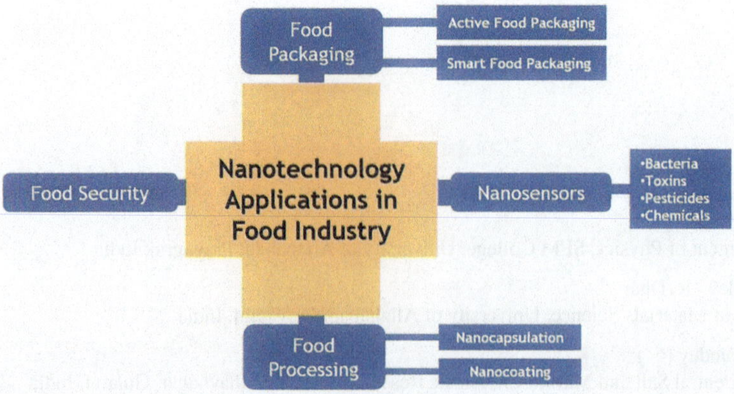

Fig. 1.1 Diverse applications of nanotechnology in the food industry

to humidity, heat, and gas exchange to enhance the shelf-life of food but also can insure the real color and flavor of the food.

1.3 Nanotechnology in Food Packaging and Security

Food packaging is another important step in the food industry in terms of food safety. Good packaging material must possess the combined qualities of strength, biodegradability, permeability to moisture, and other gases. NT-applied packaging materials confer an array of advantages over traditionally used materials for packaging in having anti-microbial properties, better mechanical strength, low weight, and alerting consumers to the safety status of the food [3, 8]. There are mainly three categories of nanofoods packaging, basic one is improved packaging; in this nanoparticles (NPs) make resistance towards temperature, humidity, and gas. However, in active food packaging, NPs act as antimicrobial material and interacts directly with the food items. Active food packaging inactivates pathogens; increases shelf-life and makes the safety of the food items. Another type of nano-packaging is known as intelligent food packaging which is designed to sense biochemical changes and development of pathogens in food items. This packaging monitors the quality of the packaged food items as well as recognizes the pathogens if present [3]. These categories of nano-packaging are based on the kind of packaging materials and applications [9].

1.3.1 Active Food Packaging Systems

Food-borne diseases create critical health problems worldwide and are responsible for the economic burden to society. Active nanocomposites used in preparing biodegradable packaging materials are not only eco-friendly but also increase the shelf-life of food items. In addition, active food packaging systems regulate moisture, absorb oxygen, carbon dioxide, ethylene, and water vapor, and also act as a thermal insulator. Active packaging systems developed by using NT are mainly utilized for the storage purpose of the food items such as meat products which need cold conditions for storage as well as distribution [10]. Such advanced packaging materials increase the shelf-life of meat foods and preserve the quality by creating needed vacuum pressure by passing gases such as oxygen and carbon dioxide, this ends the necessity of air conditioning for storage and distribution of meat food products [10].

1.3.2 Smart Food Packaging Systems

Smart packaging systems counter environmental conditions of food packaging by detecting the undesired contaminations and/or presence of microbes inside the food items. Smart food packaging warns consumers about the safety of packed food items based on the internal molecular external environment of packed food items. Many times microbiological and chemical tests are not enough to recognize the safety of the packaged food items. Whereas, smart food packaging may detect spoilage and pathogenic microorganisms in the packed food items instantly. Smart food packaging systems are equipped with various nanosensors such as moisture sensor, freshness indicator, temperature indicator, chemical toxin indicator, oxygen, carbon dioxide, and pathogen indicators. On the other hand natural dies including β-carotene, curcumin, and anthocyanins, curcumin, chlorophyll, β-carotene present in vegetables and fruits which can be used as a sensor in packaging materials by applying the technology based on NPs. Nanosensors are devices fabricated by using nanoparticles and widely used to detect food contaminants and pathogens in food items. These nanosensors can trace any modifications in the proceeded food items such as change in food's color, texture, and biological transformations. Nanosensors provide information of enzymes produced during the degradation of foodstuff and thus food products enabled with such packaging system have increased shelf-life and due to this use of preservatives becomes limited [4]. Modern smart packaging is based on selective antibody and antigen interaction to identify the pathogens in packaged food items. In this method, antibodies are conjugated with quantum dot semiconductor materials. Quantum dots are one-dimensional NPs having around 5–10 nm diameter and can absorb continuous spectra and as per their size and composition, produce narrow fluorescence emission spectra. Recently antibody attached Cds quantum dot was employed to detect *E. coli* bacteria [11]. Researchers have developed a sensor based on gold and palladium NPs labeled with an antibody that can detect *L. monogtogen* pathogen in an opened food packet. Thus smart nanofood packaging system which can detect and inhibit pathogens and unwanted microbes in packaged food items may be a milestone in successful food industry.

1.4 Mechanism of Action of Nanoparticles

Mostly, there are two factors that may affect food products safety i.e. intrinsic and extrinsic factors. The common intrinsic factors are water percentage, pH level, microbes, and enzymes which leads rate of degradation of food items while extrinsic factors such as temperature, total pressure light intensity, relative humidity and partial pressure are common factors that effectively affect the degradation rate of the foods [12]. In general, microbial growth occurs on the surface of unpreserved food items; therefore antimicrobial packaging develops but NT effectively restricts the growth of microbes in comparison to antimicrobial food additives. Some evidence-based studies were presented in Table 1.1. Studies have reported that some NPs

Table 1.1 Important nanocomposite materials and their antimicrobial activities

Sl. No.	Nanocomposite	Food items	Pathogens tested in food	Effect	Reference
1.	Chitosan/Nigella sativa extract-AgNPs	Strawberries	*B. subtilis, E. coli, P. aeruginosa*, and *S. aureus*	AgNPs showed better antibacterial activity against gram-negative bacteria in compression to gram-positive bacteria	[19]
2.	Low-density polyethylene/Ag/TiO_2NPs	Pikeperch Filets	*A. niger, C. albicans, E. coli,* and *S. aureus*	Effective against all the strains	[20]
3.	Low-density polyethylene/AgNPs	Pikeperch Filets	Apple-isolated *Penicillium expansum, E. coli, Enterococcus faecalis, Salmonella enterica* subsp. *enterica serovar Typhimurium*, and *S. aureus*	Effective against fungi and both positive and negative bacterial strains	[21]
4.	Cellulose/AgNPs	Pikeperch Filets	*E. coli* and *S. aureus*	Significant antibacterial activities against both the bacteria	[22]
5.	Low-density polyethylene/Ag^+Cu/TiO_2, NPs	Nile Tilapia Fish	*E. coli* and *L. monocytogenes*	Very effective antibacterial effect.	[23]
6.	Carboxymethyl cellulose/cellulose/AgNPs	Strawberries	*E. coli* and *S. aureus*	Remarkable anti-bacterial effects were exhibited. Quality, as well as shelf-life of packaged strawberries, were increased	[24]
7.	Chitosan-TiO_2NPs/Red apple pomace extract	Strawberries	*E. coli* and *S. aureus*	Effectiveness is better in *S. aureus* than *E. coli*	[25]
8.	Low-density polyethylene/AgNPs/TiO_2NP and low-density polyethylene/nanoclay/TiO_2NPs	Chicken meat	*E. coli* and *S. aureus*	Effective on both gram-positive and negative bacteria.	[26]

(continued)

Table 1.1 (continued)

Sl. No.	Nanocomposite	Food items	Pathogens tested in food	Effect	Reference
9.	Poly (3-hydroxybutyrate-co-3-hydroxyvalerate)/ biogenic SiO_2NPs	Chicken meat	*E. coli* and *S. aureus*	Dose-dependent antibacterial activity was shown	[27]
10.	Polyvinyl alcohol/ boiled rice starch/ AgNPs	Chicken meat	*S. aureus* and *S. typhimurium*	More effective for *S. typhimurium* than *S. aureus*	[28]
11.	Cellulose/CuNPs	Chicken meat	*E. coli* and *Bacillus* sp.	Significant antibacterial effect was shown at the dose 250 mM against *E. coli*	[29]
12.	Polylactic acid/ oligomeric lactic acid/ chitosan-AgNPs	Chicken meat	*E. coli* and *S. aureus*	Remarkable antibacterial activities were observed against both strains	[30]
13.	Carrageenan/Laponite on the oxygen plasma surface modified polypropylene film/ AgNPs	Chicken meat	*E. coli* and *S. aureus*	Remarkable antibacterial activities were observed against both strains	[31]
14.	Cellulose nanofibril/ AgNPs	Chicken meat	*E. coli*	Better effect against *E. coli* than on *L. monocytogenes*	[32]
15.	Poly(lactic acid)/3-(40-epoxyethyl-benzyl)-5,5-dimethylhydantoin/ SiO_2NPs	Chicken meat	*L. monocytogenes* *E. coli* and *S. aureus*	Remarkable antibacterial activities were observed against both strains	[33]

namely Ag, Au, Cu, CuO, MgO, Fe, and TiO_2 have shown great potential towards antimicrobial properties [13, 14]. Among metal NPs, AgNPs have shown promising results of toxicity towards various food pathogens. AgNPs increase cell membrane permeability by degrading lipopolysacharide and cell surface adherence [15]. In this way, AgNPs can penetrate the cell membrane of bacteria and damage their DNA [16]. AgNPs release Ag^+ ions inside the cell which bind with oxygen, sulfur, nitrogen, and other electron donor groups and inhibit the generation of adenosine triphosphate and DNA replication which force cell death [17]. In another way, Ag^+ can also cause shrinkage of cytoplasm, separation of cell wall, and damage of ribosome which inhibit DNA replication and lead to the death of the cell. There are three basic mechanisms accepted widely involved in microbial toxicity of metal NPs; metal ions uptake via intracellular ATP depletion, reactive oxygen species (ROS) production via oxidative cellular damage, and destruction of the bacterial membrane [18].

1.5 Other Applications of Nanotechnology in the Food Industry

1.5.1 Nanocoating on Packaging Surfaces

Nano-coating on packaging surfaces of food items is one of the other most exciting applications of NT in the food industry. The presence of O_2 inside the packaged food materials creates favorable conditions for the microbes to grow and thus compromise the quality and shelf-life of the food products. Mills (2005) has developed titanium dioxide (TiO_2)NPs conjugated with photo-indicator intelligent ink using redox-activate methylene blue dye for the detection of oxygen level. This detector is very sensitive and changes its color in response to the change in the quantity of O2 inside the packaged material [34]. Besides this, magnesium oxide (MgO) and polylactic acid biopolymer mixed nanocomposite were used to develop food package materials that effectively protect the food against bacteria bio-films [35]. Recently developed coating with zero-valent iron particles with the ability to scavenge oxygen in food packaging has shown revolutionary possibility of application of NT as nano-coating on food products to protect them from degradation [36, 37].

1.5.2 Protection from Chemical Deterioration of Foods

In food items, various important components exist and they can react chemically with external environments causing chemical deterioration to the foods. Relatively less reactive nanomaterials play a key role to inhibit unwanted chemical reactions and carry antioxidants. Nano-encapsulation of a bioactive compound using polymeric nanomaterials is a recent idea for the protection from chemical deterioration, in application, vitamins and flavonoids can be delivered directly in an acidic environment such as the stomach [38]. Nano-encapsulation ensures the functionality of the nutritional ingredients and controlled release of the core materials in the targeted site. Therefore, nano-encapsulated ingredients exhibit plentiful advantages such as unceasing delivery of consecutive delivery of varied active ingredients, enhancement of shelf-life, extension of stability, and pH-triggered controlled release [39].

1.5.3 Improvement of Physical Properties of Food and Packaging Materials

Numerous nanomaterials have been developed to prevent texture and other physical properties of food products and of packaging materials too. Polymer nanocomposite materials with layered silicate have been developed for packaging that has many good properties including high flame resistance and protection from UV rays [40,

41]. Recently many NPs have been synthesized that possess the property to protect and enrich the physical appearance such as the color of food items. Nowadays, TiO_2 NPs have been approved as coloring food additives. Besides, TiO_2 NPs, SiO_2 NPs are being widely used as anti-caking materials to maintain the flow of powdered materials as a carrier of aroma in food items [42].

1.5.4 Removal of Heavy Metals

Heavy metals Pb, Zn, Cu, Hg, etc. could get into foods in many ways including agricultural contaminated water. These heavy metals may pose severe threats to human health since they can accumulate and move across the food chain [43]. They may damage kidneys, lungs, central nervous system, and many other vital organs. In addition to being a direct threat to human health, heavy metals are hazardous to the environment since they cannot be degraded by microbes and once they enter the ecosystem, they remain for years and get transmitted to the whole food chain. These metals are reported to cause cancer and other many life-threatening diseases [44]. The removal of these heath hazardous metals from food items has drawn the attention of scientists and health workers worldwide. In such a scenario, the application of nanotechnology has gained wide attention recently. Heavy metals and their oxides such as ZnO, CuO increase intracellular ROS levels which causes DNA damage. Modified magnetic nanoparticles play a very important role to adsorb heavy metals micro-extraction and possess the capacity to measure even trace amounts of Ni, Cu, Co, Pb, Mn, and Cd from food items. Animated magnetic iron oxide nanoparticles have been demonstrated as an absorbent of heavy metal ions [45, 46]. MgO nanoparticles have elicited a very remarkable ability to track and remove heavy metal ions from contaminated food products as well water samples. Interestingly, MgO NPs have also shown significant antibacterial potential against a variety of strains. It has been reported that the high demand of MgO NPs against bacterial detection and removal of heavy metals from food and beverages is due to their low-cost, easy availability, facile preparation, and eco-friendly [47].

1.5.5 Prevention from Biofilm Formation

Biofilm is defined as a tightly packed bunch of bacterial cells which adhere to any substrate and form a polymeric extracellular matrix which is very tough to break out. The biofilm can be formed by the tight binding of free-floating microbes to the surface of food items by Vander-walls forces. AgNPs increase the metabolism of bacteria and have shown the potential to prevent biofilm formation. NiONPs having sizes 10–20 nm have been proposed for antibacterial, anti-tumor as well as activity to inhibit biofilm formation [48]. Researchers have reported that the chlorhexidine-conjugated gold nanoparticles can exhibit excellent inhibitory effects against the formation of biofilm by *Klebsiella pneumonia* [49]. It has been reported that naked (superparamagnetic iron oxide)IONs and IONs coated with 3-aminopropyltriethoxy silane The two types of (IONs), such as naked IONs and IONs coated with 3-aminopropyltriethoxy silane, have elicited significant ability to prevent biofilm formation by *B. subtilis*, their growth as well as cell viabilities [50, 51].

1.6 Nanosensors in the Food Industry

Food safety and security are very important question in the food industry since it directly affects human health and economic growth. Many old and conventional methods have been discovered and utilized to detect and monitor harmful pathogens and toxic materials in the food industry from time to time. However, these methods are time taking, costly, and unresponsive in detecting harmful pathogens and unwanted materials in food products. Therefore, the food industry needs such a sophisticated and easy-to-handle instrument/sensor which should be rapid, sensitive, and inexpensive to detect pathogens and toxic materials in a food product. In recent times NT has shown very promising potential for solving food safety and security issues in terms of identifying pathogens and toxic materials which help to improve human health and economic growth of agriculture and food industry.

A sensor is a device that detects or measures the changes in the form of physical and chemical properties of the environment for a particular system. Nanosensors have several benefits in sensitivity and specificity over traditional sensors that can be applied in the food industry very successfully. Nanosensors, functionalized with chemical and biological molecules such as antibodies to increase their specialty to detect changes in physical and chemical processes. They offer high sensitivity of detection due to the high surface-to-volume ratio of nanomaterials used in fabrication. Nanotubes and nanowires, the one-dimensional nanomaterials have shown great potential in the fabrication of nanosensors. There are numerous potential applications of nanosensors in several areas such as medicine, agriculture, food technology as well as research. Nanosensors can monitor various physical changes such as pressure, temperature, volume, concentration, gravity, velocity, electric and magnetic signal in a very accurate manner. Nanosensors that use molecularly imprinted polymer can be divided into three categories such as electrochemical, piezoelectric, and spectroscopic. Electrochemical sensors are based on electrochemical properties of sensing materials such as charge, conductivity, capacitance, electric potential (Fig. 1.2). Piezoelectric sensor either converts mechanical force into electrical or electrical force in mechanical force. Spectroscopic sensor work on light-based signal chemiliuminescent, surface Plasmon resonance, and fluorescence.

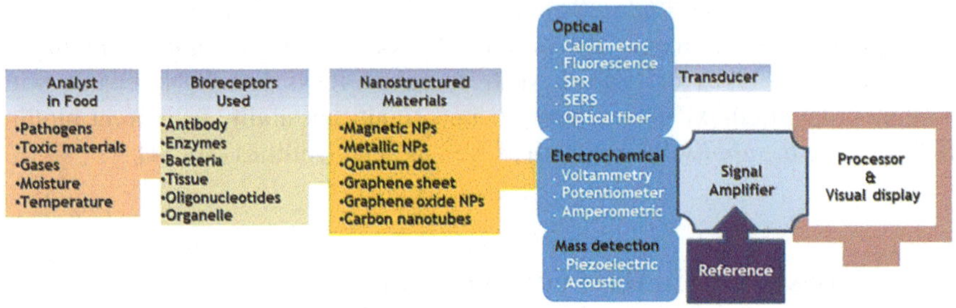

Fig. 1.2 Schematic representation of different components of nanosensors

1.6.1 Nanosensors for Detecting Food Pathogenic Bacteria

In food items, the detection of pathogenic bacteria is achieved by detecting genetic material or whole-cell of bacteria. NPs based DNA isolation from bacteria is more effective, sensitive, and fast due to which better results are expected than conventional ways. Magnetic nanoparticles such as iron oxide NPs are used for the isolation of DNA from *Listeria monocytogenes* milk bacteria [52, 53]. Several other nanoparticles are employed for the detection of pathogenic bacteria in foods, the details are listed in Table 1.2.

1.6.2 Detection of Toxins, Pesticides and Chemicals in Food

Food toxin materials such as aflatoxins, polytoxins, botulinum, mycotoxin contaminate food product and if enters in our body causes several diseases. Researchers have developed various nanosensors based on nanocomposite materials that are very efficient and sensitive in the detection of these toxic materials. Gold NPs functionalized with anti-aflatoxins have been utilized for the detection of aflatoxin. Carbon nanotubes based on electrochemical luminescent sensors are used to detect polytoxins found in mussel meat [63]. Food toxin sensors based on NPs are listed in Table 1.2.

Nowadays in the agriculture sector, excessive pesticides and fertilizers are used for more crop and fruits productions by farmers which are a serious drawback for human health. Organophosphate is a more common pesticide used in food products. Gold NPs have been functionalized as calorimetric and fluorometric sensors to detect organophosphate and carbamate pesticides in food products [67]. Other pesticide sensors are listed in Table 1.2.

Table 1.2 Nanosensors based on nanoparticles with detecting technique of food; pathogens, toxin and gases

Sl. No.	Nanosensors based on nanoparticles	Detection technique	Factors it detects	Reference
1.	Magnetic NPs	Impedance and DNA isolation based on PCR	*E. coli, L. monocytogenes* and *Salmonella typhimurium Bacterium*	[54, 55]
2.	Gold NPs	SPR and electrochemical	*E. coli, Staphylococcus aureus* and *Vibrio para Bacterium*	[56–58]
3.	Single-walled carbon nanotubes (SWCNT) Polypyrrole nanowires	FET transistor and electrochemical	*Bacillus globigii* and *Salmonella infantis Bacterium*	[59, 60]
4.	Zinc sulfite and cadmium selenide NPs	Fluorescence	*E. coli* and *Salmonella typhimurium Bacterium*	[61, 62]
5.	Gold NPs	Electrochemical	Botulinum neurotoxin type B and brevetoxins *Bacterium*	[64]
6.	Zinc oxide NPs	Cyclic voltammetry and impedance	Ochratoxin-A	[65, 66]
7.	Carbon nanotubes (CNT) and multi-walled carbon nanotubes (MWCNT)	Electroluminescent	Microcystin-LR and palytoxin	[63]
8.	Gold and Silver NPs	Fluorescence and UV-visible spectroscopy	Melamine	[68]
9.	MWCNT, graphene mixed magnetic NPs	Electrochemical	Sudan I	[69]
10.	Cobalt nitroprusside	Electrochemical	Sulfite	[70]
11.	Methylene blue–titanium dioxide	Luminescence	Oxygen	[71]
12.	Silver and gold nanorods	UV–visible	Time–temperature	[72]
13.	Gold NPs	UV–visible and electrochemical	Time–temperature	[73]
14.	Iron oxide-titanium dioxide NPs	Electrical gas sensing	Time–temperature	[74]

1.6.3 Nanosensors for Food Freshness

The packaged food materials are prone to degrade with time and temperature. The presence of oxygen and humidity in unproportioned quantities spoils the freshness of packed food products and also reduces their shelf-life. Oxygen oxidizes the nutritive vial molecules of the food consequently; this condition favors the growth of bacteria and other pathogens. Nanoparticles' color, size, and spectral response have been utilized for developing time–temperature-based indicators for food freshener. Methylene blue/titanium dioxide hybrid nanocomposite has been developed for oxygen sensor. Moreover, various freshness sensors have been developed based on nanocomposite materials which are listed in Table 1.2.

1.7 Future Remarks and Conclusion

NT has shown a possibility to play a promising role in different aspects of the food industry including packaging, storage, distribution, and enriching the quality and taste of food materials. However, the data obtained to date on the possibility of the applications of NT indicates that the utilization of NT in the food industry is in the initial stage even though it has made a revolution in the food industry through nano-processing and nano-packaging of foodstuff, further it insures the safety and quality of food products by the use of advanced nanosensors. Nano-encapsulation of bioactive ingredients facilitates targeted delivery of active ingredients as well as increases the shelf-life of the food products. Nanocoating on surfaces is not only useful in preventing the chemical and microbial deterioration of the food items but also restoring their nutritional value. Heavy metal reduction/removal by the use of NT from food and beverages is a novel and cost-effective technique to insure food quality. However, despite the revolutionary role of NT applications in the food industry, some issues need to be studied in detail. The toxic properties of some nanoparticles and less easy availability are the other important areas that must be attended to. The chapter concludes that, if managed and regulated properly, nanotechnology can play a very vital role in the food industry; improving food processing, product quality, and safety, which will benefit humans at large.

References

1. Cho YH, Jones OG. Assembled protein nanoparticles in food or nutrition applications. Adv Food Nutr Res. 2019;88:47–84.
2. Zhang H, Tikekar RV, Ding Q, Gilbert AR, Wimsatt AT. Inactivation of food borne pathogens by the synergistic combinations of food processing technologies and food-grade compounds. Compr Rev Food Sci Food Saf. 2020;19:2110–38.
3. Wahab A, Rahim AA, Hassan S, Egbuna C, Manzoor MF, Okere KJ, Walag AMP. Application of nanotechnology in the packaging of edible materials. In: Egbuna C, Mishra AP, Goyal MR, editors. Preparation of phytopharmaceuticals for the management of disorders. Cambridge: Academic; 2021. p. 215–25.
4. Powers KW, Brown SC, Krishna VB, Wasdo SC, Moudgil BM, Roberts SM. Research strategies for safety evaluation of nanomaterials. Part VI. Characterization of nanoscale particles for toxicological evaluation. Toxicol Sci. 2006;90:296–303.
5. Nile SH, Baskar V, Selvaraj D, Nile A, Xiao J, Kai G. Nanotechnologies in food science: applications, recent trends, and future perspectives. Nano-Micro Lett. 2020;12:45.
6. Peng W, Robert J, Feng K, Wu H. Electrospinning: a novel nano-encapsulation approach for bioactive compounds. Trends Food Sci Technol. 2017;70:56–68.

7. Perinelli DR, Palmieri GF, Cespi M, Bonacucina G. Encapsulation of Flavours and fragrances into polymeric capsules and cyclodextrins inclusion complexes: an update. Molecules. 2020;25:5878.
8. Singh T, Shukla S, Kumar P, Wahla V, Bajpai VK, Rather IA. Application of nanotechnology in food science: perception and overview. Front Microbiol. 2017;8:1501.
9. Hannon JC, Kerry J, Cruz-Romero M, Morris M, Cummins E. Advances and challenges for the use of engineered nanoparticles in food contact materials. Trends Food Sci Technol. 2015;43:43–62.
10. Dias MV, de Soares FF, Borges SV, de Sousa MM, Nunes CA, de Oliveira IRN, Medeiros EAA. Use of allyl isothiocyanate and carbon nanotubes in an antimicrobial film to package shredded, cooked chicken meat. Food Chem. 2013;141:3160–6.
11. Mohamadi E, Moghaddasi M, Farahbakhsh A, Kazemi A. A quantum-dot-based fluoroassay for detection of food-borne pathogens. J Photochem Photobiol B Biol. 2017;174:291–7.
12. King T, Osmond-McLeod MJ, Duffy LL. Nanotechnology in the food sector and potential applications for the poultry industry. Trends Food Sci Technol. 2018;72:62–73.
13. Spirescu VA, Chircov C, Grumezescu AM, Bogdan S, Vasile T, Andronescu E. Inorganic nanoparticles and composite films for antimicrobial therapies. Int J Mol Sci. 2021;22:4595.
14. Arora N, Thangavelu K, Karanikolos GN. Bimetallic nanoparticles for antimicrobial applications, bimetallic nanoparticles for antimicrobial applications. Front Chem. 2020;8:412.
15. Sondi I, Salopek-Sondi B. Silver nanoparticles as antimicrobial agent: a case study on *E. coli* as a model for Gram-negative bacteria. J Colloid Interface Sci. 2004;275:177–82.
16. Li Y, Tseng YD, Kwon SY, Despaux L, Bunch JS, Mceuen PL. Controlled assembly of dendrimer-like DNA. Nat Mater. 2004;3:38–42.
17. Karimi M, Sadeghi R, Kokini J. Human exposure to nanoparticles through trophic transfer and the biosafety concerns that nanoparticle-contaminated foods pose to consumers. Trends Food Sci Technol. 2018;75:129–45.
18. Pathkoti K, Manubolu M, Hwang HM. Nanostructures: current uses and future applications in food science. J Food Drug Anal. 2017;25:245–53.
19. Kadam D, Momin B, Palamthodi S, Lele SS. Physicochemical and functional properties of chitosan-based nano-composite films incorporated with biogenic silver nanoparticles. Carbohydr Polym. 2019;211:124–32.
20. Barani S, Ahari H, Bazgir S. Increasing the shelf life of pikeperch (*Sander lucioperca*) fillets affected by low-density polyethylene/Ag/TiO_2 nanocomposites experimentally produced by sol-gel and melt-mixing methods. Int J Food Prop. 2018;21:1923–36.
21. Brito SDC, Bresolin JD, Sivieri K, Ferreira MD. Low density polyethylene films incorporated with silver nanoparticles to promote antimicrobial efficiency in food packaging. Food Sci Technol Int. 2020;26:353–66.
22. Chen QY, Xiao SL, Shi SQ, Cai LP. A one-pot synthesis and characterization of antibacterial silver nanoparticle-cellulose film. Polymers. 2020;12:440.
23. Efatian H, Ahari H, Shahbazzadeh D, Nowruzi B, Yousefi S. Fabrication and characterization of LDPE/silver-copper/titanium dioxide nanocomposite films for application in Nile Tilapia (Oreochromis niloticus) packaging. J Food Meas Charact. 2021;15:2430–9.
24. He Y, Li H, Fei X, Peng L. Carboxymethyl cellulose/cellulose nanocrystals immobilized silver nanoparticles as an effective coating to improve barrier and antibacterial properties of paper for food packaging applications. Carbohydr Polym. 2021;252:117156.
25. Lan W, Wang S, Zhang Z, Liang X, Liu X, Zhang J. Development of red apple pomace extract/chitosan-based films reinforced by TiO(2) nanoparticles as a multifunctional packaging material. Int J Biol Macromol. 2021;168:105–15.
26. Lotfi S, Ahari H, Sahraeyan R. The effect of silver nanocomposite packaging based on melt mixing and sol–gel methods on shelf life extension of fresh chicken stored at 4 °C. J Food Saf. 2021;39:e12644.

27. Ojha N, Das N. Fabrication and characterization of biodegradable PHBV/SiO(2) nanocomposite for thermo-mechanical and antibacterial applications in food packaging. IET Nanobiotechnol. 2020;14:785–95.
28. Mathew S, Jayakumar A, Kumar VP, Mathew J, Radhakrishnan EK. One-step synthesis of eco-friendly boiled rice starch blended polyvinyl alcohol bionanocomposite films decorated with in situ generated silver nanoparticles for food packaging purpose. Int J Biol Macromol. 2019;139:475–85.
29. Muthulakshmi L, Rajalu V, Kaliaraj A, Siengchin GS, Parameswaranpillai J, Saraswathi R. Preparation of cellulose/copper nanoparticles bionanocomposite films using a bioflocculant polymer as reducing agent for antibacterial and anticorrosion applications. Compos B Eng. 2019;175:107177.
30. Sonseca A, Madani S, Rodríguez G, Hevilla V, Echeverría C, Fernández- García M. Multifunctional PLA blends containing Chitosan mediated silver nanoparticles: thermal, mechanical, antibacterial, and degradation properties. Nano. 2019;10:22.
31. Yadav S, Mehrotra GK, Dutta PK. Chitosan based ZnO nanoparticles loaded gallic-acid films for active food packaging. Food Chem. 2021;334:127605.
32. Yu Z, Wang W, Kong F, Lin M, Mustapha A. Cellulose nanofibril/silver nanoparticle composite as an active food packaging system and its toxicity to human colon cells. Int J Biol Macromol. 2019;129:887–94.
33. Zhao Y, Wei B, Wu M, Zhang H, Yao J, Chen X, et al. Preparation and characterization of antibacterial poly(lactic acid) nanocomposites with N-halamine modified silica. Int J Biol Macromol. 2020;155:1468–77.
34. Mills A. Oxygen indicators and intelligent inks for packaging food. Chem Soc Rev. 2005;34:1003–11.
35. Swaroop C, Shukla M. Nano-magnesium oxide reinforced polylactic acid biofilms for food packaging applications. Int J Biol Macromol. 2018;113:729–36.
36. Foltynowicz Z, Bardenshtein A, Sangerlaub S, Antvorskov H, Kozak W. Nanoscale, zero valent iron particles for application as oxygen scavenger in food packaging. Food Packag Shelf Life. 2017;11:74–83.
37. Ariyarathna IR, Rajakaruna RMPI, Nedra Karunaratne D. The rise of inorganic nanomaterial implementation in food applications. Food Control. 2017;77:251–9.
38. Pool H, Quintanar D, de Dios Figueroa J, Mano CM, Bechara JEH, Godinez LA, et al. Antioxidant effects of quercetin and catechin encapsulated into PLGA nanoparticles. J Nanomater. 2012;2012:e145380.
39. Momin JK, Jayakumar C, Prajapati JB. Potential of nanotechnology in functional foods. Emir J Food Agric. 2013;25:10–9.
40. Laoutid F, Bonnaud L, Alexandre M, Lopez-Cuesta JM, Dubois P. New prospects in flame retardant polymer materials: from fundamentals to nanocomposites. Mater Sci Eng R Rep. 2009;63:100–25.
41. Lizundia E, Ruiz-Rubio L, Vilas JL, Leon LM. Poly (L-lactide)/ZnO nanocomposites as efficient UV-shielding coatings for packaging applications. J Appl Polym Sci. 2016;133:e42426.
42. Dekkers S, Krystek P, Peters RJ, Lankveld DX, Bokkers BG, van Hoeven-Arentzen PH, et al. Presence and risks of nanosilica in food products. Nanotoxicology. 2011;5:393–405.
43. Afroze S, Sen TK. A review on heavy metal ions and dye adsorption from water by agricultural solid waste adsorbents. Water Air Soil Pollut. 2018;229:225.
44. Cocarta D, Neam TS, Deac AR. Carcinogenic risk evaluation for human health risk assessment from soils contaminated with heavy metals. Int J Environ Sci Technol. 2016;13:2025–36.
45. Fukui H, Horie M, Endoh S, Kato H, Fujita K, Nishio K. Association of zinc ion release and oxidative stress induced by intratracheal instillation of ZnO nanoparticles to rat lung. Chem Biol Interact. 2012;198:29–37.
46. McShan D, Ray PC, Yu H. Molecular toxicity mechanism of nanosilver. J Food Drug Anal. 2014;22:116–27.

47. Cai Y, Li C, Wu D, Wang W, Tan F, Wang X, et al. Highly active MgO nanoparticles for simultaneous bacterial inactivation and heavy metal removal from aqueous solution. Chem Eng J. 2017;312:158–66.
48. Shahrokh S, Emtiazi G. Toxicity and unusual biological behavior of nanosilver on Gram positive and negative bacteria assayed by microtiter plate. Eur J Biol Sci. 2009;1:28–31.
49. Saleem S, Ahmed B, Khan MS, Al-Shaeri S, Musarrat J. Inhibition of growth and biofilm formation of clinical bacterial isolates by NiO nanoparticles synthesized from Eucalyptus globulus plants. Microb Pathog. 2017;111:375–87.
50. Ranmadugala D, Ebrahiminezhad A, Manley-Harris M, Ghasemid Y, Berenjian A. The effect of iron oxide nanoparticles on Bacillus subtilis biofilm, growth and viability. Process Biochem. 2017;62:231–40.
51. Bajpai VK, Kamle M, Shukla S, Mahato DK, Chandra P, Hwang SK, Kumar P, Huh YS, Han YK. Prospects of using nanotechnology for food preservation, safety, and security. J Food Drug Anal. 2018;26:1201–4.
52. Lee NR, Lee SK, Ahn J, Shin YB, Choi HS, Lee CS, Kim S, Kim MG. High sensitivity detection of 16s rRNA using peptide nucleic acid probes and a surface plasmon resonance biosensor. Anal Chim Acta. 2018;630:168–73.
53. Yang H, Qu L, Wimbrow A, Jiang X, Sun Y. Rapid detection of listeria monocytogenes by nanoparticle-based immunomagnetic separation and real-time PCR. Int J Food Microbiol. 2007;118:132–8.
54. Amagliani G, Brandi G, Omiccioli E, Casiere A, Bruce I, Magnani M. Direct detection of Listeria monocytogenes from milk by magnetic based DNA isolation and PCR. Food Microbiol. 2004;21:597–603.
55. Varshney M, Li Y, Srinivasan B, Tung S. A label-free, microfluidics and interdigitated array microelectrode-based impedance biosensor in combination with nanoparticles immunoseparation for detection of Escherichia coli O157:H7 in food samples. Sens Actuators B Chem. 2007;128:99–107.
56. Afonso AS, Perez-Lopez B, Faria RC, Mattoso LHC, Hernandez- Herrero M, Roig-Sagues AX, Costa MM, Merkoci A. Electrochemical detection of Salmonella using gold nanoparticles. Biosens Bioelectron. 2013;40:121–6.
57. Joung HA, Lee NR, Lee SK, Ahn J, Shin YB, Choi HS, Lee CS, Kim S, Kim MG. High sensitivity detection of 16s rRNA using peptide nucleic acid probes and a surface plasmon resonance biosensor. Anal Chim Acta. 2008;630:168–73.
58. Zhao G, Xing F, Deng S. A disposable amperometric enzyme immunosensor for rapid detection of Vibrio parahaemolyticus in food based on agarose/nano-Au membrane and screen-printed electrode. Electrochem Commun. 2007;9:1263–8.
59. Villamizar RA, Maroto A, Rius FX, Inza I, Figueras MJ. Fast detection of Salmonella Infantis with carbon nanotube field effect transistors. Biosens Bioelectron. 2008;24:279–83.
60. Garcia-Aljaro C, Bangar MA, Baldrich E, Munoz FJ, Mulchandani A. Conducting polymer nanowire-based chemiresistive biosensor for the detection of bacterial spores. Biosens Bioelectron. 2010;25:2309–12.
61. Su X, Li Y. Quantum dot biolabeling coupled with immunomagnetic separation for detection of Escherichia coli O157:H7. Anal Chem. 2004;76:4806–10.
62. Yang L, Li Y. Simultaneous detection of Escherichia coli O157:H7 and Salmonella typhimurium using quantum dots as fluorescence labels. Analyst. 2006;131:394–401.
63. Zamolo VA, Valenti G, Venturelli E, Chaloin O, Marcaccio M, Boscolo S, Castagnola V, Sosa S, Berti F, Fontanive G, Poli M, Tubaro A, Bianco A, Paolucci F, Prato M. Highly sensitive electrochemiluminescent nanobiosensor for the detection of palytoxin. ACS Nano. 2012;6:7989–97.
64. Zhou Y, Pan FG, Li YS, Zhang YY, Zhang JH, Lu SY, Ren HL, Liu ZS. Colloidal gold probe-based immunochromatographic assay for the rapid detection of brevetoxins in fishery product samples. Biosens Bioelectron. 2009;24:2744–7.

65. Ansari AA, Kaushik A, Solanki PR, Malhotra BD. Nanostructured zinc oxide platform for mycotoxin detection. Bioelectrochemistry. 2010;77:75–81.
66. Norouzi P, Larijani B, Ganjali MR. Ochratoxin A sensor based on nanocomposite hybrid film of ionic liquid-graphene nanosheets using coulometric FFT cyclic voltammetry. Int J Electrochem Sci. 2012;7:7313–24.
67. Liu D, Chen W, Wei J, Li X, Wang Z, Jiang X. A highly sensitive, dual-readout assay based on gold nanoparticles for organophosphorus and carbamate pesticides. Anal Chem. 2012;84:4185–91.
68. Vasimalai N, John SA. Picomolar melamine enhanced the fluorescence of gold nanoparticles: spectrofluorimetric determination of melamine in milk and infant formulas using functionalized triazole capped gold nanoparticles. Biosens Bioelectron. 2013;42:267–72.
69. Ma X, Chao M, Wang Z. Electrochemical determination of Sudan I in food samples at graphene modified glassy carbon electrode based on the enhancement effect of sodium dodecyl sulphonate. Food Chem. 2013;138:739–44.
70. Devaramani S, Malingappa P. Synthesis and characterization of cobalt nitroprusside nano particles: application to sulfite sensing in food and water samples. Electrochim Acta. 2012;85:579–87.
71. Borisov SM, Klimant I. Luminescent nanobeads for optical sensing and imaging of dissolved oxygen. Microchim Acta. 2009;164:7–15.
72. Wang YC, Lu L, Gunasekaran S. Gold nanoparticle-based thermal history indicator for monitoring low-temperature storage. Microchim Acta. 2015;182:1305–11.
73. Nopwinyuwong A, Kaisone T, Hanthanon P, Nandhivajrin C, Boonsupthip W, Pechyen C, Suppakul P. Effects of nanoparticle concentration and plasticizer type on colorimetric behavior of polydiacetylene/silica nanocomposite as time–temperature indicator. Energy Proc. 2014;56:423–30.
74. Lou Z, Li F, Deng J, Wang LL, Zhang T. Branch-like hierarchical heterostructure (α-Fe_2O_3/TiO_2): a novel sensing material for trimethylamine gas sensor. ACS Appl Mater Interfaces. 2013;5:12310–6.

Chapter 2
Nanomaterials in Food System Application: Biochemical, Preservation, and Food Safety Perspectives

Shreya M. Hegde, Sanya Hazel Soans, Ravi Teja Mandapaka, J. M. Siddesha, Ann Catherine Archer, Chukwuebuka Egbuna, and Raghu Ram Achar

2.1 Introduction

The use of nanotechnology in the food industry has been growing consistently with time. This new technology has paved its way through the food system by proving its benefits right from cultivation to processing, packaging, storage, and transportation. Nanomaterials are in demand because of their small size (diameter <100 nm), shape, and novel properties [1]. This might provide us with the solution to food wastage which leads us to a huge food crisis. Human exposure to this new technology is increasing day by day. Hence, as a concern to public health, it is important to understand how nanotechnology works in the food system [2].

S. M. Hegde · J. M. Siddesha · R. R. Achar (✉)
Division of Biochemistry, School of Life Sciences, JSS Academy of Higher Education and Research, Mysuru, Karnataka, India
e-mail: siddeshajm@jssuni.edu.in; rracharya@jssuni.edu.in

S. H. Soans · A. C. Archer
Department of Microbiology, School of Life Sciences, JSS Academy of Higher Education and Research, Mysuru, Karnataka, India
e-mail: archerann@jssuni.edu.in

R. T. Mandapaka
Centre for Gender Studies, Nutritional Security, and Urban Agriculture, National Institute of Agricultural Extension Management (MANAGE), Hyderabad, Telangana, India

C. Egbuna
Nutritional Biochemistry and Toxicology Unit, World Bank Africa Centre of Excellence, Centre for Public Health and Toxicological Research (ACE-PUTOR), University of Port-Harcourt, Rivers State, Nigeria

Faculty of Natural Sciences, Department of Biochemistry, Chukwuemeka Odumegwu Ojukwu University, Anambra State, Nigeria

C. Egbuna et al. (eds.), *Application of Nanotechnology in Food Science, Processing and Packaging*, https://doi.org/10.1007/978-3-030-98820-3_2

Nanoparticles are extensively used in packaging and preservation thanks to their far-reaching antimicrobial activity, the generation of ROS by the nanoparticles can damage the microorganisms present on the exterior of the foodstuff. They have wide-ranging surface area per unit mass and their reduced particle dimension helps enhance the beneficial properties like biological activity, bioavailability, and solubility of the processed encapsulated food products when compared to other modes of packaging [3]. Moreover, if organic composite nanoparticles are used, the food products will have improved bioactivity. The enclosed packaging may also contain nanosensors, which are known as active/smart/intelligent food packaging which are a revolutionary method to detect the condition of food when packaged. Foodborne pathogens such as *Escherichia coli*, *Listeria*, and *Salmonella* on steel food production surfaces can be eliminated by engineered water nanostructures generated as aerosols [4]. Apart from this, nanoparticles can act as transporters for biomolecules like enzymes and antioxidants. They transport anti-browning agents, flavors, and other bioactive materials even after opening the packaging which improves their shelf life. These advantages make the food industry run smoothly without any hazards [5]. Table 2.1 describes the various components and their applications in food industries.

2.2 Biochemical Perspective for the Usage of Nanomaterials

The nanoparticles in our food may consist of both **organic** and **inorganic** constituents. The organic nanoparticles include a fabricated form of biomolecules like proteins, carbohydrates, and lipids. They can either be liquid, or they can have properties of semi-solid (gel), or solid (crystalline or amorphous) completely based upon their arrangement and modification [6]. Their activity in the human body considerably varies within various parts of the human digestive system. Nanomaterials if not digested may get solubilize, form a precipitate, or get lumped up. Digestible nanoparticles undergo enzymatic breakdown in the region of the mouth, stomach, small intestine, or colon according to their properties. Under the organic nanoparticles, the most common are lipids, proteins, and carbohydrates nanoparticles [7].

Table 2.1 Types of nanomaterials and their application in food industries

Nanocomponent	Primarily used in	Advantages
Nanoporous membranes, nanofibrils, nano aggregates, nanocrystals	Food processing	Anticaking agents Gelating agents Enzyme immobilization
Nanocoating (nanoemulsions/nanoencapsulations)	Food preservation	Encapsulation of bioactive compounds Antimicrobial activity Nutrient delivery
Nanocomposites (polymer/clay), nanotracers, nanosensors	Food packaging and food safety	Active packaging Smart packaging

2.2.1 Lipid Nanoparticles

Lipids by their nature are present within many commercial food products like soft drinks, dairy drinks, and fortified water in the form of small oil droplets dispersed in water. The major advantages of these are, they increase the bioavailability and the stability of the product and they provide optical transparency which is desirable in many foods. They can be present in the form of micelles, vesicles, oil droplets, and fat crystals. Solid lipid nanoparticles (SLN) are very commonly used under this [8].

2.2.2 Protein Nanoparticles

Nanoparticles of proteins are found naturally in the casein micelles in bovine milk and other dairy products along with phosphate ions. They usually have an aggregated cluster of proteins held by physical interactions or covalent bonds. Typically, they are spherical, but fibrous structures can also be created. The functions include increasing the bioavailability of vitamins, minerals, and nutraceuticals and also encapsulating attributes like flavor, preservatives, and colors [9].

2.2.3 Carbohydrate Nanoparticles

Digestible polysaccharides like starch and non-starch polysaccharides like cellulose, alginate, carrageenan, pectin, and xanthan are important sources of Carbohydrate nanoparticles. They are formed by breaking larger carbohydrate units like starch granules and chitosan or cellulose fibrils. The shape of the nanoparticles can be spherical or not, digestible or indigestible depending on their source. The functions are the same as that of protein nanoparticles [6].

2.2.4 Inorganic Nanoparticles

Inorganic nanoparticles are widely in use in the form of silver, and oxides and dioxides of iron, titanium, silicon, and zinc. At ambient temperature, they can possess properties of either crystalline or amorphous solids, their shape can be spherical or non-spherical, and the range of sizes depends on their modification process. Under different solution conditions with parameters like pH and ionic strength, their tendency to dissolve also varies. Their chemical reactivities also changes based on the environment around them. Inorganic substances like Ag, Cu, CuO, ZnO, Pd, Fe, and TiO_2 are commonly used in the process of packaging [10]. They are adhered via developing electrostatic, hydrogen bonding, and covalent interactions. Silver nanoparticles (AgNP) have claimed to be useful in diverse applications within the food industry according to the manufacturers because of their ability to target bacterial metabolism by attaching to its DNA, protein, and enzymes, which results in bacteriostatic effect along with destabilization of the cytoplasmic membranes [1].

Furthermore, several nanoparticles oxides and dioxides such as titanium (TiO_2), zinc oxide (ZnO), silicon dioxide (SiO_2), and magnesium oxide (MgO) also have shown to be helpful in food packaging due to their capacity to behave as UV blockers and as photo-catalytic disinfecting agents. Dioxides like silicon have proven to improve the quality of many polymer matrices by improving their physical and mechanical properties. Amongst these TiO_2 is highly explored by incorporating it into various forms of nanomaterials like nanorods, nanowires, nanotubes, etc. The exact action of this dioxide is not known, but it may be due to the alterations in the coenzyme A-dependent enzyme activity, and their ability to damage the DNA through the hydroxyl radicals [4].

2.2.5 Organic Nanomaterials Versus Inorganic Nanomaterials

The composition of the nanoparticles plays a main role in deciding their fate in the gastrointestinal tract (GIT). Generally, organic nanomaterials are considered less toxic than inorganic ones. Organic components are readily digestible by the digestive enzymes. But if they are fabricated by dietary fiber or mineral oils, they are not digested by the upper GIT. The inorganic components are also not completely digestible in the GIT, but with adjustments in pH or dilution, they can be partially digested. When the upper part of the GIT cannot digest certain components, it is passed to lower GIT where the gut bacteria conditions can be changed. Few inorganic nanoparticles components like silver can dissolve and release ions which can bring about undesirable changes by chemical reactions, whereas nanoparticles of TiO_2 are relatively inert in the lower GIT. Along with this, the size of the nanoparticles also decides the digestibility in the GIT and its toxicity. The smaller the size of the nanoparticles, the easy it is to get digested and absorbed by virtue of their wide surface area and the capacity for uptake of nanoparticles by intestinal epithelium cells through transport systems (active or passive) or through tight junctions highly depends on the particle size [6].

There are various uncertainties surrounding the mechanism and possibilities of nanoparticle toxicity, hence understanding the properties of nanoparticles and their ability to produce toxic effects are essential. Our body can consume nanomaterials in a variety of ways which can be circulated throughout the body and damage human cells by fluctuating function of an important cell organelle like mitochondria, generate reactive oxygen species, and increase the permeability of membranes, finally leading to toxic effects and chronic disease.

As this new approach of using nanomaterials through various applications in food industries is most likely here to stay, more research is needed on the development of more sustainable use of this technology with proper regulation [11].

2.3 Preservatives and Antimicrobial Nanomaterial Components

2.3.1 *Preservatives*

Food preservation refers to the treatment and handling of food to slow down decomposition and prevent loss of food quality, edibility, or nutritional content due to microorganisms [12]. Microorganisms present on certain food might be pathogenic, which produce various poisons or cell segments that are inconvenient to human wellbeing. Heat treatment is the most effective actual strategy for food preservation. However, excessive heat may alter the nutritional value of the food. As a result, obstacle innovation such as nanotechnology has been used to combat this issue [13]. The development of active food packaging that blends the characteristics of an exterior barrier with antimicrobial agents has the ability to limit or delay microbial growth, thereby reducing food deterioration and extending shelf life. Nanotechnology has a wide scope in applications related to food. Among these applications, infusing a particular kind of nanoparticle into a particular food item so that the food material can foster some of its ideal properties is the most common one. The most common nanomaterials employed in the food sector are silver nanoparticles and nanocomposites as antimicrobials [14].

2.3.1.1 Silver Nanoparticles

Silver nanoparticles mainly cause cell death by Ag ions which bind to membrane proteins, generate pits, cause various structural changes, and they play an important role in catalyzing the production of ROS in cells of bacteria, resulting in apoptosis due to the oxidative stress [15]. Zhang et al. [16] synthesized silver nanoparticles (AgNps) which exhibited strong preservative effects on *Citrus* fruit rot caused by *Penicillium italicum.* The mechanism of preservation was based on damage to the cell wall and membrane, which caused the disruption of essential intracellular substances such as DNA and proteins and also leading to leakage from the cytoplasm, causing cell distortion and death. To summarise, the study doesn't only present a fresh and effective process for the synthesis of AgNPs, but it also demonstrates how AgNPs could be utilized to preserve citrus fruits [16]. The *in vivo* antibacterial activity of Chitosan silver nanoparticles (Ch-AgNPs) against *Escherichia coli* in minced meat samples was high when compared to either controls or Ch alone, with AgNPs playing a prominent role in antimicrobial activity. The findings show that Ch-AgNPs could be employed as antibacterial agents to extend the shelf life of meat [17]. Food preservation procedures are most commonly associated with the use of chemical preservatives, which have a number of negative consequences, including a change in the composition of food, a reduction in nutritional quality, and a hazardous effect on human health [18]. Bacteriocins are small toxic proteinaceous compounds secreted extracellularly by bacteria and theyare garnering a lot of interest because of their GRAS (generally recognized as safe) status and lack of food toxicity [19]. When bacteriocins are nanoencapsulated and employed in food preservation, they are protected from gastrointestinal enzyme breakdown.

2.3.1.2 Poly-Lactic Acid Nanoparticles

PLA (Poly-lactic acid) nanoparticles with nisin were produced for potent and stable release of bacteriocins for food preservation purpose. Long-lasting antibacterial activity was discovered in Nisin-loaded polymeric nanoparticles produced using the gas precipitation process. Indeed, this formulation allowed for delayed nisin release and stability, resulting in an effective antibacterial system that can be used to preserve food and pharmaceuticals [20].

Chitosan in combination with nisin treatment appears to be a good method for preserving the quality of aquatic products during storage. Moisture loss management, volatile spoilage inhibition, total volatile basic nitrogen (TVB-N) reduction, total viable counts (TVC) growth control, color, and sensory acceptability maintenance were all achieved with 1% chitosan mixed with 0.6% nisin [21]. Li et al. [22] synthesized a new packaging made up of nanomaterials with an improved barrier of upgraded mechanical properties to preserve Chinese jujube at room temperature. Polyethylene was mixed with nanopowders like nano-Ag, kaolin, anatase TiO_2, and rutile TiO_2 to create the nano-packing material. When compared to standard packing material, the outcomes demonstrated that the new packaging which had nanomaterials had a significant positive impact on its physicochemical properties along with sensory quality [22]. Swaroop et al. [23] fostered a food packaging material with a blend of magnesium oxide (MgO) nanostructures and polylactic acid biopolymer, which showed that it had potent effects against bacterial biofilms [23].

2.3.1.3 Zinc oxide Nanoparticles

ZnO is an inorganic substance that is extensively used in everyday life, as well as a food additive. The Food and Drug Administration has classified ZnO as a GRAS material [24]. The release of Zn^{2+} antibacterial ions has been proposed as a plausible explanation for ZnO's toxicity to *Saccharomyces cerevisiae* [25]. The purpose of this research was to develop a gelatin-based nanocomposite incorporating chitosan nanofibres (CHNF) and zinc oxide nanoparticles (ZnONPs) and test its efficacy in increasing the shelf life of chicken fillets. During storage, the physical and chemical quality of chicken fillets and cheese samples were protected by the film [26]. ZnO in a homogeneous dispersion of liquid media, such as culture medium or any food, was found to be crucial for increasing microbial inhibition efficacy. In TSB (Tryptic Soy Broth) and milk, ZnONPs suppressed the development of *Escherichia coli* and *Staphylococcus aureus* in a preferential manner. Therefore nanoparticles containing ZnO formulations can be used for the area of food safety and other external reasons because of their antibacterial properties which can help to suppress biofilm formation [27].

2.3.1.4 Titanium Dioxide Nanoparticles

TiO_2-NPs are an active photocatalyst because they are known to be chemically stable, non-toxic, and cost-effective, as well as a potent antibacterial agent that has been approved by the US FDA. The usage of designed antimicrobial food packaging product decreases food product losses and maintains quality and safety during transit, storage, and sale [28].

2.3.1.5 Polymeric Nanoparticles

They are well suited to encapsulating bioactive chemicals such as vitamins and flavonoids and protect them from releasing in an acidic environment, such as the stomach [29]. Essential oils (EO) encapsulated in Polymeric poly ε-caprolactone, also known as (PCL) nanocapsules have proven to have better antibacterial action against foodborne pathogens than pure essential oils in the market, demonstrating the procedure efficacy of their encapsulation of essential oils in a nanometric structure. Those containing Thymus capitatus (Th-EO) were the most active among the *Origanum vulgare* (EO-NCs), owing to its distinct chemical makeup of bioactives [30]. The zein electrospun nanofibers are found to have low water resistance, limiting their use in food preservation. For *Agaricus bisporus* preservation, cinnamon essential oil (CEO) was encapsulated and was made to electrospun zein/ethyl cellulose (EC) hybrid nanofibers to improve the water resistance of nanofibers. The zein/EC nanofibers loaded CEO considerably reduced weight loss, maintained *Agaricus bisporus* stiffness, and improved it's quality during storage. Through a simple and customizable on-demand packaging procedure, it exhibited a considerable influence on extending the storage life of *Agaricus bisporus* [31]. To encapsulate thymol into poly lactide-co-glycolide (PLGA) because of its antibacterial behavior, coaxial electrospinning was used to generate nanofibers consisting of core-shell. Because PLGA successfully prevents thymol from volatilizing, the core-shell nanofibers containing thymol can gently evaporate into the atmosphere where the fruits and vegetables are kept. The film demonstrated outstanding antibacterial activity and fresh-keeping qualities, and it might be used in a variety of fruit preservation applications [32].

Sodium alginate (SA) and polyethyleneoxide (PEO) blended nanofibers were encapsulated with phlorotannin (Ph) using electrospinning. The ability of active nanofibers to protect chicken against *Salmonella enterica* was tested at 4 °C and 25 °C. At 4 °C, the cell count dropped from 6.20 to 3.28 Log CFU/g, and at 25 °C, it dropped from 8.80 to 2.53 Log CFU/g. Phlorotannin's method of action on bacteria was investigated, and it was discovered that the primary targets of phlorotannin was the cell membrane, along with ATP, protein, and genetic material DNA. On a whole, nanofibers increase the shelf life without compromising their sensory qualities [33].

2.3.1.6 Bimetallic and Trimetallic Nanoparticles

They are nanostructured materials with hybrid properties that improve physical properties like thermal, mechanical, and gas obstruction capabilities. Because of their environmentally friendly nature, hybrid NPs preserve the packaging and extend the shelf life of food, they also reduce the usage of regular plastics [34]. Silver-copper (Ag-Cu) alloy nanoparticles (NPs) were incorporated into a glycerol plasticized agar solution to create agar-based active nanocomposite films. The addition of this alloy to the agar films changed the color, transparency, and UV barrier qualities. The agar films with the nanocomposite exhibited strong antibacterial activity against both Gram-positive (*Listeria monocytogenes*) and Gram-negative (*Salmonella enterica sv typhimurium*) bacteria. They could be employed as food packaging materials to reduce foodborne pathogens and spoilage bacteria [35]. The fabrication and characterisation of bionanocomposite films based on fish skin gelatin (FSG) and bimetallic Ag-Cu nanoparticles (Ag-Cu NPs) were undertaken. Due to this property of the agar films, both Gram-positive and Gram-negative bacteria were eradicated. Overall, the films have the potential to be employed as active food packaging materials that control foodborne pathogens and spoilage bacteria [36].

2.4 Advantages and Issues with Nanomaterials in Food Safety

For life to exist beautifully and race to increase its progeny, food, shelter, and clothing are considered a minimum requirement. Indeed, it is actively studying food waste management through developing and understanding a questionnaire towards reducing waste management in food. Food waste management should be focused on knowing about it, why it happens, where is food being wasted, and how and when is it being wasted. Food safety and equal wisdom of food waste are no exception. One can probe food safety with the above questions [37].

Interestingly, we can think about what type of food is 'prepared,' where, why, when, and how it is 'prepared.' Researchers generally do not believe this line of probing is either low standard or insignificant. In the past, if one can see, there has been many an attempt to quantify global food waste and many an assessment on various food supply chains and envisioned to a much bigger picture. It has even been said that more than 60% of the actual cultivated food is gone waste before it can reach the waiting hands of the consumer [38].

In fact, food safety is all about using resources to ensure the safe preparation and preservation of foods and make them even safer for consumption. We should conduct active internal auditing to ensure external auditing happens with no bigger glitches. We should focus on increasing sales in foods and be ready to give away food products that are either expired or damaged. In more than a specific method, food wastage is a gripping problem in the rising scientific scenario [39].

Food safety is of national and international concern. We have to protect food from being damaged at the time of processing, handling, and distribution. This contamination can happen due to physical factors, chemical factors, and biological factors. Also, the safety, security, and longevity of food businesses and merchandise are as vital to our existence as to our environment. Moreover, in present-day life, food safety has become a severe concern due to the presence of pathogens, toxicants, pollutants, and contaminants offering to our health [1].

Nanomaterials, for that matter, are naturally occurring tiny materials created as the byproducts of combustion reactions or forced byproducts of engineering to perform a specialized function [40]. Nanotechnology in foods is very emerging and throws up a new world of opportunities in the food industry. Basic functionalities in the modern-day nanotechnology applications towards improving food packaging systems are, enhancing the hurdles of plastic materials, adding active compounds and components that deliver various functional attributes above orthodox active packaging. Packaging foods with nanomaterials will improve the shelf life of foods, improve food safety, and intimate consumers well in advance about the quality of food and its expected date of deterioration. The food is not contaminated or spoiled, packaging errors are repaired, and even preservatives are added to extend the life of the food in the package. By applying nanotechnology in the food industry, we can identify bacteria in packaging or produce more stringent flavors, improve color quality, and safety by increasing certain internal properties [41]. Nanotechnology offers enormous hope to provide benefits not just within food products but also around food products. Nanotechnology introduces chances anew for innovation in the food industry. That said, we should not neglect the uncertainty and health concerns that are simultaneously emerging and existing [1].

Additionally, the latest advancements in nanotechnology have changed the course of several scientific and industrial areas, including the food industry, and have cemented their place with the ever-increasing demand for use of nanoparticles in fields of varied scientific interests. They are food science and microbiology, processing technologies of nutrition, food packaging, functional food development, food safety, a shelf-life extension of food and food products.

Over the past years in the food industry, there has been a vast increase in the popularity of structures. Hence, research in this area should stand highly focused. Nanotechnology and its very application in food packaging and food safety are not new. In addition, promising results have been achieved in food preservation using nanomaterials to save foods from water vapor, fats, gases, bad flavors, and odors. Nanotechnology proffers excellent vehicle systems for delivering bioactive compounds to the target tissues. Despite many advances in nanotechnology every day, there are new challenges and opportunities towards enhancing and bolstering the current technology and issues about the aftermaths of nanotechnology that must address towards easing consumers' concerns. The openness of safety and environmental concerns should be a mandate while tackling nanotechnology in food particles. Hence, mandatory testing of nano foods is needed before we release them to the consumer market [29].

The following are four critical areas in the food industry where nanotechnology can be applied and explored (Fig. 2.1).

(a) Functional materials
(b) Nanoscale processing and microscale processing
(c) Development of products
(d) Food safety and biosecurity [42]

Within these areas, the possible applications of nanomaterials in the food industry are,

1. Using Nanosilver in packaging and on first-aid dressings.
2. Innovative packaging where we put detectors into packaging that provides data if the food inside has deteriorated.
3. We are facilitating the controlled release of active ingredients by controlling bioavailability and allowing flavors.
4. Improve functionality of food and micro-ingredients and novel processes in food manufacturing systems. Improved ingredient functionality [12].

Nanotechnology improves foods and, makes them more nutritious and healthier. It produces new food products, food packaging systems, and storage systems [43]. Many of these applications are at a starting stage, and we expect most of them to be high-value products in less time. There's a limit in food nanotechnology applications that are successful. We can use nanotechnology to increase flavor and texture, decrease fat content, or encapsulate nutrients like vitamins and see that they are not damaged during the shelf life of the food. In addition to this, nanomaterials can be used to keep the product inside the package fresher for a long time. By incorporating nanosensors in the nanopackaged food, we can provide consumers with information on the state of the food inside.

Food packages embedded with nanoparticles intimate consumers when a product is no longer safe for consumption. Sensors can raise an alarm before the food goes damaged and tell us of the nutritional status of the food at that point in time. Certainly, nanotechnology will reshape the fabrication of the entire packaging industry [44]. Simple packaging of food nonpackaging has been shared below for easy understanding.

Advancements in food nanotechnology offer exciting challenges for both the government and the private industries. The food processing industry should take care of the consumer confidence and acceptance of nano foods. Food regulatory bodies like the U.S. Food and Drug Administration (FDA), should author guidance concerning the criteria to evaluate the safety of food, the package of the food, and provide uses of nanomaterials with novel properties [41].

It is vital to note that nano foods originate in the laboratory. Therefore, they are not as same as conventional nano foods. Unfortunately, there has been an insufficient exploration of scientifically and naturally happening nanosystems and the advantages they provide. Thus, it is tough to make broad generalizations if nanotechnology is good, bad, or ugly. However, nanotechnology food packaging was estimated to be less troublesome than nanotechnology foods [41].

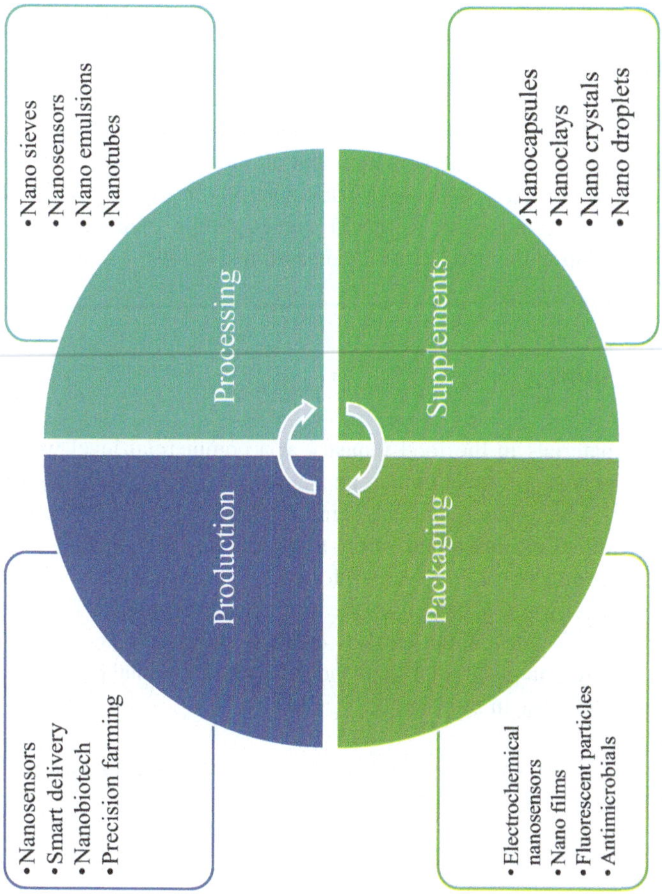

Fig. 2.1 An overview of the application of nanomaterials in the food industry

Also, nano foods are not named as they are, and consumers are not given an open option to avoid these based on their individual choices. Thus, compulsory testing of nano-modified foods is advisable before they are permitted to go into the market. New methods and standardized test procedures towards studying the value and role of nanoparticles on living cells are highly required to evaluate hazards of human exposure to nanoparticles. We expect on a wider range, that food products derived from nanotechnology will be available a lot to consumers in the coming years across the globe [29].

We are living in a fast-paced society eyeing perfection every second in technological advancements. We should focus on deducing the most suitable quarantine measures in avoiding and decreasing spoilage and wastage of food. We should only aim at producing food when it is vital, and we must ensure it's decrement of overproduction. Producing and procuring food and generating efficient waste alters the path of an internally contaminated environment of ours. We agree that there is very minimal progress in the assessments of nanomaterials in food safety. And it will impact the global food business. Research studies should focus on developing a guideline for regulating the security of nanomaterials [45, 46].

2.5 Conclusion

Applying nanomaterials in the food industry at a commercial level will increase multiple folds because of their unique and novel qualities. It will also increase human exposure to nanomaterials. That being the case, the effect of nanomaterials on food and health is of paramount worry to the public. The capacity to quantify nanomaterials throughout the life cycle of food is vital for consistency in manufacturing, safety, and prospective benefits of the final consumer product. Being accepted and acknowledged by the public depends on their safety. We should lay utmost importance on putting in place a constant and steady international regulatory framework for nanotechnology in foods.

Acknowledgments All the authors are thankful to JSS Academy of Higher Education & Research for the support towards research activities.

References

1. Singh RP. Utility of nanomaterials in food safety. In: Singh RL, Mondal S, editors. Food safety and human health [Internet]. Academic Press; 2019 [cited 2021 Oct 25]. p. 285–318. Available from: https://www.sciencedirect.com/science/article/pii/B9780128163337000114
2. Naseer B, Srivastava G, Qadri OS, Faridi SA, Islam RU, Younis K. Importance and health hazards of nanoparticles used in the food industry. Nanotechnol Rev. 2018;7(6):623–41.
3. Kumar P, Mahajan P, Kaur R, Gautam S. Nanotechnology and its challenges in the food sector: a review. Mater Today Chem. 2020;17:100332.

4. Sharma C, Dhiman R, Rokana N, Panwar H. Nanotechnology: an untapped resource for food packaging. Front Microbiol. 2017;8:1735.
5. Primožič M, Knez Ž, Leitgeb M. (Bio)nanotechnology in food science-food packaging. Nanomaterials (Basel). 2021;11(2):292.
6. McClements DJ, Xiao H. Is nano safe in foods? Establishing the factors impacting the gastrointestinal fate and toxicity of organic and inorganic food-grade nanoparticles. NPJ Sci Food. 2017;1:6.
7. Bergin IL, Witzmann FA. Nanoparticle toxicity by the gastrointestinal route: evidence and knowledge gaps. Int J Biomed Nanosci Nanotechnol. 2013;3(1–2): 163–210.
8. da Silva Santos V, Badan Ribeiro AP, Andrade Santana MH. Solid lipid nanoparticles as carriers for lipophilic compounds for applications in foods. Food Res Int. 2019;122:610–26.
9. Głąb TK, Boratyński J. Potential of casein as a carrier for biologically active agents. Top Curr Chem (Cham). 2017;375(4):71.
10. Sahoo M, Vishwakarma S, Panigrahi C, Kumar J. Nanotechnology: current applications and future scope in food. Food Front. 2021;2(1):3–22.
11. Sharifi S, Behzadi S, Laurent S, Forrest ML, Stroeve P, Mahmoudi M. Toxicity of nanomaterials. Chem Soc Rev. 2012;41(6):2323–43.
12. Pradhan N, Singh S, Ojha N, Shrivastava A, Barla A, Rai V, et al. Facets of nanotechnology as seen in food processing, packaging, and preservation industry. BioMed Res Int. 2015;2015:365672.
13. Nanotechnology in the food industry|Encyclopedia [Internet]. [cited 2021 Oct 25]. Available from: https://encyclopedia.pub/4228
14. Ahari H, Soufiani SP. Smart and active food packaging: insights in novel food packaging. Front Microbiol. 2021;12:657233.
15. Dakal TC, Kumar A, Majumdar RS, Yadav V. Mechanistic basis of antimicrobial actions of silver nanoparticles. Front Microbiol. 2016;7:1831.
16. Zhang J, Si G, Zou J, Fan R, Guo A, Wei X. Antimicrobial effects of silver nanoparticles synthesized by Fatsia japonica leaf extracts for preservation of citrus fruits. J Food Sci. 2017;82(8):1861–6.
17. Tan C, Han F, Zhang S, Li P, Shang N. Novel bio-based materials and applications in antimicrobial food packaging: recent advances and future trends. Int J Mol Sci. 2021;22(18):9663.
18. Food Preservative—an overview|ScienceDirect Topics [Internet]. [cited 2021 Oct 25]. Available from: https://www.sciencedirect.com/topics/food-science/food-preservative
19. Sharma N, Kapoor G, Neopaney B. Characterization of a new bacteriocin produced from a novel isolated strain of Bacillus lentus NG121. Antonie Van Leeuwenhoek. 2006;89(3–4):337–43.
20. Fahim HA, Khairalla AS, El-Gendy AO. Nanotechnology: a valuable strategy to improve bacteriocin formulations. Front Microbiol. 2016;7:1385.
21. Hui G, Liu W, Feng H, Li J, Gao Y. Effects of chitosan combined with nisin treatment on storage quality of large yellow croaker (Pseudosciaena crocea). Food Chem. 2016;203:276–82.
22. Li H, Li F, Wang L, Sheng J, Xin Z, Zhao L, et al. Effect of nano-packing on preservation quality of Chinese jujube (Ziziphus jujuba Mill. var. inermis (Bunge) Rehd). Food Chem. 2009;114:547–52.
23. Swaroop C, Shukla M. Nano-magnesium oxide reinforced polylactic acid biofilms for food packaging applications. Int J Biol Macromol. 2018;113:729–36.
24. Siddiqi KS, Ur Rahman A, Tajuddin null, Husen A. Properties of zinc oxide nanoparticles and their activity against microbes. Nanoscale Res Lett. 2018;13(1):141.
25. Kasemets K, Ivask A, Dubourguier H-C, Kahru A. Toxicity of nanoparticles of ZnO, CuO and TiO_2 to yeast Saccharomyces cerevisiae. Toxicol In Vitro. 2009;23(6):1116–22.
26. Amjadi S, Emaminia S, Nazari M, Davudian SH, Roufegarinejad L, Hamishehkar H. Application of reinforced ZnO nanoparticle-incorporated gelatin bionanocomposite film with chitosan nanofiber for packaging of chicken fillet and cheese as food models. Food Bioprocess Technol. 2019;12(7):1205–19.
27. Mirhosseini M, Firouzabadi F. Antibacterial activity of zinc oxide nanoparticle suspensions on food-borne pathogens. Int J Dairy Technol. 2013;66:291–5.

28. Daneshniya M, Maleki MH, Jalilvand Nezhad H, Jalali V, Behrouzian M. In: Application of titanium dioxide (TiO_2) nanoparticles in packaging and coating of food products; 2021.
29. Singh T, Shukla S, Kumar P, Wahla V, Bajpai VK, Rather IA. Application of nanotechnology in food science: perception and overview. Front Microbiol. 2017;8:1501.
30. Granata G, Stracquadanio S, Leonardi M, Napoli E, Consoli GML, Cafiso V, et al. Essential oils encapsulated in polymer-based nanocapsules as potential candidates for application in food preservation. Food Chem. 2018;269:286–92.
31. Niu B, Zhan L, Shao P, Xiang N, Sun P, Chen H, et al. Electrospinning of zein-ethyl cellulose hybrid nanofibers with improved water resistance for food preservation. Int J Biol Macromol. 2020;142:592–9.
32. Encapsulation of thymol in biodegradable nanofiber via coaxial eletrospinning and applications in fruit preservation. J Agric Food Chem [Internet]. [cited 2021 Oct 25]. Available from: https://pubs.acs.org/doi/10.1021/acs.jafc.8b06362
33. Duraiarasan S, Cui H, Lin L. Encapsulation of Phlorotannin in alginate/PEO blended nanofibers to preserve chicken meat from Salmonella contaminations. Food Packag Shelf Life. 2019;21:100346.
34. Bimetallic and trimetallic nanoparticles for active food packaging applications: a review|SpringerLink [Internet]. [cited 2021 Oct 25]. Available from: https://link.springer.com/article/10.1007/s11947-019-02370-3
35. Arfat YA, Ahmed J, Jacob H. Preparation and characterization of agar-based nanocomposite films reinforced with bimetallic (Ag-Cu) alloy nanoparticles. Carbohydr Polym. 2017;155:382–90.
36. Pan S, Goudoulas TB, Jeevanandam J, Tan KX, Chowdhury S, Danquah MK. Therapeutic applications of metal and metal-oxide nanoparticles: dermato-cosmetic perspectives. Front Bioeng Biotechnol. 2021;9:724499.
37. Paritosh K, Kushwaha SK, Yadav M, Pareek N, Chawade A, Vivekanand V. Food waste to energy: an overview of sustainable approaches for food waste management and nutrient recycling. BioMed Res Int. 2017;2017:2370927.
38. Vågsholm I, Arzoomand NS, Boqvist S. Food security, safety, and sustainability—getting the trade-offs right. Front Sustain Food Syst. 2020;4:16.
39. Schanes K, Dobernig K, Gözet B, Food waste matters - A systematic review of household food waste practices and their policy implications. Journal of Cleaner Production, 2018;182(1): 978-991. https://doi.org/10.1016/j.jclepro.2018.02.030
40. Jeevanandam J, Barhoum A, Chan YS, Dufresne A, Danquah MK. Review on nanoparticles and nanostructured materials: history, sources, toxicity and regulations. Beilstein J Nanotechnol. 2018;9:1050–74.
41. Sekhon BS. Food nanotechnology—an overview. Nanotechnol Sci Appl. 2010;3:1–15.
42. Functional materials in food nanotechnology—Weiss—2006—J Food Sci—Wiley Online Library [Internet]. [cited 2021 Oct 26]. Available from: https://ift.onlinelibrary.wiley.com/doi/10.1111/j.1750-3841.2006.00195.x
43. Wahab A, Rahim AA, Hassan S, Egbuna C, Manzoor MF, Okere KJ, Walag AMP. Application of nanotechnology in the packaging of edible materials. In: Egbuna C, Mishra AP, Goyal MR, editors. Preparation of phytopharmaceuticals for the management of disorders. Cambridge: Academic; 2021. p. 215–25.
44. Kumar V, Guleria P, Mehta SK. Nanosensors for food quality and safety assessment. Environ Chem Lett. 2017;15(2):165–77.
45. McClements DJ. Nanotechnology approaches for improving the healthiness and sustainability of the modern food supply. ACS Omega. 2020;5(46):29623–30.
46. Ameta SK, Rai AK, Hiran D, Ameta R, Ameta SC. Use of nanomaterials in food science. In: Ghorbanpour M, Bhargava P, Varma A, Choudhary DK, editors. Biogenic nano-particles and their use in agro-ecosystems [Internet]. Singapore: Springer Singapore; 2020 [cited 2021 Oct 26]. p. 457–88. Available from: http://link.springer.com/10.1007/978-981-15-2985-6_24

Chapter 3
Use of Nanotechnology for the Improvement of Sensory Attributes of Foods

Neelesh Kumar Nema, Nayana Rajan, Sachithra Sabu, Swapnil Devidas Khamborkar, Smitha Sarojam, Linson Cheruveettil Sajan, Marin Babu, Aeena Peter, Baby Kumaranthara Chacko, and Viju Jacob

Abbreviations

CNC	Cellulose nanocrystals
CNF	Cellulose nanofiber
CNs	Cellulose nanomaterials
IFIC	International Food Information Council
LNC	Lipid nanocapsules
MNPs	Metal based-nanoparticles

3.1 Introduction

Food has a multifaceted range of meanings without a single definition. However, according to the US Department of Health and Human Services, an item considered food has to have a minimal nutritional value. More precisely, "food items are physical entities comprised of biological components that give energy inside the body and are employed to repair and improve body functions [1]. Food protects the body from disease and regulates body functions as well. Over the past few decades, science-driven food research and newer technologies have revolutionized the food sector in addressing the challenges associated with a balanced diet by 'being edible' and 'being nutritive', which are the key requirements in the food processing industries. Food research is a systematized and applied tool for investigating and

N. K. Nema (✉) · N. Rajan · S. Sabu · S. D. Khamborkar · S. Sarojam · L. C. Sajan · M. Babu · A. Peter · B. K. Chacko · V. Jacob
Nutraceuticals Division, C.V.J. Creative Centre-Bioingredients, Synthite Industries Pvt. Ltd., Ernakulum, Kerala, India
e-mail: neeleshk@synthite.com; nayana@synthite.com; sachithra@synthite.com; swapnil@synthite.com; smithas@synthite.com; linson@synthite.com; merinbabu@synthite.com; aneena@synthite.com; babykc@synthite.com; viju@synthite.comm

C. Egbuna et al. (eds.), *Application of Nanotechnology in Food Science, Processing and Packaging*, https://doi.org/10.1007/978-3-030-98820-3_3

compiling information about foods and their components with the help of multidisciplinary fields, e.g. chemistry, biochemistry, nutrition, microbiology, and engineering, while food technology is the applicable food science knowledge for selecting, processing, preserving, packaging, and distributing safe food without distressing its nutritive values [2]. Selection of good food and its processing to make quality products are two exercises that can be improved using applied technologies. These technologies can improve the product's characteristics, nutritive value, shelf-life, modification of food structures and textures, controlling of biochemical, microbiological, chemical change, and traceability of the food products, which are foremost requirements in the food industry.

Nanotechnology helps to develop products like nanoparticles, nanoemulsions, nanofibers, nanocomposites, nanolaminates, nanocapsules, and similar nanocapsulation structures. It is one of the fast-emergent multidisciplinary technologies to develop food ingredients, food additives, carriers for nutrients/supplements, and food contact materials at a molecular level. This technology offers a wide array of benefits in terms of quality to the customer, including novel characteristics and improved sensory attributes, e.g. improved appearance, original tastes, novel textures, fresh sensations, minimum extent of fat, enhanced nutrients absorptivity, upgraded packaging to ensure microbiological safety, etc. [3–6]. The word "nano" comes from the Latin word "nanus" and the Greek word "ν'ανς,", which indicate a person of very low height, i.e., a dwarf. However, today it is used at the level of a nanometer (nm) or an atomic level. Atom(s) or molecule(s) of some elements, compounds, and/or macromolecules aggregate in the matrix at a specific environment, and form nanomaterials of a size ranging from 1–1000 nm and acquire some physical and chemical characteristics that are quite dissimilar to those of an individual molecule or bulk material [7].

Nano-sized materials (characterized dimensionally from a geometrical standpoint) such as quantum dots, nanoparticles (Zero dimension-0D), nanofibers, nanotubes (One dimension-1D), nano-thin films (Two dimension-2D order), and nanopatterned bulk structures (Three dimension-3D) [5] have unique properties to modulate physicochemical properties such as physical appearance, color intensity, and odor characteristics.

The sensory attributes are often used in practice by consumers to evaluate food quality with reference standards while purchasing any food product. According to the International Food Information Council (IFIC, 2021), taste, price, and healthfulness are the top-ranked purchase drivers and/or characteristics when selecting any food substance [3, 8]. Therefore, it is very vital to utilize sensory evaluation methods, when developing any new food products or introducing any of the processing technologies [9]. Sample size, product type, and type of products should be taken into account when selecting test methods [10] to evaluate sensory characteristics with the help of analytical tools [11].

The chapter also summarizes how nanotechnology plays a significant role at the molecular level in improving the quality of food products using their sensory acceptance and major gaps in knowledge that require further research.

3.2 Quality of Foodstuffs and Sensory Attributes

Quality attributes such as physical, chemical, microbiological, and organoleptic (sensory) evaluations are the foremost procedures to evaluate food quality. Sensory evaluation can be performed through sensory receptors that are broadly classified into four categories and respond to primary stimuli:

1. Light (photoreceptors)
2. Chemicals (chemoreceptors)
3. Pressure (mechanoreceptors)
4. Temperature (thermoreceptors)

Each sensory receptor is tuned to detect specific characteristics of food by performing a specific sensory function (Fig. 3.1). A scientific discipline is used to evoke, measure, analyze, and interpret reactions to those characteristics of food and materials as they are perceived by the senses of sight, smell, taste, touch, and hearing [12]. In the course of vision, photoreceptors activate the rod and cone and respond to the brain for light intensity, color, clarity (visual range of light wavelengths), geometric shape (square, circular, etc.), and size (length, height, thickness, width), etc. During smell, olfactory receptors recognize molecular features of floating particles through chemoreceptors and help detect the aroma. In hearing, eardrums detect vibrations and recognize the sound with mechanoreceptors, which help perceive different sounds that are characteristic of foodstuffs. On the other hand, mechanoreceptors that are present in the skin and other tissues also respond to touch and countless pressures and further indicate consistency, e.g., roughness, evenness, and surface

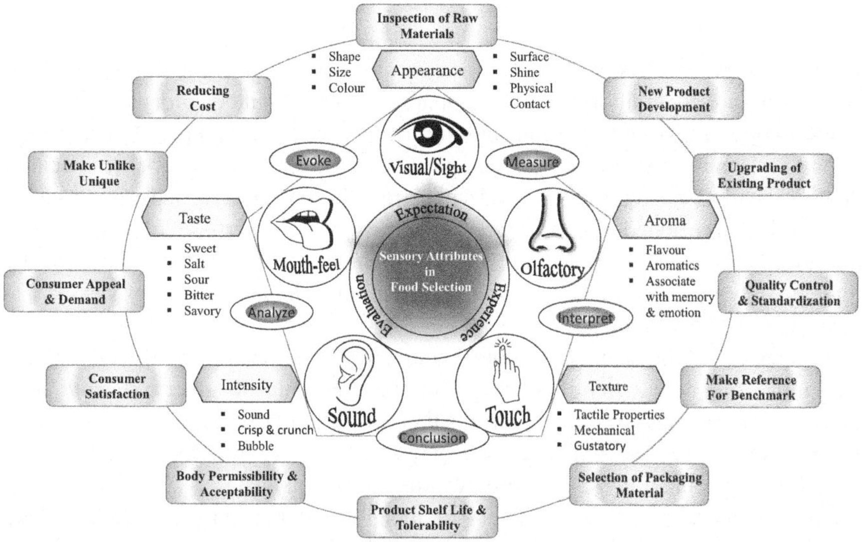

Fig. 3.1 Sensory receptors and detection of specific characteristics of food

characteristics (hardness, cohesiveness, adhesiveness, denseness, springiness); in the same region, thermoreceptors help to recognize temperature gradations. In the taste perspective, sensory and motor nerves available in taste buds detect the potentiality of gustatory perceptions including sweetness, bitterness, sourness, saltiness, and umami (savory taste) with various tastes, e.g. spice heat/pungency, cooling/chilling, sharp astringency, and metallic/irony.

3.2.1 Food Sensory Analysis

A wide range of sensory techniques, including analytical sensory testing, discrimination testing, descriptive testing, effective analysis, preference testing, acceptance testing, and conjoint analysis, are available for evaluating the quality of foodstuffs that are based on the principles of experimental design and statistical analysis.

3.2.2 Sensory Attributes Improved by Nanotechnology

Currently, nanostructured principles are very much used in food applications, especially nano-ingredients to make food formulations and pack food items.

Some of the examples of sensory attributes are presented in Table 3.1.

3.3 Nanoparticles

Particles in the size range of 1–100 nm on a nanometric scale with one or more dimensions are generally known as nanoparticles. They could be organic, inorganic, or carbon-based particles that improve surface properties as compared to the larger sizes of respective materials. The nanoparticles show enhanced properties because of their small size, such as high reactivity, strength, surface area, sensitivity, stability, etc. They are added to many foods to improve flow properties, color, and stability during processing or to increase shelf life. The nanoparticles are synthesized by three main types of methods, i.e., physical, chemical, and mechanical processes that have improved over time.

Hesperetin, a flavanone, is a potent antioxidant that cannot be directly mixed with milk or applied with milk to make milk beverages due to adverse sensory attributes. Nanostructured hesperetin encapsulated with lipid carriers is a possible solution that overcomes the associated sensory issues in dairy beverages by increasing solubility, reducing bitterness, and preventing color change [13].

In food packaging, nanopolymers or metal-based nanoparticles (M-NPs) are used in general practice instead of conventional materials. Food packaging materials made of M-NPs do not allow moisture and oxygen into the food-stuff and

Table 3.1 Nanotechnology with improving sensory attributes of foods

Product category	Characteristics	Application in food processing	Example	Benefits	Ref.
Nanoparticles	– Tiny materials – Nanoscale size having ranged from 1 to 100 nm – Zero dimension – Possess unique physical and chemical properties – High surface area	– Gelation and viscosifying agent – Food additive/ supplement – Used as sensors to detect contamination and standardize the food environment	– The nano-selenium-enriched Nano tea (Shenzhen Become Industry & Trade Co., Ltd. China)-Supplement for Selenium – Nanoslim beverages with liquid suspended nanoparticles for low calories diet – Food-grade ZnO and TiO_2 nanoparticles (maintain color; whitening and texture modifier in candies and ice-creams) – Gold nanoparticles AuNPs maintain food texture and longevity	– Enhanced gastrointestinal uptake of metal – Nanoslim beverages offer a low-calorie diet – Silver – Nanoparticles-Antibacterial wheat flour (WPI ACC NO: 2006-489267/200650) by Park KH, South Korea	[49, 50]
Nanoencapsulation	– Core or active, within a matrix or shell to form nanocapsules – Encapsulation at the nano level	– Offer better absorption – shelf life and target delivery – As flavor enhancer – Used in food fortification	– Nanoencapsulated fish oil can be used for fortification of low fat probiotic fermented milk – Utilized in fortifying yogurt to reduce acidity, syneresis and peroxide value upon storage – Rutin (dietary flavonoid) encapsulated with ferritin nanocages improved the thermal, UV radiation, and firmness properties	– Protection of products from oxidation – Preservation until intestinal absorption	[14, 51, 52]

(continued)

Table 3.1 (continued)

Product category	Characteristics	Application in food processing	Example	Benefits	Ref.
Nanoemulsions	– Isotropic dispersed systems of two non-miscible liquids – Colloidal systems – Forming droplets of nanometric sizes – Oily system dispersed in an aqueous system, or an aqueous system dispersed in an oily system	– Offer better availability and dispersion – Optically clear – In the edible coatings to enhance the functionality and shelf life of the food products – Reduces the amount of required surfactant – Modulate product texture	– Nanoemulsions in the form of carbohydrates or proteins – Have improved the texture that aids in ice cream uniformity – Nanoemulsion-based fortified beverages with nutrients – Nanoemulsions of lemongrass oil and sodium alginate, edible coatings in fresh-cut apples	– Improving digestibility – Nanoemulsions containing natural antimicrobials, such as EOs – Easy absorptivity	[53, 54]
Nanocapsule	– Vesicular systems are made of a polymeric membrane that encapsulates an inner liquid core – Shell made from a nontoxic polymer – Size at the nanoscale level	– Masks the off flavor	– Nanocapsules containing tuna fish oil in "Tip-Top" Up bread	– Source of omega 3 fatty acids – The microcapsules are designed to break open only when they reach the stomach, thus avoiding the unpleasant taste of the fish oil	[55, 56]

Nanofibers	– Fibers with diameters less than 100 nm – Made up of carbon and polymers – Electrospinning process	– Act as a nanocarrier system – Improves shelf life – Improve physical appearance – Stabilizing food emulsions – Inhibiting syneresis or oil separation	– Stabilization of food emulsions – Gelatin nanofiber© – Thermally stable nanocarrier e.g.: Zein nanofibers used as a nanocarrier for β-carotene, a potential food colorant – Offer aroma retention in encapsulated products e.g.: Saffron extract encapsulated in zein/tragacanth core-shell to make thermally stable products	– Increasing the viability of the probiotics and facilitating their handling – Making functional foods	[5, 38]
Nanocomposites	– A combination of nanoparticles with polymers	– Offer high barrier properties – Improves shelf life	– Used in carbonated beverages bottles as gas barriers so that carbon dioxide leakage can be minimized – Keeping beverages fresh till consumption	– Minimize environmental impact	[57]
Nanolaminates	– Thin films formed by two or more layers of food-grade materials – Deposited on a substrate through the layer-by-layer assembly technique	– Used in edible films – Improve the textural properties – Serve as carriers of colors, flavours, antioxidants, nutrients, and antimicrobials	– Significantly – Good to protect food from moisture – Edible films are present in a wide variety of foods; fruits, vegetables, meats, chocolate, candies, baked goods, and French fries – Helps to extend the shelf life of cheese	– It is possible to design nano-laminated coatings that can control the digestibility of encapsulated lipids. These coatings could be used to maintain – Decrease or control the release of bioactive lipophilic components within the human GI tract	[2, 58, 59]

protectvthe food products from oxidation while retaining their food properties. Apart from packaging materials, M-NPs are now also integrated with some types of favors, antioxidants, antimicrobial, and preserving agents. Laminated and coated M-NPs packing materials reduce food losses due to preventing different microbial growth and infections and, sometimes, detection of food spoilage organisms, UV protection activities. The gold-coated NPs increase the visualization proficiency to obtain shiny and quality effects. Other particles like silver (Ag), zinc (Zn) oxide, iron (Fe) oxide, magnesium (Mg) oxide, and zerovalent iron have been exposed for their antimicrobial properties, which protect the food quality and keep the food healthy with a fresh texture [14]. Titanium dioxide (TiO2-common food whitener and brightener additive) and silicon (Si) dioxide are used in a variety of applications in the food industry for sensory improvement. Alumino-silicate nano-materials are commonly used as anti-caking agents in granular or powdered processed foods [15]. Ag, TiO2 (1%) coated on low-density polythene, maintains the quality and shelf life. The shelf life of fresh-cut carrots can be prolonged by controlling both microbial and sensory quality during refrigerated storage.

3.4 Nanoencapsulation

The unique technology embeds small particles or miniatures referred to as "core" or "bioactive" compounds (BACs) of solid, liquid, or gas into the secondary material termed as the matrix, shell, or inert material. Finally, it forms a small capsule at the nano level to protect or contrary to deteriorating environmental conditions such as high temperatures, oxidation, photosensitivity, light impact, moisture effect, pH variations, interactions with associated compounds, etc. known as nanoencapsulation technology [16, 17].

The technology delivers core or active in a controlled manner with an appropriate rate at a specific targeted site through partitioning, dissolution, osmosis, diffusion, swelling, and erosion methods. As the matrix protects the core, it improves the stability of ingredients and enhances the shelf-life of finished food products. The encapsulation technique can improve the nutritional content of food without affecting sensory attributes or enhancing the organoleptic performance, e.g. taste, aroma, the texture of food, masking off-flavors, etc. [16–20].

This technique may also transform liquid ingredients into free-flowing powders for easy handling, and incorporation into dry foodstuffs. Carbohydrate, protein, or lipid-based alternatives such as chitosan, peptide–chitosan, and β-lactoglobulin nanoparticles (NPs), or emulsion biopolymer complexes are the most popular choices used in food and nutraceutical applications. Technologies can be categorized into five groups: those used to make nanocapsules in the food industry; lipid-based techniques; nature-inspired techniques; specialized equipment techniques, biopolymer-based techniques; and other miscellaneous techniques [17, 18, 21]. In lipid-based techniques, nanoencapsulation is classified prominently into nano-emulsions, nano-structured phospholipid carriers, and nano-lipid carriers.

Rutin is considered a dietary flavonoid with numerous significant pharmacological properties. Its use in the food industry is limited due to its low solubility.

3.4.1 Nano-emulsions

To make nanoencapsulation, nanoemulsions are considered ideal vehicles either in the form of oil in water emulsions (O/W) that carry lipophilic bioactive agents such as plant sterols, carotenoids (like β-carotene), α-tocopherol, dietary fats, nutritional oils, or a combination of these actives surrounded by a hydrophilic matrix (oil phase dispersed in an aqueous phase) or water in oil emulsions (W/O) that encapsulate hydrophilic compounds, e.g. xanthophyll, vitamins, and polyphenols surrounded by an aqueous system dispersed in an oil system) [22, 23].

They are isotropic colloidal kinetically stable systems where the particle size typically ranges from 50 to 500 nm [24]. Food nanoemulsions have an extensive application in the food manufacturing industry because of their high ability to digest, improved bioavailability, and enhanced stability with extended shelf life. Nanoemulsions could be a fat replacer to make low-fat products with a creamy texture such as mayonnaise, spreads, and ice cream [9, 25]. These are examples of improving the texture and physical appearance. For example, nanoemulsions can potentially be used in muscle food products to improve the quality of muscle food while inhibiting lipid oxidation and reducing physiological changes during long-term storage with extended shelf life [25]. Chicken breast fillets coated with nanoemulsion-based edible sodium caseinate and ginger essential oil (3 and 6%) reduced the color difference and cooking loss [26]. Flaxseed oil was pre-emulsified and encapsulated with cross-linker genipin, placed on PET trays, and overwrapped with PVC film before being stored at 4 °C in the dark for 3 days, changing the structure of the meat matrix and texture. It also influenced the physical characteristics of chicken sausages [27]. In another example, olive oil-based nanoemulsions at various concentrations (15, 30, or 45%) modified the sensory, chemical, and microbiological quality of rainbow trout (*Oncorhynchus mykiss*) fillets. The nanoemulsions masked the strong fishy odor of rainbow trout in storage and improved the organoleptic properties such as taste, odor, and texture. Out of three, 30 and 45% of olive oil nanoemulsions extended the shelf life by 6 days with improved physical characteristics [28]. In another study, the fishy odor of rainbow trout fillets was also reduced to a maximum extent while applying the nanoemulsions, which contained orange, grapefruit, and mandarin essential oils, and kept in a refrigerated condition at 4 ± 2 °C.

Aside from the aforementioned, a nanoemulsion made of chitosan solution (1% w/v) and citrus essential oil applied to silvery pomfret (*Pampus argenteus*) fillets and stored in air-proof polyethylene pouches at 4 ± 0.3 °C for 16 days increases the shelf life from 12 to 16 days while improving sensory scores [29]. Improved organoleptic quality, reduced biochemical changes (TVB-N, FFA, and PV), and improved shelf life were observed for up to 16 days [30]. Oil-in-water nanoemulsion prepared with canola oil and coated on pork patties improved the physicochemical and sensory characteristics, especially the tenderness [31]. Turkey breast meat coated with nano emulsified mustard (*Brassica juncea*) essential oil (NME) combined with gelatin/hydroxypropyl-β-cyclodextrin, improved sensory attributes, especially the odor and overall consumer acceptability. The nano emulsified coating also improved overall quality and extended the shelf life of turkey meat [32] and provided a better sensory experience with healthier food options for the customers. Despite the above-revealed muscle food applications, nanoemulsions can also be used in vegetable preparations to improve sensory attributes. Desirable sensory attributes can be achieved by replacing objectionable ingredients with nanoemulsions and getting the expected values. Nanoemulsions can manage lipid constancy, stability, oxidation rancidity, and also offer an extended shelf-life of the products. This novel technology, with its small droplet size and associated large droplet surface area [33] opens up new possibilities for the food industry.

3.4.2 Nano-Lipid Carriers

Novel vitamin D_3-loaded lipid nanocapsules (LNC) are a great example of a nanostructured molecule, prepared with nanotechnology for fortification of milk without disturbing any of the sensory properties. LNCs were made of three main components: an oily phase, an aqueous phase containing NaCl, and a non-ionic surfactant. In this study, the lipid nanoparticles extended the release pattern of vitamin D_3 in fortified milk [34].

3.5 Nanofibers

Nanofibers are categorized as any slender, elongated, threadlike object or structures (natural or synthetic) having at least one dimension characteristic with a diameter of between 50 and 300 nm. They provide physical, chemical, biological, and mechanical properties, depending on the size, shape, and composition of the material. Nanofibers are commonly made from polymers, carbon, and semiconductor materials for a variety of applications [35, 36]. Bi-component extrusion, phase separation, template synthesis, drawing, melt blowing, electrospinning, and centrifugal spinning are several techniques available for fabricating nanofibers [37, 38]. Electrospun nanofibers, on the other hand, are popular due to their ability to exhibit a variety of interesting properties such as high porosity, interconnectivity, large surface area per unit mass, high gas permeability, and small interfibrous pore size [39]. Concentration, electrical conductivity, surface tension, rheological properties, applied voltage, and feed rate are the important parameters by which desirable properties for specific purposes, e.g., regular shape, more stable, and diffusive nanofibers can be achieved. Due to their rheological properties, natural polymers such as collagen, gelatin, chitosan, and hyaluronic acid are generally used to enhance the elasticity, stability, and consistency of food products [38].

Cellulose nanomaterials (CNs) including cellulose nanofibers (CNF), cellulose nanocrystals (CNC), and chitin, are the most common nanostructures used in food applications. Cellulose nanofibers are frequently used in fruit coatings for improving barriers, retaining moisture content, stability, and other functional properties. In one study, it was found that cellulose nanofiber (CNF) and soy protein isolate (SPI) make a stable protein-polysaccharide complex that improves the rheological and textural properties of the individual and is offered as a fat replacer (viable fat substitute) in ice-cream industries with low-calorie, low-fat, and anti-melting texture. It was also noticed that an appropriate combination of natural cellulose nanofiber with soy protein isolate does replace an equivalent amount of cream in ice-cream making and improves the strength in terms of hardness, adhesiveness, springiness, cohesiveness, gumminess, chewiness, and resilience when evaluated by instrumental texture profile analysis (TAXT2i). The resultant SPI–CNF mixture is a great example of a fat replacer (replacing 10% of the cream) in ice-cream preparation, where nanofibers mimic the instrumental texture profile by reducing fat content [40]. Similarly, cellulose nanofibers were utilized as fat substitutes in emulsified sausage [41]. In the sausage, a total of 30% and 50% of the original fat were substituted by palm oil Pickering emulsion (CPOE-pre-emulsified by 1 wt% CNF) alone, independently and in combination with cellulose nanofiber (CNF-1% aqueous dispersion) at a ratio of 1:1 (water: oil, v/v) in both percentages separately and observed for sensory evaluation for overall acceptability. Compared to a full-fat product, there was no significant difference in both the products for color intensity, whereas the flavor sensory attribute was found to have changed significantly. Compared to full-fat, 30% fat sausages substituted by CPOE showed good texture, which was further upgraded by the addition of cellulose nanofibers (CNF-1%) to achieve better hardness, springiness, and chewiness. According to the rheology and scanning electron microscope results, 50% substituted fat by a combination of CPOE and CNF-1% showed an increased elastic and compact effect, but overall acceptability and higher sensory scores were favored for 30% substituted fat with cellulose nanofiber and would be a good candidate for low-fat meat products [9, 41]. Chitin, a long-chain polymer composed of N-acetyl-d-glucosamine groups linked by β (1→4) linkages, is the second most abundant polysaccharide in nature after cellulose polysaccharide, formed into chitin nanofibers (CNFs) with the help of protein matrices. It's a biodegradable, biocompatible, non-toxic, and environmentally friendly biomaterial. Ultrasonication is one of the most effective methods to form it as compared to electrospinning, TEMPO-mediated oxidation, acid hydrolysis, grinding, the starburst system, dynamic high-pressure homogenization, and microfluidization [42]. These fibers absorb the chloride ion from the NaCl matrix and increase the level of free sodium to change the taste in the mouth and improve sensory attributes associated with salty taste and flavor. High salt is not good for the healthy body and, therefore, salt-intake reduction methods are gradually in high demand, e.g., using substitutes such as KCl, $CaCl_2$, and $MgSO_4$, or else by adding flavor enhancers such as citric acid, spices, and monosodium glutamate. In continuation, CNFs could be a good alternative to enhance the saltiness and viscosity of products like salted mackerel, soy sauce, seasoning of instant food, etc. [42].

Dextran (a bacterial polysaccharide consisting of R-l, 6 linked d-glucopyranose residues with some R-1, 2-, R-1, 3-, or R-1, 4-linked side chains) is another molecule that makes nanofibers with hydrophobic molecules where vitamin E is uniformly distributed within the polymer fiber matrix. In the Dextran nanofibers, the amorphous structure of dextran fibers entraps vitamin E without any chemical interactions inside. This matrix acts as a novel carrier and helps deliver vitamin E in cheese fortification. Sensory analysis of the fortified product shows that this carrier has better acceptability and texture adequacy in cheese making as compared to blank and direct fortified samples. These nanofibers can also hold water and help to increase cheese firmness. In conclusion, dextran ultrathin nanofibers could be a good candidate for entrapping hydrophobic compounds by the electrospinning method and utilizing the design of novel functional foods with high sensory texture, homogeneity, and firmness for long shelf life [39].

The above examples highlight that nanofibers have important implications in the food industry and can be alternative sources to improve the sensory attributes in terms of texture, taste, and flavor.

3.6 Polymer Nanocomposites (PNCS)

Apart from nanoparticles, nanocomposites and nanolaminates are also extensively used for the packaging of food to maintain texture, physical appearance, and freshness. A composite is a commercially prepared polymer mixture made by joining or combining two or more geometries substances (fibers, flakes, spheres, particulates) with a nanometric range (nano-polymers or nanoparticles) of at least one-dimensional shape to obtain tailored physical properties than the individual ingredients. Exhibit different barrier properties. Nanocomposites resist thermal stress, support transportation, and increase the shelf life while storing food products [14]. Clays and other silicate materials, for example, montmorillonite, kaolinite, hectorite and saponite [43], polymer/Carbonaceous nanocomposite [44, 45], and polymer/silica [poly(3-hydroxybutyrate-co-3- hydroxyl-valerate) (PHBV) nanocomposite fused with silica] [46], graphene [47] are the principal PNCs composite used in food packing for flame resistance, improved strength, migration issues, recyclability aspects, toxicity, and consumer acceptability [48]. Magnesium oxide (MgO)/PLA composite improves 25% gas barrier properties and improves the mechanical properties to make the food fresh, and healthy. Currently, nanocomposites are also used in bottles of beer to enhance the shelf life of up to 6 months [48].

3.7 Conclusion

Currently, nanostructured principles are very much used in food applications, especially nano-ingredients to make food formulations, and pack food items. Food nanotechnology plays a prominent role in obtaining high-quality foods of desired sensory characteristics and functionality without distressing nutritive values. Indispensable sensory evaluation, such as taste, smell, and texture, as well as a physical appearance with the combination of multiple tests, are key disciplines by which consumers interpret food quality about the standards reference. Nanostructured foods, e.g. nanoparticles, nanoemulsions, nanoencapsulation, and nanofibers, are the pathfinding solutions for the food manufacturing industries to make quality food that meets the customer's requirements and delivers unique sensory experiences without affecting the nutritive values. These technologies can be an alternate way out as compared to conventional food processes for making long-lasting nutritious food with a high self-life without compromising the sensory effect desired by consumers. In upcoming days, these future generation technologies can also be involved in all the food handling steps from beginning to end, including the selection of food ingredients, processing, conclusive recipe-formulations, decoration, packing, and stability, including transportation and storage, etc.

Acknowledgments The authors acknowledge the Management of Synthite Industries Pvt. Ltd., India for their support and inspiration.

References

1. Borghini A, Piras N. Food and foods. Toward a definition. Riv Ital Filos Linguaggio. 2019:384–92. https://doi.org/10.4396/SFL2019ES06.
2. Sekhon BS. Food nanotechnology—an overview. Nanotechnol Sci Appl. 2010;3:1–15.
3. Chaudhry Q, Scotter M, Blackburn J, Ross B, Boxall A, Castle L, et al. Applications and implications of nanotechnologies for the food sector. Food Addit Contam A Chem Anal Control Expo Risk Assess. 2008;25:241–58. https://doi.org/10.1080/02652030701744538.
4. Yu H, Park JY, Kwon CW, Hong SC, Park KM, Chang PS. An overview of nanotechnology in food science: preparative methods, practical applications, and safety. J Chem. 2018; 2018: 1-10. https://doi.org/10.1155/2018/5427978.
5. Leena MM, Yoha KS, Moses JA, Anandharamakrishnan C. Nanofibers in food applications. Innovative Food Processing Technologies,A Comprehensive Review, Amsterdam: Elsevier; 2021, p. 634-50. https://doi.org/10.1016/b978-0-08-100596-5.22952-1
6. Rashidi L, Khosravi-Darani K. The applications of nanotechnology in food industry. Crit Rev Food Sci Nutr. 2011;51:723–30. https://doi.org/10.1080/10408391003785417.
7. Trotta F, Mele A. Nanomaterials: classification and properties. Nanosponges: Synthesis and Applications, 1st ed. Germany: Wiley-VCH Verlag GmbH & Co. KGaA; 2019, p. 1-26. https://doi.org/10.1002/9783527341009.ch1.
8. International Food Information Council. Perceptions on processed: consumer sentiment and purchasing habits; 2021.
9. Sun R, Lu J, Nolden A. Nanostructured foods for improved sensory attributes. Trends Food Sci Technol. 2021;108:281–6. https://doi.org/10.1016/j.tifs.2021.01.011.

10. Hein KA, Jaeger SR, Tom Carr B, Delahunty CM. Comparison of five common acceptance and preference methods. Food Qual Prefer. 2008;19:651–61. https://doi.org/10.1016/j.foodqual.2008.06.001.
11. Marsili R. Sensory-directed flavor analysis, vol. 148. Boca Raton: CRC Press; 2006. https://doi.org/10.1201/9781420017045.
12. Sidel JL, Stone H. The role of sensory evaluation in the food industry. Food Qual Prefer. 1993;4:65–73. https://doi.org/10.1016/0950-3293(93)90314-V.
13. Molina G, Pelissari IFM, Asiri AM. Food applications of nanotechnology. 1st ed. Boca Raton: CRC Press; 2019. https://doi.org/10.1201/9780429297038.
14. Shafiq M, Anjum S, Hano C, Anjum I, Abbasi BH. An overview of the applications of nanomaterials and nanodevices in the food industry. Foods. 2020;9:1–27. https://doi.org/10.3390/foods9020148.
15. Samal D. Use of nanotechnology in food industry: a review. Int J Environ Agric Biotechnol. 2017;2:2270–8. https://doi.org/10.22161/ijeab/2.4.90.
16. Augustin MA, Hemar Y. Nano- and micro-structured assemblies for encapsulation of food ingredients. Chem Soc Rev. 2009;38:902–12. https://doi.org/10.1039/b801739p.
17. Assadpour E, Jafari SM. Nanoencapsulation: techniques and developments for food applications. Nanomaterials for food applications. Amsterdam: Elsevier; 2019. p. 35–61. https://doi.org/10.1016/B978-0-12-814130-4.00003-8.
18. Jafari SM, Faridi Esfanjani A, Katouzian I, Assadpour E. Release, characterization, and safety of nanoencapsulated food ingredients. Elsevier Inc.; 2017. https://doi.org/10.1016/b978-0-12-809740-3.00010-6.
19. Yada RY, Buck N, Canady R, Demerlis C, Duncan T, Janer G, et al. Engineered nanoscale food ingredients: evaluation of current knowledge on material characteristics relevant to uptake from the gastrointestinal tract. Compr Rev Food Sci Food Saf. 2014;13:730–44. https://doi.org/10.1111/1541-4337.12076.
20. Huang Q, Yu H, Ru Q. Bioavailability and delivery of nutraceuticals using nanotechnology. J Food Sci. 2010;75:50–7. https://doi.org/10.1111/j.1750-3841.2009.01457.x.
21. Jafari SM, Katouzian I, Rajabi H, Ganje M. Bioavailability and release of bioactive components from nanocapsules. Nanoencapsulation technologies for the food and nutraceutical industries, Washington: Academic Press; 2017, p. 494–523. https://doi.org/10.1016/B978-0-12-809436-5.00013-6.
22. Jafari SM, Paximada P, Mandala I, Assadpour E, Mehrnia MA. Encapsulation by nanoemulsions. Nanoencapsulation technologies for the food and nutraceutical industries, Washington: Academic Press; 2017, p. 36–73. https://doi.org/10.1016/B978-0-12-809436-5.00002-1.
23. Zuidam NJ, Shimoni E. Overview of microencapsulates for use in food products or processes and methods to make them. Encapsulation technologies for active food ingredients and food processing. New York: Springer; 2010. p. 3–29. https://doi.org/10.1007/978-1-4419-1008-0_2.
24. Ranjan S, Dasgupta N, Chakraborty AR, Melvin Samuel S, Ramalingam C, Shanker R, et al. Nanoscience and nanotechnologies in food industries: opportunities and research trends. J Nanopart Res. 2014;16:2464. https://doi.org/10.1007/s11051-014-2464-5.
25. Cushen M, Kerry J, Morris M, Cruz-Romero M, Cummins E. Nanotechnologies in the food industry—recent developments, risks and regulation. Trends Food Sci Technol. 2012;24:30–46. https://doi.org/10.1016/j.tifs.2011.10.006.
26. Noori S, Zeynali F, Almasi H. Antimicrobial and antioxidant efficiency of nanoemulsion-based edible coating containing ginger (Zingiber officinale) essential oil and its effect on safety and quality attributes of chicken breast fillets. Food Control. 2018;84:312–20. https://doi.org/10.1016/j.foodcont.2017.08.015.
27. Bolger Z, Brunton NP, Monahan FJ. Impact of inclusion of flaxseed oil (pre-emulsified or encapsulated) on the physical characteristics of chicken sausages. J Food Eng. 2018;230:39–48. https://doi.org/10.1016/j.jfoodeng.2018.02.026.

28. Durmuş M, Özoğul Y. The effects of nanoemulsions on the fatty acid profiles of sea bass fillets during storage at 2 ± 2 °C. Ege J Fish Aquat Sci. 2018;35:227–35. https://doi.org/10.12714/egejfas.2018.35.3.01.
29. Wu C, Wang L, Hu Y, Chen S, Liu D, Ye X. Edible coating from citrus essential oil-loaded nanoemulsions: physicochemical characterization and preservation performance. RSC Adv. 2016;6:20892–900. https://doi.org/10.1039/C6RA00757K.
30. Durmus M. The effects of nanoemulsions based on citrus essential oils (orange, mandarin, grapefruit, and lemon) on the shelf life of rainbow trout (Oncorhynchus mykiss) fillets at 4 ± 2 °C. J Food Saf. 2020;40:1–10. https://doi.org/10.1111/jfs.12718.
31. Lee J, Kim H, Choi MJ, Cho Y. Improved physicochemical properties of pork patty supplemented with oil-in-water nanoemulsion. Food Sci Anim Resour. 2020;40:262–73. https://doi.org/10.5851/kosfa.2020.e11.
32. Milani MA, Dana MG, Ghanbarzadeh B, Alizadeh A, Afshar PG. Effect of novel bioactive coating enriched with nanoemulsion of mustard essential oil on the quality of Turkey meat. J Food Nutr Res. 2020;59:71–80.
33. Das AK, Nanda PK, Bandyopadhyay S, Banerjee R, Biswas S, McClements DJ. Application of nanoemulsion-based approaches for improving the quality and safety of muscle foods: a comprehensive review. Compr Rev Food Sci Food Saf. 2020;19:2677–700. https://doi.org/10.1111/1541-4337.12604.
34. Kiani A, Fathi M, Ghasemi SM. Production of novel vitamin D3 loaded lipid nanocapsules for milk fortification. Int J Food Prop. 2017;20:2466–76. https://doi.org/10.1080/10942912.2016.1240690.
35. Tan EPS, Lim CT. Mechanical characterization of nanofibers—a review. Compos Sci Technol. 2006;66:1102–11. https://doi.org/10.1016/j.compscitech.2005.10.003.
36. Kenry LCT. Nanofiber technology: current status and emerging developments. Prog Polym Sci. 2017;70:1–17. https://doi.org/10.1016/j.progpolymsci.2017.03.002.
37. Almetwally AA, El-Sakhawy M, Elshakankery MH, Kasem MH. Technology of nano-fibers: production techniques and properties—critical review. J Textile Assoc. 2017;78:5–14.
38. Okutan N, Terzi P, Altay F. Affecting parameters on electrospinning process and characterization of electrospun gelatin nanofibers. Food Hydrocoll. 2014;39:19–26. https://doi.org/10.1016/j.foodhyd.2013.12.022.
39. Fathi M, Nasrabadi MN, Varshosaz J. Characteristics of vitamin E-loaded nanofibres from dextran. Int J Food Prop. 2017;20:2665–74. https://doi.org/10.1080/10942912.2016.1247365.
40. Sun L, Chen W, Liu Y, Li J, Yu H. Soy protein isolate/cellulose nanofiber complex gels as fat substitutes: rheological and textural properties and extent of cream imitation. Cellulose. 2015;22:2619–27. https://doi.org/10.1007/s10570-015-0681-4.
41. Wang Y, Wang W, Jia H, Gao G, Wang X, Zhang X, et al. Using cellulose nanofibers and its palm oil pickering emulsion as fat substitutes in emulsified sausage. J Food Sci. 2018;83:1740–7. https://doi.org/10.1111/1750-3841.14164.
42. Jiang WJ, Tsai ML, Liu T. Chitin nanofiber as a promising candidate for improved salty taste. LWT Food Sci Technol. 2017;75:65–71. https://doi.org/10.1016/j.lwt.2016.08.050.
43. Mansoori Y, Sanaei SS, Zamanloo MR, Imanzadeh G, Atghia SV. Synthesis and properties of new polyimide/clay nanocomposite films. Bull Mater Sci. 2013;36:789–98. https://doi.org/10.1007/s12034-013-0542-4.
44. Kuan CF, Kuan HC, Ma CCM, Chen CH. Mechanical and electrical properties of multi-wall carbon nanotube/poly(lactic acid) composites. J Phys Chem Solids. 2008;69:1395–8. https://doi.org/10.1016/j.jpcs.2007.10.060.
45. Wu CS, Liao HT. Study on the preparation and characterization of biodegradable polylactide/multi-walled carbon nanotubes nanocomposites. Polymer. 2007;48:4449–58. https://doi.org/10.1016/j.polymer.2007.06.004.
46. Khankrua R, Pivsa-Art S, Hiroyuki H, Suttiruengwong S. Thermal and mechanical properties of biodegradable polyester/silica nanocomposites. Energy Procedia. 2013;34:705–13. https://doi.org/10.1016/j.egypro.2013.06.803.

47. Borriello C, De Maria A, Jovic N, Montone A, Schwarz M, Antisari MV. Mechanochemical exfoliation of graphite and its polyvinyl alcohol nanocomposites with enhanced barrier properties. Mater Manuf Process. 2009;24:1053–7. https://doi.org/10.1080/10426910903022346.
48. Sarfraz J, Gulin-Sarfraz T, Nilsen-Nygaard J, Pettersen MK. Nanocomposites for food packaging applications: an overview. Nano. 2020;11:10. https://doi.org/10.3390/nano11010010.
49. Wei Y, Yan B. Nano products in daily life: to know what we do not know. Natl Sci Rev. 2016;3:414–5. https://doi.org/10.1093/nsr/nww073.
50. Pradhan N, Singh S, Ojha N, Shrivastava A, Barla A, Rai V, et al. Facets of nanotechnology as seen in food processing, packaging, and preservation industry. Biomed Res Int. 2015;2015 https://doi.org/10.1155/2015/365672.
51. Moghadam FV, Pourahmad R, Mortazavi A, Davoodi D, Azizinezhad R. Use of fish oil nanoencapsulated with gum arabic carrier in low fat probiotic fermented milk. Food Sci Anim Resour. 2019;39:309–23. https://doi.org/10.5851/kosfa.2019.e25.
52. Nedovic V, Kalusevic A, Manojlovic V, Levic S, Bugarski B. An overview of encapsulation technologies for food applications. Proc Food Sci. 2011;1:1806–15. https://doi.org/10.1016/j.profoo.2011.09.265.
53. Hogan SA, McNamee BF, Dolores O'Riordan E, O'Sullivan M. Microencapsulating properties of sodium caseinate. J Agric Food Chem. 2001;49:1934–8. https://doi.org/10.1021/jf000276q.
54. Acevedo-Fani A, Soliva-Fortuny R, Martín-Belloso O. Nanostructured emulsions and nanolaminates for delivery of active ingredients: improving food safety and functionality. Trends Food Sci Technol. 2017;60:12–22. https://doi.org/10.1016/j.tifs.2016.10.027.
55. Singh J, Sengar R. Nanotechnology in agriculture and food. Ann Hortic. 2020;13:14. https://doi.org/10.5958/0976-4623.2020.00015.8.
56. Chakotiya AS, Sharma RK. Phytoconstituents of zingiber officinale targeting host-viral protein interaction at entry point of sars-COV-2: a molecular docking study. Def Life Sci J. 2020;5:268–77. https://doi.org/10.14429/dlsj.5.15718.
57. Bitinis N, Hernandez M, Verdejo R, Kenny JM, Lopez-Manchado MA. Recent advances in clay/polymer nanocomposites. Adv Mater. 2011;23:5229–36. https://doi.org/10.1002/adma.201101948.
58. Jalilzadeh A, Tunçtürk Y, Hesari J. Extension shelf life of cheese: a review. Int J Dairy Sci. 2015;10:44–60. https://doi.org/10.3923/ijds.2015.44.60.
59. McClements DJ. Design of nano-laminated coatings to control bioavailability of lipophilic food components. J Food Sci. 2010;75:R30–R42. https://doi.org/10.1111/j.1750-3841.2009.01452.x.

Chapter 4
Nano and Microencapsulation of Foods, Vitamins and Minerals

Dunya Al-Duhaidahawi

Abbreviations

AAO	Aryl-alcohol oxidase
BHA	Butylated hydroxyanisole
BHT	Butylated hydroxytoluene
CD	Cyclodextrin
CMC	Critical micellar concentration
CPO	Chloroperoxidase
CTAB	Cationic surfactant cetyl trimethyl ammonium bromide
DFS	Dual fortified salt
EPC	Egg phosphatidylcholine
$FeCl_3$	Ferric chloride
$FeSO_4$	Ferrous sulphate
FFAs	Free fatty acids
FIP WP	Ferri polyphosphate whey protein
GA	Gum Arabic
GIT	Gastrointestinal tract
IDFA	International Dairy Foods Association
L-Ca	Encapsulated calcium in liposome
LMFS	Large microencapsulated $FeSO_4$
MD	Maltodextrin
Mm	Micrometer
MRI	Magnetic resonance imaging
nm	Nanometer
SDS	Sodium dodecyl sulfate
SLNs	Solid-liquid nanoparticles
SMFS	Small microencapsulated $FeSO_4$

D. Al-Duhaidahawi (✉)
Department of Pharmacognosy, College of Pharmacy, University of Kufa, Al-Najaf, Iraq
e-mail: dunyal.mohammed@uokufa.edu.iq

C. Egbuna et al. (eds.), *Application of Nanotechnology in Food Science, Processing and Packaging*, https://doi.org/10.1007/978-3-030-98820-3_4

TBA	Thiobarbituric acid
TBARS	TBA reactive substance
TFS	Triple fortified salt
USDA	United State Department of Agriculture
WPI	Whey protein isolate

4.1 Introduction

Nanoparticles delivery approach has obtained constant recognition from the food industry in the last few decades for applications that range from the slow-release vitamin and mineral formulations and micronutrients encapsulated and stabilized for nutritional supplement utilization. Because of the current growth and prominence of "niche" subgroups of useful ingredients, the colloidal delivery approach (dealing mainly with the micro and nanoencapsulations) has been considered as a useful resource for resolving multiple difficulties associated with food fortification. Encapsulation can be defined as either layering a droplet, gas, or solid particle center with a solid polymeric matrix or integrating active components into polymer matrices to reduce the interaction of the encapsulated substances with the surrounding environment [1]. Colors, flavors, preservatives, and other bioactive components are added to packaged foods to improve their aesthetic and shelf life. Furthermore, fortified and value-added items frequently require specific health-promoting bioactives, such as micronutrients and nutraceuticals. There are numerous challenges in food fortification; among which, durability, matrix compatibility, and organoleptic changes are common difficulties [2]. Fortification is a difficult process due to difficulties in product processing and issues related to in vivo bioavailability, controlled release, and bio accessibility. Nano- and microencapsulation can address some or all of these problems. The gastro-intestinal (GI) tract stability of the encapsulated materials could also be improved to enable controlled drug release at suitable GI targets [3]. The present chapter discusses a variety of colloidal systems (micro and nano) and their applications in vitamin and mineral encapsulation and delivery.

4.2 Foods and Nutraceuticals Vitamins

4.2.1 Vitamins: Dietary and Biological Needs

Probiotics should include vitamins necessary for human growth and development (like niacin and choline). The advantages of nutraceuticals are more than simply preventing illness. Oil and water-soluble vitamins function as antioxidants and hormone regulators. Vitamin B-complex is beneficial for vision, immunity, bones, teeth, and skin. Vitamin E improves cognitive function.

4.2.2 Vitamins: Stability and Formulation Concerns

Vitamins encounter a number of challenges, including oxidative stability. Other additives may also impede water-soluble vitamins, impacting texture, appearance, and storage. Vitamin A (retinoids): When exposed to air and heat, trace metal ions Fe^{2+} and Cu^{2+} increase the oxidation sensitivity of vitamin A.

Thiamine, a heat and UV-sensitive vitamin B, is a potent Maillard reactive due to its chemistry. Baking, pasteurizing, or boiling thiamine-fortified meals can reduce their content by up to 50%. B2 vitamin: Riboflavin is a brilliant yellow pigment soluble in water. Visible spectrum light below 500–520 nm can degrade riboflavin. **Vitamin B3:** Niacin is highly stable when exposed to heat, light, air, and alkalis. Niacin can be destabilized by other micronutrients, including minerals.

Vitamin B5: Pantothenic acid decomposes when heated under slightly acidic to neutral conditions but stays stable within the pH range of 5–7. It is used as calcium and sodium salts to enhance stability in the gastrointestinal pH. **Vitamin B6**: Pyridoxine is heat stable, but not pyridoxal or pyridoxamine. **Vitamin B6** is degraded by oxidation, UV radiation, and alkaline conditions. **Vitamin B7** appears to be reasonably stable; however, heat treatment causes relatively minor losses. It is stable in air at neutral and acidic PH, but susceptible to alkaline pH. **Vitamin B9:** Folic acid is a reddish, tasteless, and odorless vitamin, which is water soluble at 12.5 mg/mL. Light, air, acid, and alkali deactivate it. **Vitamin B12**: In addition to being heat stable, light, oxygen, acid, and alkali cause vitamin B12 to lose its activity. **Vitamin C**: Ascorbic acid is easily degraded by metals such as copper and iron during production and storage. In neutral or alkaline conditions, it is oxidized by air. Metals, particularly copper and iron, and enzymes expedite vitamin C breakdown (particularly enzymes containing copper or iron, such as ascorbic acid oxidase). **Vitamin D**: D2 and D3 are crystalline compounds white to yellow in color. They are insoluble in vegetable oils and water but can be dissolved in organic solvents. Light, oxygen, and acid quickly dissociate both these vitamins, especially in solution and powder forms. However, they tend to isomerize in oily solutions. The crystalline compounds have good thermal stability. **Vitamin E**: Food processing and storage reduce vitamin E levels due to light, oxygen, and heat. Some foods can lose half their flavor after just 2 weeks. Vegetable oils are destroyed by frying. During storage, tocopherol esters (acetate and succinate of tocopherol) resist oxidation, which makes them useful in nutritional supplements. **Vitamin K:** Light and alkaline substances decompose K1 into a yellow-oily liquid. Vitamin K2 is a crystalline substance that is resistant to heat and reducing agents but susceptible to acid, alkali, light, and oxidizing agents.

4.3 The Encapsulation of Vitamins (Nano and Micro) Colloidal Form

4.3.1 Dispersions of Solids in Liquids

4.3.1.1 Microparticles

In recent years, microencapsulation has become more popular in pharmaceuticals. It has been found that functional ingredients can be delivered in microstructures of size between 1 and 1000 μm. In contrast to microspheres, the microcapsule core is enclosed in a well-formed polymer shell. Encapsulated vitamins are protected from the external environment. Vitamin C is a well-studied model vitamin for encapsulation. Vitamin C powder is stable but not in the solution form as light and oxygen readily destroy vitamin C in the solution. Trace metal ions and formulation can also affect vitamin C stability. Thus, processing can easily degrade vitamin C, causing discoloration. Vitamin C can also interact with other substances, changing the product's flavor or color. Microencapsulation has been widely used for the stabilization and processing of vitamin C in food items. A wall-forming material, ethyl cellulose, and carnauba wax were used to study the characteristics of ascorbic acid and four different encapsulation methods. Carnauba wax releases slower than ethyl cellulose. Core-shell microcapsules encapsulating vitamin C (2–5 μm) was prepared using a new interfacial/emulsion reaction. Folic acid was encapsulated in alginate and pectin microparticles [4]. Depending on the experiment, the microparticles were of 300–650 nm in size. Less trapping efficiency means less folic acid release. Microencapsulation has also been utilized for light-sensitive vitamins like A and D. Gelatin-acacia coacervates contain vitamin A palmitate. Acacia was mixed with vitamin A palmitate in maize oil, and the pH was adjusted to make microcapsules. This research focused on the effects of hydrocolloid mixing ratio, hardening agent concentration, and drying technique. An average diameter of 25 μm and encapsulation efficiency of 83.43% were achieved. Alginate and chitosan were used by Albertini et al. to create a double-layer microcapsule containing vitamin A palmitate [10]. Encapsulation increased vitamin A palmitate storage stability compared to the pure form. A novel sort of microcapsule was recently reported by solidifying heated emulsion droplets. Tocopherol, a liposoluble vitamin, was effectively encapsulated into microcapsules. Sunflower oil and water were emulsified at 85 °C, and then diluted in cold water to create micro-capsule-size particles. The sudden change in temperature led the oil droplets to freeze. Making microcapsules with a core-shell structure is as simple as adding calcium chloride to the cold water phase and stirring. Firstly, α-tocopherol was dissolved in the oil phase, and then encapsulation was accomplished as previously described. The antioxidant tests showed that encapsulation has no influence on α-tocopherol antioxidant activity [5].

4.3.1.2 Polymeric Nanoparticles

The size of colloidal particles is in the sub-micron range. This category comprises polymer and solid lipid nanoparticulate systems. Adding vitamins and minerals to a "clear" food beverage that makes fat-based foods taste better without changing their overall taste, or creating a bioavailable formula relies on tiny particles that are easily dissolved. Vitamins encapsulated in colloidal particles can be used in products like fortified milk, juices, and cooking oils. Liposoluble actives can be encapsulated by antisolvents, or nanoprecipitation, and co-precipitating them with hydrophilic polymers makes it convenient to load them in colloidal particles [6]. This approach has been used to encapsulate liposoluble vitamins like A, D, E, and K. Vitamin E (tocopherol) was co-precipitated with wheat gluten vegetable protein fractions (gliadins). They were formed as colloidal particles with an encapsulation effectiveness of 77% or higher and dual release features. The anti-solvent technique works well with zein, corn protein rich in proline. Zein interacts with particles composed of charged biopolymers, such as caseinate and chitosan. Submicron nanoparticles were synthesized using the ionic zein and α-tocopherol with hydrophobic interactions [7]. These particles were poly dispersible and ranged from 300–1000 nm in size. The zein-chitosan combination generated substantially smaller and homogenous sized particles (400–500 nm). When chitosan was added to the preparation process, it improved the control of the encapsulated vitamin's release. A liposoluble vitamin, cholecalciferol, was coated with zein-carboxymethyl chitosan nanoparticles [8].

Phase separation was used to produce monodispersed nanoparticles as small as 120 nm. The controlled release of vitamin D3 from nanoparticle increased its photostability. These solid lipid nanoparticles (SLNs) have been employed to deliver lipophilic bioactives for food applications. The initial stage of making an oil-soluble vitamin loaded SLNs is dispersion in the lipid phase. For high melting lipid carriers, dispersion of vitamins in lipid carriers takes place at high temperature (about 80 °C). Aqueous surfactant solution is subsequently mixed with the melt, resulting in the spontaneous production of SLNs [9]. SLNs are made from glyceryl behenate and tripalmite encapsulated vitamin A and its palmitate. Factors affecting size and shape were assessed, such as manufacturing process, surfactant system, and lipid type, which have an influence on experimental outcomes. The interplay between these variables affects the internal and external architecture of SLNs, and therefore the distribution of vitamins inside SLNs (in the lipid matrix or in the outer shell).

4.3.2 Liquid-in-Liquid Dispersions

4.3.2.1 Emulsions, Microemulsions and Nanoemulsions

In food science, emulsions are the most explored colloidal system, forming the basis of a wide range of edible goods, like ice creams, sauces, spreads, whipped creams, etc.

A mixture is made up of dispersed and continuous liquids. The use of surface active amphiphilic emulsifiers stabilizes the internal phase. Emulsions oil/water (o/w) or water/oil (w/o) are commonly used in food science and technology. For margarines and butter, a continuous phase of solid fat gives the finished product a structural appearance. This is because the water continuous phase is less viscous. However, changing the internal phase can affect emulsion rheology in general, it is either gelled or increases the proportion of internal phase (or both). Double or multiple emulsions generate an o/w emulsion, which is further emulsified into an oil phase or water phase [10]. Size and stability of droplets can also be considered when classifying emulsions. Emulsions having droplet sizes greater than 1 μm are termed as macroemulsions. The droplets of size 1 μm/1000 nm or smaller fall. When homogenization is performed, high-pressure or high-shear homogenizers are used, which might produce nanoemulsions or mini emulsions. The droplet size of microemulsions is typically smaller than 100 nanometers and they are formed spontaneously within liquids. Surfactant and co-surfactant synergy reduces interfacial tension, resulting in spontaneous production [11].

Despite their popularity, macroscale emulsions are thermodynamically unstable and tend to cream and coalesce with time. Emulsions are frequently employed to encapsulate and supply both liposoluble and water-soluble vitamins. This enhances water dispersibility and bioavailability while preventing oil-soluble vitamins from deterioration in the product condition and the human body condition. Numerous studies have used the most active form of vitamin E that is α-tocopherol emulsion, which is food-grade [12]. It was determined that the distribution of α-tocopherol in macroemulsion systems may be enhanced by emulsions of α-tocopherol using a pseudo phase model for thermodynamically stable microemulsions. As a result of α-tocopherol's surfactant-like properties (polar phenolic headgroup with hydrocarbon tail), they identified it mostly in the interfacial region. High pressure homogenization encapsulated α-tocopherol in o/w nanoemulsions with droplet sizes typically smaller than 500 nm [13]. The protein coating the tocopherol-filled fat droplets protected them from oxidation. Vitamin E-loaded nanoemulsions were recently described. This novel nanoemulsion contains vitamin E with >99% efficiency and particles are on average 78 nm in size. Microemulsion consisting of water, vitamin E, ethyl butyrate, EL-35, and ethanol was also studied. A regulated vitamin E released from microemulsions was shown having droplet size of 20 nm. Vitamins A and D can also be encapsulated in emulsions. Encapsulated trans retinoic acid/retinol is required in a variety of emulsion processes [14]. Besides high loading, encapsulation of vitamin A must address retention of loaded vitamins. Researchers employed an oil gelling agent, and coated oil droplets with solid fat as an alternative to oil to enhance vitamin retention in the oil phase thus encapsulating the vitamin. By means of a high- pressure homogenizer, 10% oil/90% water emulsion was created with an average particle size of 220 nm.

4.3.3 Self-Assembled Colloidal Dispersions

An organized structure or pattern is produced when pre-existing, disordered components interact. In an aqueous medium, an amphiphilic compound generates self-assembled colloids. The transport of hydrophobic and hydrophilic bioactive can be achieved by microparticles, procolloidals, lipid bilayers, liquid crystal phases, and mesophases [15].

4.3.3.1 Micelles

Molecules containing lipids are micelles that form spheres in aqueous solutions. Fatty acids have an amphipathic nature, which contain both water-loving (polar head groups) and water-hating sites, causing micelle formation (the long hydrophobic chain). A micelle's polar head group creates a surface, and the nonpolar hydrophobic tail group is located inside and away from water. Single hydrocarbon chains make micelle-forming fatty acids round, reducing steric hindrance. Micelles are too bulky for the two hydrophobic chains in the fatty acids found in glycolipids and phospholipids and they prefer "lipid bilayers" [16]. Micelles arise spontaneously due to hydrophobic interactions between molecules. From simple casein micelles in milk to sophisticated digesting dietary mixed micelles, fat-soluble vitamins (A, D, E, and K) enter the bloodstream via GIT. Micelles also act as mechanisms for transporting fat-soluble vitamin from the digestive tract to the blood. Externally manufactured micelles include lipid-soluble vitamins [17]. In order to supply lipid-soluble vitamin D2, casein micelles from milk were used. Nano delivery systems such as casein micelles are found in milk to stabilize and transport minerals like calcium and vitamin D_2 which are encapsulated in 155 nm sized particles .In addition to stabilizing the encapsulated vitamins, the assembled micelles also protect them from photochemical destruction. In a recent study, an ultra-high-pressure homogenization was employed to encapsulate vitamin D3 into reassembled casein micelles. A double-blind placebo-controlled clinical research in 87 human volunteers reported that vitamin D3 infused in reassembled casein micelles was highly bioavailable. The vitamin-encapsulated micelles formed large aggregates after they interacted with bile. Unlike empty micelles, the encapsulated vitamins influenced aggregation. As polymeric micelles cannot afford vitamin K absorption, the researchers concluded that free bile acid is essential for bioavailability. Micellar systems have also been used to analyze that vitamin C is water-soluble, in addition to liposoluble vitamins. In a surfactant solution, the degree of ascorbic acid oxidation depends on the concentration of surfactant. Critical micellar concentration (CMC) is the concentration at which ascorbic acid oxidation is at its peak, which is then declined by the surfactant concentration in both SDS and CTAB.

4.3.3.2 Liposomes

Liposomes in aqueous solutions are self-assembled vesicles made up of lipids and lipid-like molecules. Liposomes may contain both hydrophilic and hydrophobic actives. Their internal pH may be altered, allowing them to contain otherwise unstable chemicals. Liposomes can be made in many forms and sizes. Phospholipids can be organized to produce numerous or single vesicles. The dairy industry uses vitamins A, D, and E so that vitamins are stabilized and delivered to enhance nutritional quality. Banville et al. [18] found that liposome-encapsulated vitamin D recovered 62% more vitamin D than vitamin D produced commercially (43%) and cream contains 41% solubilized vitamin D. Vitamins were protected from degradation by liposomal encapsulation. The liposome approach protects water-soluble vitamins and has been studied showing beneficial effects on water-soluble vitamins (vitamin C). Soy phosphatidylcholine liposomes protected heat-labile vitamin C [19]. They were also evaluated with orange juice. Some researchers have found that poly phosphatidyl choline (60:40) liposomes can improve the storage stability of vitamin C. Encapsulated water-soluble and liposoluble vitamins are shown in Fig. 4.1.

4.3.3.3 Procolloidal System

Incorporating an aqueous phase dilutes a procolloidal system. Water-soluble colloids are formed when these concentrated systems are diluted with liquid crystal phases that form microphases when diluted. A fine oil-in-water emulsion is formed when oil, a surfactant, a co-surfactant, and a co-solvent are gently mixed together. Spontaneously produced aqueous microemulsions (50–1000 nm droplet size) are developed in the GI tract after ingestion of soft or firm gelatin capsules. When a surfactant interacts with a co-surfactant, spontaneous dispersion is produced on the interface, and as a result of co-surfactant assistance, the surfactant film formed has a low interfacial energy and a low bending stress. The incorporation and encapsulation of vitamin E in procolloidal systems, especially those that self-emulsify, have been extensively studied, due to its powerful nutraceutical and pharmaceutical properties [20]. Self-emulsifying and self-micro emulsifying vitamin E systems notably enhance bioavailability. Self-emulsifying systems increased tocotrienol bioavailability by two to three times in six healthy individuals who participated in the study. Food products have been extensively studied for water- and liposoluble vitamins delivered through liposomes. However, mass-production of these systems in a reproducible manner remains unsolved. The use of proliposomes is one of the methods for scaling up liposome production. Proliposomes are phospholipid particles formulated with functional ingredients, which are

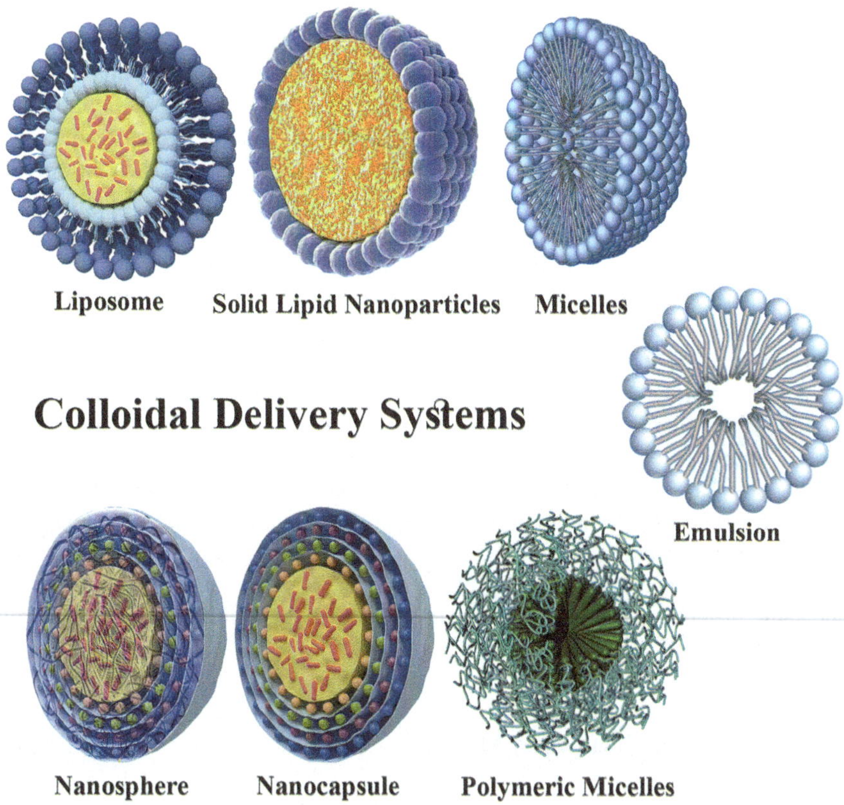

Fig. 4.1 Illustrations of colloidal delivery systems for water- and liposoluble vitamins

soluble in water and when hydrated above the transition temperature and agitated form liposomes. The addition of beta-carotene to a proliposomal formulation containing hydrogenated soy phosphatidylcholine improved chemical stability. The shelf life was reported to be 60 days when refrigerated [21]. Moreover, proliposomal formulations of glyceryl dioleate-PEG12 and glyceryl dipalmitate-PEG23 at 35% and 20% weight, respectively, were effectively combined with Vitamins D and E.

4.3.4 Dry Matrix Encapsulation

Industrial feasibility and scale-up include spray-drying, extruding, coating with pans, and using fluidized beds. These methods can easily produce microcapsules with a consistent particle size distribution.

4.3.4.1 Spray-Drying

The food industry's most popular encapsulation method is spray-drying. The process is cheap, easy, and produces consistent particles. Spray-drying is thought to be cheaper than most encapsulation processes. There are usually two steps in spray-drying: Firstly, a coating or core material solution is prepared, and then the solvent is evaporated, which is termed as dispersion. Secondly, a layer of coating material is applied over the core material or a coating of polymer matrix containing a core material is distributed. Food manufacturers have used spray-drying encapsulation since the late 1950s to protect flavor oils from deterioration and to better handle liquids. However, one drawback of spray-drying is the requirement of good water solubility of the coating material. The availability of coating materials is limited due to their water-based nature; therefore, several biopolymers have been used to coat water-soluble vitamins (vitamin C) such as gum Arabic, pea protein, and chitosan [22]. A microcapsule of less than 10 nanometers in size was produced using pea protein, chitosan, and gum Arabic, while rice starch was used to produce microcapsules of 5–40 μm. We found that microencapsulating vitamin C microcapsules containing gum Arabic and starch improved its oxidative stability, and that microcapsules of vitamin C prepared using pea protein and chitosan showed controlled release. The same group used pea protein concentrate liposoluble vitamins (α-tocopherol) for encapsulation [23].

4.3.4.2 Extrusion (in a Glassy Matrix)

Extrusion-based microencapsulation is a relatively newer method of drying in contrast to spray-drying, which is performed by forcing a series of dies, molten polymer and core material into a dehydrating liquid. Discrete microcapsules, drying the material, and pulverizing have also been accomplished with other materials such as gelatin, sodium alginate, carrageenan, gum acacia, fats and fatty acids, waxes, etc.; however, with this method, the coating material completely encapsulates the core material, providing excellent stability against oxidative and other chemical degradation. Maltodextrin and gelatin were encapsulated in a glassy matrix containing tocopherol in a 3:1:1 ratio [24]. The glassy matrix also protected the encapsulated α-tocopherol from oxidation, while encapsulating carotene in amorphous trehalose delayed degradation time.

4.3.5 Vitamins Encapsulation in Cyclodextrins

Cyclodextrins are cyclic oligomers formed by enzyme conversion of starch. The D-glucopyranose unit has six, seven, or eight (-CD) units joined together by α-(1,4) bonds. Their structure allows them to form molecular complexes with many hydrophobic molecules. Encapsulation of hydrophobic drugs protects volatile aromatic compounds from degradation. The delivery of functional ingredients and bioactives, including vitamins, is the major advantage of CDs [25]. A CD's structure resembles a cage with a hydrophobic cavity. Hydrophobic compound inclusion complex is formed when less polar guest molecules replace the polar hydrophobic cavity, which is lined with water molecules. Though the exact cause of the inclusion complex formation is unknown, the following mechanism is thought to be involved: A polar interaction between a guest molecule and the hydrophobic cavity releases molecules of cavity i.e., water, and bulk water replaces the energetically disfavored polar interactions between the water molecules in the cavity and the included water. True inclusion complexes can only be formed by lipid soluble vitamins in cyclodextrins. Thus, the creation of inclusion complexes to encapsulate liposoluble vitamins has been studied extensively. The retinoid-CD complexes were made in aqueous alcohol at ambient temperature. Many studies have shown that hydroxyl propyl-CD and -CD inclusion complexes increase the solubility of vitamin A. Other studies have shown that CD complexation stabilizes vitamin A by inhibiting photoisomerization and photodegradation of retinoids [26]. Beyond retinoids-CD complexes, most studied vitamin is vitamin D as CD complex. The low water solubility of vitamin D affects its biological activity. Vitamin D and CD inclusion complexes can be made using ethanol as a solvent. A study on the stability of vitamin D after cyclodextrin complexation has also been reported. Menadione is another liposoluble vitamin (or vitamin K3). Benzene is incorporated into the cavity of a CD, while the quinine moiety is inserted outside the cavity, thus improving its aqueous solubility [27].

4.4 The Food Industry and Encapsulated Minerals

Encapsulated minerals can be integrated into the formulations of food and medicine to provide unique benefits and sensory features to the physically and chemically improved functional products. Encapsulated mineral salts have advantages such as reduced discoloration, reduced taste and smell by masking-off flavor, production and storage processes that control the release of mineral components, and improved taste and smell. Encapsulation of minerals into dairy products and table salt has been studied extensively during the last two decades. $C_4H_8FeN_2O_4$, $NH_4Fe(SO_4)_2$, and $C_6H_{10}FeO_6$ are some examples of electrolytic-Fe (Table 4.1) [28]. Soy-yogurt and soy milk are fortified with encapsulated tricalcium phosphate ($Ca_3(PO_4)_2$) and calcium citrate ($Ca_3(C_6H_5O_7)_2$) salts. KI and KIO_3 are the most common encapsulated iodine fortification salts, while $C_4H_2FeO_4$ and $FeSO_4$ are the most common encapsulated Fe fortification salts.

Table 4.1 Bioavailability, toxicity, and technological aspects of several mineral nanoparticles

Mineral	Shape/size	Technical aspects	Bioavailability/ toxicology aspect	References
Se	Spherical/10–50 nm	Eliminates six food-borne pathogens from growing and forming biofilms	No toxicity to Artemia larvae at 100 mg/mL	[29]
Se	Spherical/80 nm	When heat is applied to Se-NPs, their thermo-stability varies with their size: Smaller Se-NPs are more resistant to transforming into nanorods than larger ones	Se-deficient mice are less likely to experience phase II enzyme induction	[30]
Se	Spherical/~36	Comparison of dietary supplements with Se and Se-NPs	At supranutritional levels, induce phase II enzymes without affecting their ability to up-regulate seleno-enzymes	[31]
Se	Spherical/20-30 nm	Introducing elemental Se-NPs to lambs through probiotic bacteria resulted in enriching lamb meat with Se (1) as an indirect method of improving the nutritional value and functionality of meat products.	The activity of glutathione peroxidase (GPx), thioredoxin reductase (TrxR), and glutathione-S-transferase (GST) is equally increased by Se-NPs	[32]
Iron oxide (Fe_2O_3)	NR/<50 nm	The application of food with quercetin and Fe_2O_3 nanoparticles	Cytotoxicity induced by Fe_2O_3-NPs and apoptosis and quercetin (50*mol/L) protects against them	[33]

(continued)

Table 4.1 (continued)

Mineral	Shape/size	Technical aspects	Bioavailability/ toxicology aspect	References
Magnesium oxide(MgO)	Hexagonal/10–50 nm	Combination of Mg-NPs and nisin in milk, resulting in a leak in the bacterial cell membrane, leads to death. A protocol developed for reducing pasteurizing milk temperature and controlling pathogens required MgO-NP levels	MgO-NPs may cause DNA damage and cell death. They cause oxidative damage in consumer products	[34]
Zinc oxide(ZnO)	Spherical/10–30 nm	This engineered film of Ca alginate loaded with ZnO-NPs exhibited antimicrobial activity against two pathogens (*S. typhimurium* and *S. aureus*) 10 days at 8 °C produced zero log drop in poultry meat	In adult male Wistar rats, ZnO-NPs more than 50 mg/kg significantly altered hepatic enzymes, oxidative stress, and renal tissue	[35]
Ferric phosphate (FePo4)	Spherical/10.7 nm	Food and beverage formulations can be fortified with stable nanosized FePO4 colloid additives	FePO4-NPs are soluble and bioavailable. A nanoscale coating (e.g., $FePO_4$) might optimize Fe uptake and bioavailability	[36]

(continued)

Table 4.1 (continued)

Mineral	Shape/size	Technical aspects	Bioavailability/ toxicology aspect	References
Calcium phosphate($Ca_3(PO_4)_2$	Spherical/<100 nm	$Ca_3(PO_4)_2$-NPs have great potential industry of poultry, particularly in feed management and waste minimization	$Ca_3(PO_4)_2$-NPs are 200% bioavailable in broiler chickens	[37]
Calcium carbonate and citrate $CaCO_3$/ $Ca_3(C_6H_5O_7)_2$	NR/≤50 nm	Due to its high MIC in broth, $CaCO_3$-NP has proven to be useful antimicrobial agent, which can also be used in food and agriculture industries	Studies reveal that taking calcium carbonate or citrate NPs is more effective than administering micro-NPs to enhance serum calcium level in order to maintain bone mineral density	[38]

4.4.1 *Dairy Products Fortified with Encapsulated Minerals*

4.4.1.1 Milk

Despite being one of the world's most vital foods, milk is low in Fe. Thus, fortifying milk with encapsulated iron might be beneficial. Increasing Fe intake, for example, $FeSO_4$ microencapsulated in milk and its bioavailability was studied using a lecithin liposome method. The fortified milk's Fe bioavailability did not change after 6 months of storage and heat treatment. A similar bioavailability rate of Fe to high-bioavailable $FeSO_4$ absorption was also recorded. Coatings made from PGMS were used to microencapsulate Fe to fortify milk. While just 3% of Fe was released in vitro (pH 6.0), increasing pH from 5.0 to 8.0 dramatically boosted the rate of Fe release from 12.3 to 95.7% .The sensory study found no significant variations in most sensory features between the control and microencapsulated-Fe samples after 3 days of storage [39]. Moreover, the unencapsulated Fe sample had greater TBA than the encapsulated Fe sample. It was discovered that free Fe-fortified milk had a higher TBA value. Supplemented TBA of milk with high microencapsulated-Fe loadings was not altered. Microencapsulation slowed down lipid oxidation by 60%, while masking milk's metallic taste. However, comparing astringency and bitterness to control milk, microencapsulated Fe milk produced similar results. The fortification ability of Fe microcapsules in milk and the factors that affect the physical, chemical, and sensory properties of the final product were investigated. Fortification of low-level milk with Fe microcapsules (0.1 0.3% w/v) did not increase TBA levels. It was found that 0.1% (w/v) Fe-microcapsule powder was optimal for producing fortified milk. Researchers observed that microencapsulated $FeSO_4$ in a powdered milk diet had better Fe bioavailability than the control diet. One study selected three Fe microcapsules fortified with EE from four distinct techniques (62.97% 74.85%) [40]. To compare the organoleptic ratings of Fe-fortified milk with the control milk, they used modified starch and sodium alginate microcapsules (10 mg/L Fe). GA, MD, and modified starch were mixed and vaporized to create microcapsules of Fe with an average size of 15.54 μm [41]. When compared to fortified milk containing Fe microcapsules, panelists preferred salt-fortified milk over microcapsule-fortified milk. In addition, the microcapsules of Fe in fortified milk had better bioavailability than those in control and Fe salt enriched milk. The sensory qualities of microcapsules of Fe from three Fe salts were also studied. All treatments had the same appearance and flavor. The flavor of the control and $C_6H_{10}FeO_6$ enriched milks did not differ significantly. This product was the finest option for fortifying pasteurized liquid milk. In comparison to cow milk (120 mg/100 g), soy milk contains a lower level of calcium (12 mg/100 g) [42]. As a result, many researchers have tried to improve the nutritional value of soy milk by adding calcium. It was discovered that adding Ca salts ($Ca_3(PO_4)_2$ and $Ca_3(C_6H_5O_7)_2$) to soy beverage was ineffective due to unfavorable Ca protein interactions, coagulation, and precipitation. Using a lecithin liposome structure could fortify soy milk in a better way. Researchers added Ca to soy milk (110 mg Ca/100 g soy milk) to match the level to that in cow's milk. The fortified soy milk was stable and had high Ca bioavailability for 1 week at 4 °C. $Ca_3(PO_4)_2$ in the soy milk was 2000 mg per liter, and to prevent Ca protein interaction and enhance the stability of soy milk, researchers added 30 g/L potassium citrate to samples. However, combining $Ca_3(PO_4)_2$ and $C_6H_5K_3O_7$ reduced soy milk stability.

4.4.1.2 Yogurt

The main consumers of yogurt are children, menstruating, pregnant or nursing women, and teenagers. FeSO4 and a microencapsulated compound of Fe whey protein were added to yogurt for 7 days at −70 °C to assess lipid oxidation and sensory properties. The three oxidation rates (TBA and peroxide) did not change significantly. Iron-enriched yogurt tasted rusted and metallic. The sensory panel favored the Fe whey protein complex microencapsulated yogurt since the overall flavor and quality were better [43]. In recent years, researchers have enriched probiotic yogurt with *Lactobacillus acidophilus* and iron microcapsules derived from whey protein complexes. Compared to unfortified yogurt, fortified yogurt had less oxidized flavors. TBA levels were elevated due to interactions among iron and casein proteins, prooxidant activity of oxygen species, and stimulated lipid oxidation. In oxidative conditions, free fatty acids get accumulated. Three commercial Fe-salt microcapsules were added to yogurt to improve its organoleptic quality. Both the control and

$FeSO_4$ microcapsule samples passed the sensory test after 7 days of storage. $C_4H_8FeN_2O_4$ and $C_6H_{10}FeO_6$ yogurts were clearly distinguished from the control. TBA and peroxide levels were the highest in $C_4H_8FeN_2O_4$ yogurt, whereas the control and $C_6H_{10}FeO_6$ supplemented samples showed the lowest TBA and peroxide level. Microcapsules of $FeSO_4$ were found to be useful and ideal for fortifying yogurt. On 0, 1, 2, and 3 weeks, the starter bacteria count did not differ between the control and Fe-fortification samples. The quantity of these bacteria decreased with extended storage in both control and Fe-fortified yogurt samples. The added Fe had no impact on yogurt bacterial viability. Free Fe enriched yogurts had lower TBA values than encapsulated Fe enriched yogurt [44]. After 21 days, there were significant variations in astringent, oxidized, and overall acceptability assessments. To improve Fe-bioavailability, it was advised to add 80 mg Fe-WP/L to yogurt without affecting its appearance or organoleptic properties. This yogurt drink had Fe microencapsulated in it. Unencapsulated Fe was added, pH was decreased and titratable acidity was increased, while the microbial count of the yogurt was found to be unaffected by unencapsulated Fe. TBA kinetics in encapsulated therapies was also noted. Unencapsulated Fe changed the astringency and bitterness of yogurt drink, whereas encapsulated Fe concealed Fe taste and flavor. $FeSO_4$ $7H_2O$, 12 mg Fe/L, and calcium $(Ca_3C_6H_5O_7)_2$, 600 mg Ca/L, enhanced the stability of soy yogurt. The investigation was conducted for 28 days at 10 °C. Although the fortified sample had low viscosity, extending the storage period had no significant effect on this parameter. Moreover, there were no significant variations in all sensory characteristics as well as acidity levels during the storage. Fe-fortified yogurt stored for 14 days exhibited both physicochemical and sensory qualities. The yogurts were fortified with Cold-set WPI gel powder encapsulating Fe-encapsulated in $FeSO_4$ solution (20–60 mg Fe/kg). Compared to the unfortified control sample, WPI-Fe particles fortified yogurt (up to 60 mg of Fe/kg) exhibited identical qualitative characteristics (color and flavor). A physicochemical and sensory evaluation demonstrated that samples fortified with WPI-Fe may retain quality criteria for 2 weeks regardless of the Fe concentration employed [45]. A Fe-enriched usual yogurt (B225 g) was likewise reported to give up to 60% of a woman's daily Fe requirement, with no evident color or flavor differences from the control yogurt. This feat would be unachievable with even minimal levels of Fe fortification in yogurt. To maximize the effects of three independent factors on EE and Hunter Lab color features, an application of a second-order polynomial model using the RSM-central composite design was used. The optimal pH, WPI content, and Fe concentration for pre-paring WPI-Fe particles were 7.0, 6.8%, and 18.8 mM, respectively. This is because the particles were released into the intestinal environment at a high rate (95%), WPI-Fe particles were generated under optimal conditions, and a site-specific network was an ideal delivery option of Fe (Pancreatin was added, pH 7.5). However, it was found that iron encapsulated in only 28% of powders could be released in the stomach environment (pepsin, pH 1.2).

4.4.1.3 Cheese

Cheese, like milk, is low in Fe. Most research on Fe encapsulation in cheese has been focused on Cheddar. According to IDFA (2015), 28.5% of all cheese consumed in the US was cheddar. Researchers reported that $FeCl_3$, Fe casein, FIP WP, and Fe WP complex constitute the Fe content of cheddar cheese. The study used $FeSO_4$ as a Fe source to fortify cheese, which showed a somewhat higher TBA level than unfortified cheese. TBA value and oxidized off-taste had no correlation with Fe level. Another link was between off-taste or flavor of the oxidized cheese and TBA value. Aged cheese had no effect on TBA levels, oxidized off-taste, or cheese flavor scores. The greatest bioavailability of Fe was 85% for $FeCl_3$, 71% for Fe casein, 73% for FIP-WP, and 72% for Fe WP. Feeding equivalent fortified Fe for 10 and 14 days showed similar basal Fe level, while the bioavailability sources were 5%, 4%, 6%, and 3% ($P \leq 0.05$) [46]. Nonetheless, $FeSO_4$ had 85% maximum and 5% basal bioavailability, and it accelerated lipid oxidation. In addition to WP $FeCl_3$, researchers employed the same fortification method as before. $FeSO_4$ produced oxidation and off flavors in aged cheddar. Sensory perception was better in samples having Fe WP complexes. The chemical and sensory qualities of cheddar cheese were examined with encapsulated-Fe. Encapsulated cheese ripened slower than unencapsulated cheese. During ripening, compounds with short-chain FFAs and neutral volatiles were generated insignificantly. The samples with maximum encapsulated Fe had the most bitterness, astringency, and sourness. However, it did not influence the quality of cheddar cheese, whether LMFS (large microencapsulated $FeSO_4$, 700, 1000 m) or SMFS (small microencapsulated $FeSO_4$, 220, 422 m) enriched cheddar cheese [47]. After 90 days, LMFS (66%) and SMFS (91%) recovered Fe. No significant differences in lipid oxidation rate or chemical makeup (for example, dry matter, fat, ash, magnesium, zinc, and calcium) were observed. Microencapsulating $FeSO_4$ in cheddar cheese did not overpower the flavor, odor, or color of Fe. Less SMFS-fortified cheddar cheese was found to have better Fe retention and sensory attributes. Mozzarella cheese was fortified with ferric chloride, WP, and casein sodium iron salts (25–50 mg Fe/kg). Mozzarella cheese with 50 mg Fe/kg exhibited an off-odor, metallic or oxidized flavor, and more unattractive qualities. Consumers despised all mozzarella due to metallic and off flavors. Cheddar cheese was fortified with $FeCl_3$ and Fe protein components, indicating that it is not possible to generally apply Fe fortification to every other cheese. To plain feta cheese, 80 mg/kg of compounds containing iron ($FeSO_4$, $FeCl_3$, and microencapsulated $FeSO_4$) was added. The chemistry and FFAs of the control and Fe or vitamin C samples were not different. Fe-supplemented cheese compared to unsupplemented cheese revealed a significant difference in Fe TBA and organoleptic ratings, which were lowest in feta cheese containing microencapsulated iron (80 mg) and vitamin C (150 mg) [48].

4.4.2 *Fortification of Salt by Encapsulating Iron and Iodine*

4.4.2.1 Dual Fortified Salt

Since Fe compounds have a greater stability than iodine compounds, encapsulating iodine to provide a protective partition was explored. $C_4H_2FeO_4$ or $FeSO_4$ (1 g/kg Fe) was blended with FBC to make DFS from iodine (KI and KIO_3) microcapsules. Later, the stability of several fortifiers was studied in relation to temperature and relative humidity during storage. A good result was obtained for encapsulating iodine into dextrin, indicating a wall-to-core ratio of 1:200. After 30 days of storage at 40 °C and 60% relative humidity, DFS was enhanced with $FeSO_4$ and KI lost 100% of its iodine concentration, whereas KIO_3 lost 93% at 40 °C and 60% relative humidity; DFS retained 79.7% iodine and KIO_3 retained 80.9% iodine. $C_4H_2FeO_4$ and KI microcapsules demonstrated the greatest iodine retention (98.4%, 101.9%) at temperature of 40 °C and a relative humidity of 60% for 12 months. Despite improvements in iodine chemical stability, fine microcapsules generated by the technique prevented encapsulation [49]. Also, iodizing salt with KIO_3 is expensive. Condensing Fe-compounds in salt formations, iodine and Fe DFS were created and tested for storage stability in highland (Nairobi) and coastal (Mombasa) cities in Kenya. Nairobi retained 87.3% Fe and only 92% iodine in encapsulated DFS with a KI and $C_4H_2FeO_4$ premix, whereas Mombasa retained 92% Fe and 90% iodine. DFS encapsulated in $C_4H_2FeO_4$ resisted color fluctuation and increased bioavailability. It was found that encapsulating Fe in DFS reduced unpleasant sensory characteristics and iodine loss, while maintaining considerable bioavailability. Researchers studied 19 Fe-containing compounds (e.g., fumarate, sulfate, pyrophosphate, and elemental Fe) bound to indigenous Moroccan and Cote d'Ivoire salts. After 6 months of storage, samples were examined for iodine concentration and color. According to the findings, encapsulation did not prevent most chemicals from ionic losses and undesired organoleptic alterations. Iron oxide (B2.5 and 0.5 m) particles were used to enhance salt. In the presence of Fe ($C_4H_2FeO_4$) and iodine (KI and KIO_3), the stability of table DFS was measured in the mountains and coasts of Kenya. Also, high iodine (90%) was preserved after 3 months of storage. Encapsulated $C_4H_2FeO_4$ was combined with salt-encapsulated-KI or iodate salt to make DFS. Because of the significant retention of Fe21, a little quantity of Fe21 was oxidized to Fe31 (17%). During distribution and retail, both iodine and Fe21 were preserved, making DFS with iodine and encapsulated $C_4H_2FeO_4$ extremely stable for 3 months. In vitro and in vivo studies have been focused on Fe bioavailability in iodine DFS and encapsulating $C_4H_2FeO_4$; initially, researchers made a $C_4H_2FeO_4$ premix for salt integration by agglomerating $C_4H_2FeO_4$ into salt-size particles and then encapsulation was performed [50]. The in vitro bioavailability of various DFS and encapsulated $C_4H_2FeO_4$ premixes generated by numerous techniques was then examined. The materials and techniques used in encapsulating $C_4H_2FeO_4$ marginally affected the in vitro Fe bioavailability. The researchers discovered that adding TiO_2 to the coating materials as a masking agent effectively concealed the reddish-brown color of $C_4H_2FeO_4$ and improved overall sensory acceptance and iodine stability. Bioavailability of $C_4H_2FeO_4$ encapsulated in a lipid structure was determined indicating that $FeSO_4$ was 95% bioavailable compared to $FeSO_4$ in rats. Extrusion-based encapsulation of Fe ($C_4H_2FeO_4$) enhanced the iodine stability in iodized salt. The optimal formulation contained $C_4H_2FeO_4$ (10%), binders (30%), TiO_2 (25%), and Methocel, a water-soluble polymer (10%). Uncoated Fe particles, on the other hand, lost 50% of their DFS iodine content during extrusion. The usual salt distribution mechanism will be used to supply iodine and iron to DFS. Li et al. [51] investigated whether it was possible to encapsulate $C_4H_2FeO_4$ in DFS using the cold-forming extrusion process. Extrudable dough containing substantial levels of $C_4H_2FeO_4$ was thus possible using grain flours [51]. Additionally, all extruded Fe particles (300,700 nm) were analyzed indicating high in vitro Fe digestibility. Yadava et al. [52] enhanced masking colors and coating polymers phases utilizing extrusion agglomeration to prepare a fortification of salt with Fe premix. After mixing iodine and Fe21, stability was tested at 35 °C and 60% relative humidity for 90 days after Fe premix with iodized salt. Water-soluble polymeric coatings showed a good efficacy to retain micronutrients at 10% encapsulation capacity. They also had great bioavailability and pleasing sensory characteristics. To add encapsulated Fe ($C_4H_2FeO_4$) to coarse iodized salt, Romita et al. [53] used spray drying. A 10% w/v suspension was employed, and 1.2% of the suspension was dissolved using biopolymers for 6 months. DFS was monitored in iodine stability at room temperature (20% RH) and under harsh conditions (40 °C and 60% RH). Sixty percentage of unencapsulated samples exhibiting reduced iodine stability was not surprising.

4.4.2.2 Triple Fortified Salt

A formulation containing iron, iodine, and vitamin A, three essential micronutrients in a single food product is a low-cost method of fortifying foods. Strong metabolic interactions among these components can treble salt fortification. A mixture of micronized $Fe_4O_{21}P_6$, KIO_3, and retinyl palmitate, was incorporated into HPF to create TFS with 30 g of I, 2 mg of iron, and 60 g of vitamin A in each gram of salt. After 6 months of storage, the color stable TFS lost iodine and vitamin A (B12 15%) [54]. At 10 months, the TFS group had much less vitamin A and Fe anemia deficit than the iodized salt group. In addition to $Fe_4O_{21}P_6$, KI, and retinyl palmitate (vitamin A), HPF nano capsules were also used to formulate TFS. TFS's color changes during storage were acceptable, as was the loss of iodine, which was similar to that in iodized salt. Retinyl palmitate was found to be highly stable, with only 12% loss after 6 months. The overall sensory tolerability of TFS and iodized salt was not found to be changed.

4.5 Iron Encapsulation in Cereals and Bakery Products

$FeSO_4$ was incorporated into mixers that mix baking flour, and a fine-sized white powder with freely flowing characteristics was obtained which could be stored without quality loss for 6 months [55]. Noodles encapsulated with Fe (B5 mg Fe per serving) got comparable sensory scores to those containing inorganic fortification agents such as ferric sodium ethylenediamine tetraacetate (NaFeEDTA) and other ingredients. A three-month storage at ambient temperature did not influence physical (color) or oxidative (peroxide value) parameters. Biofortified wheat flours were made by Biebinger et al. with HPF wall material [56]. The prepared microcapsules may be able to overcome undesirable sensory changes and Fe shortage in biscuits made with wheat flour. Average iodine loss during spray-cooling was B25%, but no measurable iodine losses occurred during baking. Fortified wheat-based biscuits were compared with non-fortified controls in a feeding trial randomized with Kuwaiti young women. Biscuits that were fortified with Fe microcapsules and those that were unfortified were the same in taste and color. However, the fortified ones lowered the prevalence of Fe-deficiency by 50%. The fortified samples group had higher serum ferritin and urine iodine levels. The encapsulated $FeSO_4$ Fe absorption was measured. Iron deficiency in Thai women was overcome by consuming foods based on wheat reinforced with unencapsulated $FeSO_4$, electrolytic Fe, or elemental Fe reduced by hydrogen. After 20 weeks, about 11% of the prescribed Fe dosage was absorbed. Encapsulation could not impact Fe bioavailability. Low-extraction wheat flours can be supplemented with $FeSO_4$ or $C_4H_2FeO_4$ encapsulated (less than 0.8% ash) at the same amounts as without encapsulation [57]. In terms of quality parameters, Fe ($Fe_4O_{21}P_6$, $FeSO_4$, NaFeEDTA, and decreased Fe) and gluten have significant effects. Both elements also influenced the sensory attributes such as odor and the color of wheat bread. The color, stiffness of the crust, cell count, odor, and metallic taste of the crust were found to significantly differ in unfortified gluten-free loaves than Fe-fortified loaves. The encapsulation also failed to protect Fe from oxidation, since the color, taste, and texture of breads fortified with $FeSO_4$ varied from those of unfortified breads. Children in Sao Paulo childcare centers previously rated bread supplements encapsulated with $FeSO_4$ as acceptable as reported by Souto et al. [58]. Breads that were fortified with Fe were less acceptable to children than those that were not. In young children, fortified bread may offer some protection against Fe deficiency anemia. Using a paste made from rice flour, iron compounds (0.5 and 1.0 g Fe/100 g), and 25% water, Moretti et al. [59] produced a replica of rice grains. In order to achieve a final Fe content of 5 mg per 100 g of rice, Fe components (1:100 or 1:200) were combined with rice grains. For this purpose, dispersible micronized $Fe_4O_{21}P_6$ (B0.5 m) was cold extruded, and reagent-grade $Fe_4O_{21}P_6$, $Fe_4O_{21}P_6$ (B2.5 and B20 m), $FeSO_4$ encapsulated in a liposome and HPF, as well as rice grains containing $FeSO_4$ (positive control) were utilized [60]. The rice seeds were simulated and dried overnight to a 10% water content. Uncooked and cooked rice grains had very different colors due to all Fe components except for micronized $Fe_4O_{21}P_6$ (0.5 20 m). The enriched rice seeds with $Fe_4O_{21}P_6$ and elemental Fe had the lowest Fe loss)2%) during the 30 min pre-washing at 30 °C. Researchers in India hope to develop Rice seeds fortified with iron and outstanding sensory properties and enhanced bioavailability rates using micronized $Fe_4O_{21}P_6$. Li and colleagues created Ultra Rice with $C_4H_2FeO_4$, mixtures of micronized $Fe_{41}O_{21}P_6$, FeNaEDTA, thiamine, and other antioxidants. Fe was not bioavailable or stable when added to Ultra Rice. Rice seeds that were encapsulated with Fe_4O_{21}P6 exhibited an acceptable creamy-yellow color, but 50% less thiamine [61]. Unencapsulated $C_4H_2FeO_4$ exhibited a reddish-brown tint. By combining FeNaEDTA with encapsulated $C_4H_2FeO_4$, in vitro bioavailability was improved. Using colloids or particles of $Fe_4O_{21}P_6$ could reduce thiamine losses while maintaining physical and organoleptic properties for 32 weeks at 60% RH and 40 °C. Dietary thiamine loss was minimal; however, oxidative rancidity was observed. BHA, BHT, hydrophilic citric acid, and sodium hexa meta phosphate were found to be the most effective storage forms of uncoated and encapsulated Fe containing free-radical scavenging constituents. Women who consumed cooked Ultra rice containing micronized $Fe_4O_{21}P_6$ daily reduced their anemia by 80% and Fe deficiency by 29%. Beinner, Velasquez-Melendez, Pessoa, and Greiner also reported that the consumption of Ultra Rice encapsulated with micronized $Fe_4O_{21}P_6$ (3.14 μm) reduced Fe deficiency and anemia in Brazilian youngsters (6–24 months) [62].

4.6 Minerals Encapsulation in Other Foods Fortification

The effects of injecting L-Ca with EPC into rabbits before slaughter were studied to determine its impact on meat aging. An infusion of L-Ca into rabbit meat could lower meat's ageing without contaminating it or causing any physical trauma. Water in oil mulsion was prepared using an emulsifier such as Tween 60 to make a water-in-corn-oil emulsion, indicating that the TBARS generation was low, while the Fe-EE (99.75%) was high [63]. To assess the effect of Fe on fish oil stability, fish oil-based emulsion was created and combined with droplets of the initial emulsion. TBA values increased when Fe interacted with fish oil and activated oxidation processes. A spray drying process was conducted to encapsulate natural green pigments from Pandan leaves with three coatings (GA, MS, and MD). For powders encapsulated using Ms. wall material, SEM micro-graphs showed spherical and smooth particles, while powders were generated by the encapsulants. MD and GA surface shrinking Greenness, total chlorophyll, and antioxidants were determined to be best in Zn chlorophyll microcapsules, which comprised of 30% Ms. The produced powder has a predicted half-life (462 days) longer than GA (330 days) and MD (330 days) powders. Fe microparticles were studied in black beans in 2011, which contains 5–10 mg encapsulated Fe per spoon [64]. Using samples supplemented with $FeSO_4$ micro-encapsules could help prevent or control anemia. Peach and apple juice were used to determine Fe status. $Fe_4O_{21}P_6$ was the core and lecithin was the wall of spray-dried Fe microcapsules.

4.7 Application of Nanoparticles in Minerals

In recent years, scientists have become interested in reducing mineral inclusion levels and enhancing mineral absorption through nanoparticles. Mineral NPs are generally spherical and under 100 nm in size. They are frequently employed in agriculture, food, and pharmaceutical industries. Natural mineral nanoparticles (NMNs) have numerous industrial and medical applications. Se-NPs are the most commonly used NPs in the food industry to inhibit bacterial growth and biofilm formation. Melatonin, PEG, adenosine triphosphate, polysaccharides, and epigallocatechin-3-gallate are employed to make Se-NPs with minimal toxicity [65]. A nutritional supplement containing Se-NPs may help improve the body's function and bio accessibility, as Se offers several benefits for the body's health, including enhancing the immune system and reducing cancer cell proliferation. Meal supplementation with Se-NPs lowered bioavailability in Se-deficient mice. We are also aware of the antibacterial properties of inorganic nanometal oxides such magnesia, zirconium, and calcium oxide (CaO) NPs. In order to kill bacteria, metal oxide nanoparticles produce reactive oxygen species. Surface superoxide and H_2O_2 can cause intracellular leakage and cell death. In vitro studies of MgO-NPs conducted on human liver (HepG2), kidney (NRK-52E), gut (Caco-2), and lung (A549) cell lines indicated oxidative stress, DNA damage, and cell death, according to the USDA (21CFR184.1431) [66]. Thus, their use in consumer goods should be monitored for safety reasons. Several studies have shown the efficacy of MgO-NPs in treating various ailments. In U87 human astrocytoma, MgO-NPs were less cytotoxic than ZnO and TiO_2 NPs. Boubeta et al. [67] employed MRI agents made of Fe/MgO nanoshells. Iron/MgO-NPs mediated magnetic hyperthermia in cancer therapy. Nano cryosurgery for cancers and Iron oxide (Fe_2O_3) nanoparticles are biocompatible and less toxic than other metals [67]. A scientist reported the use of Fe_2O_3-NPs as mineral supplements, controlled release medicinal and nutraceutical biomaterials, and colorants in cosmetics. The USDA (21CFR182.8991) also considers ZnO, one of the five Zn compounds, to be safe. By adding antimicrobial ZnO-NPs to active packaging films, especially for meat formulas, we investigated ways to prevent food-borne infections [68]. It was discovered that these NPs are not only antimicrobial but also a Zn source for numerous meals. Although they possess antibacterial properties, $Ca_3(PO_4)_2$-, $CaCO_3$-, and $Ca_3(C_6H_5O_7)_2$-NPs have also shown promise in agriculture, poultry, and food processing. In addition to improving serum Ca levels and bone mineral density, Ca-based nanoparticles are highly bioavailable in the body.

4.8 Conclusion

In summary, the information provided herein can help in the encapsulation of difficult-to-formulate liposoluble vitamins, reactive water-soluble vitamins, and chemically sensitive water-soluble and liposoluble vitamins. Inventors select encapsulation techniques and micro/nanostructures based on resource availability, material qualities, process economics, and end-use applications. These techniques should also be safe to use, have a long shelf life, and require no dangerous chemicals or toxic solvents in their production. Recent research has focused on how to provide active substances like micronutrients in foods. In this perspective, most food fortification and nutraceutical R&D should concentrate on nano- and microencapsulation. To avoid bottlenecks, the hunt for food-grade protective coatings, shorter development times, better understanding of encapsulating techniques, and better understanding of in vivo behavior of encapsulates should be continued.

References

1. Desai KGH, Park HJ. Encapsulation of vitamin C in tripolyphosphate cross-linked chitosan microspheres by spray drying. J Microencapsul [Internet]. 2005;22(2):179–92. https://doi.org/10.1080/02652040400026533
2. Fang Z, Bhandari B. Encapsulation techniques for food ingredient systems. In: Food materials science and engineering. Oxford: Wiley-Blackwell; 2012. p. 320–48.
3. Champagne CP, Fustier P. Microencapsulation for the improved delivery of bioactive compounds into foods. Curr Opin Biotechnol [Internet]. 2007;18(2):184–90. https://doi.org/10.1016/j.copbio.2007.03.001.
4. Madziva H, Kailasapathy K, Phillips M. Alginate pectin microcapsules as a potential for folic acid delivery in foods. J Microencapsul. 2005;22:343–51.
5. Patel AR, Remijn C, Heussen PCM, Adel R, Velikov KP. Novel low-molecular-weight-gelator-based microcapsules with controllable morphology and temperature responsiveness. ChemPhysChem. 2013;14:305–10.
6. Velikov KP, Pelan E. Colloidal delivery systems for micronutrients and nutraceuticals. Soft Matter [Internet]. 2008;4(10):1964. https://doi.org/10.1039/b804863k.
7. Duclairoir C, Orecchioni AM, Depraetere P, Andnakache E. Alpha-tocopherol encapsulation and in vitro release from wheat gliadin nanoparticles. J Microencapsul. 2002;19:53–60.
8. Luo Y, Teng Z, Wang Q. Development of zein nanoparticles coated with carboxymethyl chitosan for encapsulation and controlled release of vitamin D3. J Agric Food Chem [Internet]. 2012;60(3):836–43. https://doi.org/10.1021/jf204194z.
9. Gonnet M, Lethuaut L, Boury F. New trends in encapsulation of liposoluble vitamins. J Control Release [Internet]. 2010;146(3):276–90. https://doi.org/10.1016/j.jconrel.2010.01.037.
10. Tcholakova S, Vankova N, Denkov ND, Danner T. (2007). Emulsification in turbulent flow: 3. Daughter drop-size distribution. Journal of Colloid and Interface Science. 310(2), 570–589. http://dx.doi.org/10.1016/j.jcis.2007.01.097
11. Ziani K, Fang Y, Mcclements DJ. Encapsulation of functional lipophilic components in surfactant-based colloidal delivery systems: vitamin E, vitamin D, and lemon oil. Food Chem. 2012;134:1106–12.
12. Feng JL, Wang ZW, Zhang J, Wang ZN, Liu F. Study on food-grade vitamin E microemulsions based on nonionic emulsifiers. Colloids Surf A Physicochem Eng Asp. 2009;339:1–6.

13. Hukat R, Relkin P. Lipid nanoparticles as vitamin matrix carriers in liquid food systems: on the role of high-pressure homogenisation, droplet size and adsorbed materials. Colloids Surf B: Biointerfaces. 2011;86:119–24.
14. Eskandar NG, Simovic S, Prestidge CA. Chemical stability and phase distribution of all-trans-retinol in nanoparticle-coated emulsions. Int J Pharm [Internet]. 2009;376(1–2):186–94. https://doi.org/10.1016/j.ijpharm.2009.04.036.
15. Patel AR, Velikov KP. Biopolymeric colloidal particles loaded with polyphenolic antioxidants. In: Antioxidant polymers. Hoboken, NJ: Wiley. 2012; 427–58.
16. Semo E, Kesselman E, Danino D, Livney Y. Casein micelle as a natural nano-capsular vehicle for nutraceuticals. Food Hydrocoll [Internet]. 2007;21(5–6):936–42. https://doi.org/10.1016/j.foodhyd.2006.09.006.
17. Haham M, Ish-Shalom S, Nodelman M, Duek I, Segal E, Kustanovich M, et al. Stability and bioavailability of vitamin D nanoencapsulated in casein micelles. Food Funct [Internet]. 2012;3(7):737–44. https://doi.org/10.1039/c2fo10249h.
18. Banville C, Vuillemard JC, Lacroix C. Comparison of different methods for fortifying cheddar cheese with vitamin D. Int Dairy J. 2000;10:375–82.
19. Marsanasco M, Márquez AL, Wagner JR, del V. Alonso S, Chiaramoni NS. Liposomes as vehicles for vitamins E and C: an alternative to fortify orange juice and offer vitamin C protection after heat treatment. Food Res Int [Internet]. 2011;44(9):3039–46. https://doi.org/10.1016/j.foodres.2011.07.025.
20. Alayoubi A, Satyanarayanajois SD, Sylvester PW, Nazzal S. Molecular modelling and multi simplex optimization of tocotrienol-rich self-emulsified drug delivery systems. Int J Pharm. 2012;426:153–61.
21. Moraes M, Carvalho JMP, Silva CR, Cho S, Sola MR, Pinho SC. Liposomes encapsulating beta-carotene produced by the proliposomes method: characterisation and shelf life of powders and phospholipid vesicles. Int J Food Sci Technol [Internet]. 2013;48(2):274–82. https://doi.org/10.1111/j.1365-2621.2012.03184.x.
22. Pierucci APTR, Andrade LR, Baptista EB, Volpato NM, Rocha-Leão MHM. New microencapsulation system for ascorbic acid using pea protein concentrate as coat protector. J Microencapsul [Internet]. 2006;23(6):654–62. https://doi.org/10.1080/02652040600776523.
23. Pierucci AP, Andrade LR, Farina M, Pedrosa C, Rocha-Leuo MH. Comparison of α-tocopherol microparticles produced with different wall materials: pea protein a new interesting alternative. J Microencapsul. 2007;24:201–13.
24. Farias MC, Moura ML, Andrade L, Leão MHMR. Encapsulation of the alpha-tocopherol in a glassy food model matrix. Mater Res [Internet]. 2007;10(1):57–62. https://doi.org/10.1590/s1516-14392007000100013.
25. Patel AR, Vavia PR. Effect of hydrophilic polymer on solubilization of fenofibrate by cyclodextrin complexation. J Incl Phenom Macrocycl Chem. 2006;56:247–51.
26. Yap KL, Liu X, Thenmozhiyal JC, Ho PC. Characterization of the 13-cis-retinoic acid/cyclodextrin inclusion complexes by phase solubility, photostability, physicochemical and computational analysis. Eur J Pharm Sci [Internet]. 2005;25(1):49–56. https://doi.org/10.1016/j.ejps.2005.01.021.
27. Zielenkiewicz W, Terekhova IV, Koźbiał M, Poznanski J, Kumeev RS. Inclusion of menadione with cyclodextrins studied by calorimetry and spectroscopic methods. J Phys Org Chem [Internet]. 2007;20(9):656–61. https://doi.org/10.1002/poc.1223.
28. Majeed H, Jamshaid Qazi H, Safdar W, Fang Z. Microencapsulation can be a novel tool in wheat flour with micronutrients fortification: current trends and future applications—a review. Czech J Food Sci. 2013;31:527.
29. Khiralla GM, El-Deeb BA. Antimicrobial and antibiofilm effects of selenium nano-particles on some foodborne pathogens. LWT—Food Sci Technol. 2015;63(1001):1007.
30. Zhang J, Taylor EW, Wan X, Peng D. Impact of heat treatment on size, structure, and bioactivity of elemental selenium nanoparticles. Int J Nanomed [Internet]. 2012;7:815–25. https://doi.org/10.2147/IJN.S28538.

31. Wang D, Taylor EW, Wang Y, Wan X, Zhang J. Encapsulated nanoepigallocatechin-3-gallate and elemental selenium nanoparticles as paradigms for nano-chemoprevention. Int J Nanomed. 2012;7:1711–21.
32. Ungvári É, Monori I, Megyeri A, Csiki Z, Prokisch J, Sztrik A, et al. Protective effects of meat from lambs on selenium nanoparticle supplemented diet in a mouse model of polycyclic aromatic hydrocarbon-induced immunotoxicity. Food Chem Toxicol [Internet]. 2014;(64):298–306. https://doi.org/10.1016/j.fct.2013.12.004.
33. Sarkar A, Sil PC. Iron oxide nanoparticles mediated cytotoxicity via PI3K/AKT pathway: role of quercetin. Food Chem Toxicol [Internet]. 2014;(71):106–15. https://doi.org/10.1016/j.fct.2014.06.003.
34. Mirhosseini M, Afzali M. Investigation into the antibacterial behavior of suspensions of magnesium oxide nanoparticles in combination with nisin and heat against Escherichia coli and Staphylococcus aureus in milk. Food Control. 2016;68(208):215.
35. Akbar A, Anal AK. Zinc oxide nanoparticles loaded active packaging, a challenge study against Salmonella typhimurium and Staphylococcus aureus in ready-to-eat poultry meat. Food Control [Internet]. 2014;38:88–95. https://doi.org/10.1016/j.foodcont.2013.09.065.
36. Rossi L, Velikov KP, Philipse AP. Colloidal iron(III) pyrophosphate particles. Food Chem [Internet]. 2014;151:243–7. https://doi.org/10.1016/j.foodchem.2013.11.050.
37. Vijayakumar MP, Balakrishnan V. Evaluating the bioavailability of calcium phos- phate nanoparticles as mineral supplement in broiler chicken. Indian J Sci Technol. 2014;7(1475):1480.
38. Ataee RA, Derakhshanpour J, Mehrabi Tavana A, Eydi A. Antibacterial effect of calcium carbonate nanoparticles on Agrobacterium tumefaciens. Iran J Mil Med. 2011;13(65):70.
39. Kwak HS, Yang KM, Ahn J. Microencapsulated iron for milk fortification. J Agric Food Chem [Internet]. 2003;51(26):7770–4. https://doi.org/10.1021/jf030199.
40. Gupta C, Chawla P, Arora S, Tomar SK, Singh AK. Iron microencapsulation with blend of gum arabic, maltodextrin and modified starch using modified solvent evapora- tion method-Milk fortification. Food Hydrocoll. 2015;43:622–8.
41. Chang YH, Lee SY, Kwak H-S. Physicochemical and sensory properties of milk fortified with iron microcapsules prepared with water-in-oil-in-water emulsion during storage. Int J Dairy Technol. 2016;69(3):452–9.
42. Saeidy S, Keramat J, Nasirpour A. Physicochemical properties of calcium- fortified soymilk with microencapsulated and chelated calcium salt. Eur Food Res Technol. 2014;238(105):112.
43. Jayalalitha V, Balasundaram B, Palanidorai A, Naresh KC. Fortification of encapsulated iron in probiotic yoghurt. Int J Agric Res Rev. 2012;2(2):80–4.
44. Subash R, Elango A. Microencapsulated iron for fortification in yoghurt. Food Sci Res J [Internet]. 2015;6(2):258–62. https://doi.org/10.15740/has/fsrj/6.2/258-262.
45. Cavallini DCU, Rossi EA. Soy yogurt fortified with iron and calcium: stability during the storage. Aliment Nutr. 2009;20(7):13.
46. Kwak HS, Ju YS, Ahn HJ, Ahn J, Lee S. Microencapsulated iron fortification and flavor development in cheddar cheese. Asian Australas J Anim Sci. 2003;16(1205):1211.
47. Arce A, Sc AM. Effect of micro-encapsulated ferrous sulfate on Cheddar cheese composition, divalent cation balance and acceptability. East Lansing: Michigan State University; 2016.
48. Jalili M. Chemical composition and sensory characteristics of Feta cheese fortified with iron and ascorbic acid. Dairy Sci Technol. 2016;96(4):579–89.
49. Diosady LL, Alberti JO, Venkatesh Mannar MG. Microencapsulation for iodine stability in salt fortified with ferrous fumarate and potassium iodide. Food Res Int [Internet]. 2002;35(7):635–42. https://doi.org/10.1016/s0963-9969(01)00166-1.
50. Diosady L, Yusufali R, Oshinowo T, Laleye L. A study of storage and distribution of double fortified salts in Kenya. J Food Eng. 2006;76:547–56.
51. Li YO, Diosady LL, Wesley A. Iron in vitro bioavailability and iodine storage stability in double fortified salt (DFS). Food Nutr Bull. 2009;30(327):335.

52. Yadava D, Olive Li Y, Diosady LL, Wesley AS. Optimisation of polymer coating process for microencapsulating ferrous fumarate for salt double fortification with iodine and iron. J Microencapsul [Internet]. 2012;29(8):729–38. https://doi.org/10.3109/02652048.2011.651493.
53. Romita D, Cheng Y-L, Diosady LL. Microencapsulation of ferrous fumarate for the production of salt double fortified with iron and iodine. Int J Food Eng [Internet]. 2011;7(3):5. https://doi.org/10.2202/1556-3758.2122.
54. Wegmüller R, Zimmermann MB, Bühr VG, Windhab EJ, Hurrell RF. Development, stability, and sensory testing of microcapsules containing iron, iodine, and vitamin A for use in food fortification. J Food Sci. 2006;71(2):S181–7.
55. Kongkachuichai R, Kounhawej A, Chavasit V, Charoensiri R. Effects of various iron fortificants on sensory acceptability and shelf-life stability of instant noodles. Food Nutr Bull. 2007;28(2):165–72.
56. Biebinger R, Zimmermann MB, Al-Hooti SN, Al-Hamed N, Al-Salem E, Zafar T, et al. Efficacy of wheat-based biscuits fortified with microcapsules containing ferrous sulfate and potassium iodate or a new hydrogen-reduced elemental iron: a randomised, double-blind, controlled trial in Kuwaiti women. Br J Nutr. 2009;102:1362–9.
57. Kiskini A, Kapsokefalou M, Yanniotis S, Mandala I. Effect of different iron compounds on wheat and gluten-free breads. J Sci Food Agric. 2010;90:1136–45.
58. Souto TS, Brasil ALD, Taddei JADC. Acceptability of bread fortified with microencapsulated iron by children of daycare centers in the south and east regions of Sao Paulo City, Brazil. Rev Nutr. 2008;21:647–57.
59. Moretti D, Lee T-C, Zimmermann MB, Nuessli J, Hurrell RF. Development and evaluation of iron-fortified extruded rice grains. J Food Sci. 2005;70(5):S330–6.
60. Hotz C, Porcayo M, Onofre G, García-Guerra A, Elliott T, Jankowski S, et al. Efficacy of iron-fortified Ultra Rice in improving the iron status of women in Mexico. Food Nutr Bull. 2008;29(2):140–9.
61. Li Y, Diosady LL, Jankowski S. Stability of vitamin B1in Ultra Rice® in the presence of encapsulated ferrous fumarate. Int J Food Sci Nutr [Internet]. 2008;59(1):24–33. https://doi.org/10.1080/09637480701554103.
62. Beinner MA, Velasquez-Meléndez G, Pessoa MC, Greiner T. Iron-fortified rice is as efficacious as supplemental iron drops in infants and young children. J Nutr [Internet]. 2010;140(1):49–53. https://doi.org/10.3945/jn.109.112623.
63. Choi SJ, Decker EA, McClements DJ. Impact of iron encapsulation within the interior aqueous phase of water-in-oil-in-water emulsions on lipid oxidation. Food Chem [Internet]. 2009;116(1):271–6. https://doi.org/10.1016/j.foodchem.2009.02.045.
64. Ferreira BS, Cardoso BT, Pereira HVR, Pierucci AP, Pedrosa C, Citelli M. Acceptability of black beans (Phaseolus vulgaris L.) fortified with iron microparticles. Rev Ceres. 2011;58:548–53.
65. Zhang Y, Li X, Huang Z, Zheng W, Fan C, Chen T. Enhancement of cell permeabilization apoptosis-inducing activity of selenium nanoparticles by ATP surface decoration. Nanomedicine [Internet]. 2013;9(1):74–84. https://doi.org/10.1016/j.nano.2012.04.002.
66. Mahmoud A, Ezgi Ö, Merve A, Özhan G. In vitro toxicological assessment of magnesium oxide nanoparticle exposure in several mammalian cell types. Int J Toxicol [Internet]. 2016;35(4):429–37. https://doi.org/10.1177/1091581816648624.
67. Boubeta CM, Balcells L, Cristofol R, Sanfeliu C, Rodrıguez E, Weissleder R, et al. Self-assembled multifunctional Fe/MgO nanospheres for magnetic resonance imaging and hyperthermia. Nanomedicine. 2010;6(2):362–70.
68. Akbar A, Anal AK. Zinc oxide nanoparticles loaded active packaging, a challenge study against Salmonella typhimurium and Staphylococcus aureus in ready-to-eat poultry meat. Food Control. 2014;38:88–95.

Chapter 5
Nanoemulsions in Food Industry

Goutam Kumar Jena, Rabinarayan Parhi, and Suvendu Kumar Sahoo

5.1 Introduction

Currently, severe health-related issues are arising worldwide, which create increasing demands for relatively healthy, safe as well as economical food products [1]. Due to the introduction of functional food, the importance of food is extensively increased. These are ever-increasing demands that can be met either by adding new bioactive components or by increasing the production of existing biologically active molecules [2]. Hence food products are significant not only for their nutritional values but also for their health promotion as well as disease prevention values [3]. The benefits of health-promoting food products are limited due to their low bioavailability as well as stability issues [4]. The applications of nanotechnology in food science have received tremendous attention in the last few decades. Nanoemulsions have been drawn significant attention in the delivery system in the food industry [5]. Nanoemulsion-based delivery systems not only improve the bioavailability of the encapsulated bioactive components but also increase food stability. Nanoemulsions are kinetically stable but thermodynamically unstable transparent or translucent colloidal systems with very small particle sizes within 100 nm [6]. As nanoemulsions have the unique property of small size, increased surface area, and sensitivity to physical and chemical changes, they are ideal candidates for the food industry [7]. The significance of food-grade nanoemulsions

G. K. Jena (✉)
Department of Pharmaceutics, Roland Institute of Pharmaceutical Sciences, Berhampur, Odisha, India

R. Parhi
Department of Pharmaceutical Sciences, Susruta School of Medical and Paramedical Sciences, Assam University, Silchar, Assam, India

S. K. Sahoo
Department of Pharmaceutics, GITAM University, Visakhapatnam, India

C. Egbuna et al. (eds.), *Application of Nanotechnology in Food Science, Processing and Packaging*, https://doi.org/10.1007/978-3-030-98820-3_5

is to enhance encapsulation efficiency, digestibility, bioavailability, and targeted delivery. Due to the aforesaid merits of nanoemulsions as compared to conventional emulsions, they play a major role in the food industry. Some important stabilizers such as emulsifiers, texture modifiers, ripening retarders, weighting agents as well as ripening retardants can be incorporated to improve the Stability of nanoemulsions. Texture modifiers, substances that increase the viscosity such as proteins (whey protein isolate, gelatin or soy protein isolate), sugars (high-fructose corn syrup or sucrose), polysaccharides (carrageenan, xanthan, pectin, alginate), and polyols (sorbitol or glycerol) can also be used as stabilizers. The widely used emulsifiers in the food industry for formulation of nanoemulsions are small-molecule surfactants such as Tweens and Spans, amphiphilic polysaccharides such as gum Arabic, or modified starch, phospholipids such as soy, egg, or diary lecithin, and amphiphilic proteins such as caseinate or whey protein isolate [8]. The weighting agents such as dense lipophilic materials such as brominated vegetable oils, sucrose acetate, isobutyrate, ester gums can be employed for balancing the densities of liquids emulsions [9].

The current chapter will highlight the method of preparation, types, and applications of nanoemulsions in food science.

5.2 Components of Nanoemulsions

Generally, a nanoemulsion consists of three components, i.e., an oil phase, aqueous phase, and an emulsifier. An emulsifier is a very important component of nanoemulsion, whose presence in little quantity helps in fabricating nanoemulsions by reducing the interfacial tension between water and oil phases and also enhancing stabilization of nanoemulsions [2]. The three aforesaid components critically attribute affect the fabrication as well as stabilization of nanoemulsions.

5.2.1 Oil Phase

The various nonpolar molecules, which are used as oil phase for fabrication of food-grade nanoemulsions, consists of free fatty acids, monoacylglycerols, diglycerol, triglycerols, waxes, mineral oils, or various lipophilic nutraceuticals. Soybean, safflower, corn, flaxseed, sunflower, olive, algae, and fish are the richest source of triglycerols oils, which are usually employed in nanoemulsions due to their less cost and more nutritional value [3]. The properties of nanoemulsions mainly depend on the physical and chemical properties of the oil phase, such as water solubility, viscosity, density, polarity, refractive index as well as interfacial tension whereas, the chemical stability has a great impact on the different properties of nanoemulsions.

5.2.2 *Aqueous Phase*

The variety of polar molecules, such as carbohydrates, proteins, acids, minerals, or alcoholic cosolvents in water, can be used as a food-grade aqueous phase for the preparation of nanoemulsion. The physicochemical properties of nanoemulsions are greatly influenced by the selection of appropriate aqueous phases [10].

5.2.3 *Stabilizers*

The long-term Stability of nanoemulsions is influenced by stabilizers. Hence, the production of nanoemulsions depends on the selection of appropriate stabilizers. Stabilizers may be texture modifiers, emulsifiers, ripening retarders, weighting agents, and emulsifiers.

5.3 Types of Nanoemulsions

Nanoemulsions are classified based on the ratio of oil and water. On the basis of the ratio of oil and water, there are three types of nanoemulsions, [1] oil-in-water [2] water-in-oil and [3] multiple nanoemulsions (oil-in-water-in-oil and water-in-oil-in-water). All types of nanoemulsions are stabilized by using a mixture of surfactants and/or cosurfactants. In the case of the O/W type of nanoemulsion, the oil phase (discontinuous phase) is dispersed in the aqueous phase (continuous phase), which is stabilized and favored by solubilizing surfactants in water whereas, in the case of the W/O type of nanoemulsion water phase is dispersed in oil phases, which are stabilized and favored by oil solubility of the surfactants. A multiple nanoemulsion is formed by mixing an equal volume of oil and water phases in which both phases are inter-dispersed.

5.4 Fabrication of Nanoemulsions

The energy is required for the fabrication of nanoemulsions as these are thermodynamically unstable systems. Based on the energy requirement, the method of fabrication of nanoemulsions is basically two types as follows [3].

5.4.1 *High Energy Methods*

In this method, a variety of devices is generally applied in order to create huge mechanical energy for producing small-sized droplets in nanoemulsions. Mechanical devices are used to break oil and water phases to produce small nanosized droplets. The devices used for preparing nanoemulsions in the case of high-energy methods are high-pressure homogenizers, microfluidizers, sonicators, etc.

5.4.2 *Low Energy Methods*

Nanoemulsions with small-sized droplets are fabricated by employing less amount of energy. The physical and chemical properties of the surfactants, cosurfactants as well as oil phase required to produce nanoemulsions affect the desired size of nanoemulsions. The low energy methods for fabricating nanoemulsions are spontaneous emulsification, phase inversion technique, membrane emulsification methods, emulsion inversion point method as well as solvent displacement method.

5.5 Characterization of Nanoemulsions

Like other nanocarriers, nanoemulsion undergoes characterization to fulfill specific objectives of managing particle size, surface properties, drug release, and also Stability. Therefore, the characterization of nanoemulsion is very crucial to control theirs in vitro and in vivo behavior. Nanoemulsions are characterized by various techniques, which are discussed below.

5.5.1 *Particle Size, Polydispersity Index, and Zeta Potential*

Particle size is a crucial parameter that differentiates nanoemulsion from other emulsion systems. There are a few methods to determine droplet size of emulsion, i.e., dynamic light scattering (DLS) or photon correlation spectroscopy (PCS), laser diffraction spectroscopy (LD), microscopic techniques such as electron microscopy, atomic force microscopy (AFM). By controlling the droplet size of the nanoemulsion, the appearance of the nanoemulsion can be tuned from transparent to milky white [11]. The polydispersity index (PDI) is determined by the PCS technique and is a measure of the broadness of the size distribution of emulsion droplets [12]. The surface charge of the droplets is determined by the zeta potential method. It also predicts dispersion stability, and its value depends on various physicochemical properties. Zeta potential is determined by the Malvern Zetasizer instrument, and its value of ±30 mV is considered to be sufficient to ensure physical Stability [13].

5.5.2 Encapsulation Efficiency and Loading Efficiency

Encapsulation efficiency (EE%) is the quantity of drug encapsulated per unit weight of nanoemulsion estimated after removal of the free drug. Separation of free drugs can be performed with methods such as ultracentrifugation, ultrafiltration, dialysis, etc. In another way, EE% can be assessed by dissolving nanoemulsion into a suitable solvent and then subsequently analyzed for drug content. On the other hand, loading efficiency is calculated as the amount of the drug entrapped into the particles divided by the total quantity present in the formulation [13].

5.5.3 Morphological Characterization

The surface morphology of nanoemulsion is generally ascertained employing electron microscopies such as scanning electron microscopy (SEM) and transmission electron microscopy (TEM). A 3D image of the globules is obtained by SEM. An image analysis software is usually used to generate automatic analysis results of shape and surface morphology. On the other hand, TEM gives higher resolution images of the disperse phase. A digital image processing program can be used to qualitatively measure the sizes and size distribution of TEM micrographs [12]. Most recently, AFM and cryo-electron microscopy has been employed to explore the structure and behavior of nanoemulsion [12].

5.5.4 Measurement of Structural Modifications

Structural modifications of components in nanoemulsion are analyzed with techniques such as Fourier-transform infrared spectroscopy (FTIR), differential scanning calorimetry (DSC), and thermogravimetric analysis (TGA). FTIR analysis is generally performed to assess the presence of drug excipient interaction, cross-linking, or polymerization. DSC involves heat exchanges such as heat uptake (melting curve) or heat emission (crystallization curve), which occurs under controlled temperature that gives a physical quality of the material. However precise nature of the thermal transition is disclosed by TGA [13].

5.5.5 *Viscosity Measurement*

Viscosity is a very important parameter to be determined in the case of nanoemulsion as it ensures Stability and efficient drug release from it. The viscosity of nanoemulsion is a function of types and concentrations of oil and aqueous phases, and the surfactant. For instance, increasing the water content of the nanoemulsion leads to a decrease in viscosity, whereas decreasing the amount of surfactant increases the viscosity of nanoemulsion [12].

5.5.6 *Drug or Food Supplement Release Study*

The release of drug or food supplements from nanoemulsion is a very important aspect as it determines the way and level to which the therapeutics molecules are released. There are a few methods to determine in vitro release pattern of the encapsulated ingredients from nanoemulsion, including (i) diffusion through dialysis bag, (ii) reverse dialysis bag method, (iii) compartmental diffusion cells with artificial or biological membrane, and centrifugal filtration method. The drug release study is conducted under controlled stirring and centrifugation [12, 13].

5.5.7 *Stability Studies and Self-Life Determination*

Stability studies are carried out to assess the Stability of the encapsulated ingredients when exposed to various environmental factors such as temperature and humidity. The stability study of nanoemulsion is performed as per International Conference on Harmonization guidelines such as short-term, long-term, or accelerated stability studies. Accelerated stability studies are conducted to determine the self-life of a nanoemulsion [13].

5.6 Applications of Nanoemulsions in the Food Industry

The applications of nanotechnology in the food industry progressed through limitless boundaries with invent of various nano-delivery systems, including nanoparticles, nanoemulsions, and most recently, nanocomposites. It is important to mention here that nanotechnology is not only applied to food processing but also food preservation. In this context, nanoemulsion is a key nano-vehicle that has been used in the food industry to improve physical properties, the efficacy of foodstuffs, enhance food safety by appropriate packaging and preservation of qualities of food materials with antibacterial, antifungal, and antioxidant agents, and most importantly delivery of various food components. Various applications of nanoemulsion in the food industry are illustrated in Fig. 5.1 and discussed with case studies [14].

5.6.1 Enhancing Physical Properties of Food

5.6.1.1 Coloring Activity

Color additives intended to be used in food must comply with the approved uses and specifications laid down by the Office of Cosmetics and Colours in the Center for Food Safety and Applied Nutrition of the US FDA. With the entry of nanotechnology in the food industry, many nanoscale color additives have been manufactured and investigated. However, only a few nanoemulsion-based color additives are being used in the food industry. In one such investigation, nanoemulsion of curcumin was developed using milk protein followed by their use in ice cream formulations. The developed nanoemulsion showed good physicochemical properties in terms of particle size (333.8 nm), encapsulation efficiency (96.9%), and surface charge (−44.1 mV). In addition, considerable Stability was observed under simulated conditions with good sensory scores [15].

5.6.1.2 Increasing Food Efficacy

Enhancing the Digestibility Quality of Food

The digestibility quality of food is an indication of the amount of food absorbed through the gastrointestinal tract into the bloodstream. It can be enhanced with the use of nanoemulsion because they are capable of increasing the surface area of food materials due to their small droplet size. Thus, nanoemulsion of foodstuffs serves as a functional behavior for food materials. However, a suitable emulsifier is necessary for its Stability during processing and storage [16].

There are several studies performed on the effect of nanoemulsion of foodstuffs on digestibility. In one study, curcumin was incorporated in the oil phase of nanoemulsion, and then its digestibility was compared with curcumin consumed directly. The result demonstrated that curcumin in nanoemulsion had higher digestibility due to the easy lipid digestion step of nanoemulsion in the GI tract [17]. In another study, vitamin D3 was incorporated into the oil core of O/W nanoemulsion prepared with polysorbate 20, soybean lecithin, and their mixtures as emulsifiers, followed by their use in the fortification of dairy emulsions. The droplets size was found to be less than 200 nm, and the resulting fortified whole-fat milk showed Stability for at least 10 days [18]. Same vitamin D3 was also encapsulated in the oil phase comprised of Kolliphor and caprylic-/capric triglyceride and sodium chloride solution as the aqueous phase. The resulted nanoemulsion showed promises for the fortification of beverages [19].

Fig. 5.1 Schematic diagram of an application of nanoemulsion in the food industry

Increasing the Bioavailability

The use nanomaterials, more specifically nanoemulsion, has been used as a carrier to increase the bioavailability of bioactive compounds as nutritional supplements. The maximization of bioavailability of bioactive in nanoemulsion can be achieved either through bioaccessibility or absorption. The former has been achieved by altering the particle size leading to improved solubility due to an increase in area-to-volume ratio. For example, a novel lipid-free nano CoQ10 (lipophilic bioactive) system demonstrated its improved solubility and bioavailability after oral administration [20]. The absorption of bioactive can be improved by enhancing their permeation across the GI wall. This can be achieved by selecting a proper surfactant. In one study, hydrochloric acid as a surfactant was found to enhance the permeation of green tea catechin from O/W nanoemulsion stabilized with soy protein [21].

5.6.2 *Increase Food Safety*

Food safety intends to ensure the quality of food remains and is safe for consumption. It can be achieved either by employing packaging or by edible coating. Both provide physical protection to the food materials and help in food preservation.

5.6.2.1 By Packaging

Packaging is a way of physically protecting food products from external environmental conditions such as temperature, oxygen and other gas, and moisture. Nonetheless, it also protects food materials from microbes, which otherwise may lead to food spoilage. Nanotechnology, particularly nanoparticles in packaging) offers a better avenue compared to traditional methods of food preservation such as freezing, drying, and canning by ensuring a longer self-life. Active packaging provides an inert barrier to the above mention external conditions regarding the packaging systems which respond to the environmental changes. Active packaging may also be incorporated with a desirable molecule to improve food stability, including antimicrobial or antioxidant agents. Active packaging can act as control-release packaging, thereby allowing the migration of food supplements such as probiotics, vitamins, and minerals [22].

5.6.2.2 Edible Coating/Packaging

Materialistic lifestyle demands ready-to-eat foods due to their convenience, better sensorial qualities, and fresh-like appearance, leading to its tremendous growth in the market. Further, its importance increased for minimally processed products such as vegetables, fruits, and dairy or meat. Application of nanoemulsion in the form of edible coating onto these food products not only to prolong their self-life but also to prevent microbial growth on the food surface resulting from peeling and cutting operation [23]. The nanoemulsion used to form edible coating comprised of the continuous phase, disperse phase, and a suitable emulsifier to stabilize both phases. A variety of coating-forming materials such as polysaccharides (e.g., starch, pectins, alginates, chitosan) and proteins (e.g., soy protein, whey protein) [24]. For instance, Otoni et al. [25] prepared edible film from nanoemulsion comprised of pectin as a film former, cinnamaldehyde (essential oil), and papaya puree. The resulting edible film was found to be effective in inactivating different bacteria [25]. In another attempt, the nanoemulsion-based edible coating was developed with alginate as a film-forming agent and essential oils such as thyme, lemongrass, or sage oil. This edible film provides different functions as well as physical properties [26].

Fruits and Vegetables

The excellent nutritional value of fruits and vegetables is due to the presence of abundant fibers, vitamins, and minerals. However, these are susceptible to faster deterioration due to the presence of high moisture content and subsequently higher water activity. In addition, they are prone to microbial contamination owning to living tissue in them, leading to enzymatic browning, off-flavors, and texture breakdown. Therefore, fruits and vegetables are coated with edible films to circumvent the above issues [27].

Fresh-cut fuji apples were better preserved with edible nanoemulsion coating with alginate, lemongrass oil, and Tween 80 from *E. coli* compared to the conventional emulsion. With the same emulsion, fresh-cut apples were preserved for 2 weeks from the microbes without impacting their quality attributes [28]. In a recent study, the self-life of grape berries was found to be increased with the edible nanoemulsion coating prepared with lemongrass oil, chitosan, and Tween 20. The mentioned coating provides exceptional protection against *Salmonella* species. In addition, there was no significant difference in sensorial effect between edible-coated and uncoated grape berries [29].

In one study, nanoemulsion-based edible coating comprised of lemon oil, chitosan as coating forming polymer, and surfactants such as glycerol monooleate and Tween 20 was prepared for the improvement of self-life of rucola leaves. The edible coating was found to extend the self-life of the rucola leaves for 7 days during cold storage [30]. In another study, chitosan and carvacrol-based nanoemulsion were found to completely inhibit inoculated *E. coli* growth on fresh green beans for 11 days in cold storage [31]. The self-life of other fruits and vegetables such as plums, arugula, and lettuce has been extended employing nanoemulsion-based edible coating [32].

Meat and Poultry

Meat and chicken, either in fresh or cured form, fulfill the animal protein requirement for humans. However, pathogens such as *Listeria monocytes* pose immense challenges for the preservation of ready-to-eat meat products [33]. The Application of nanomaterials in the meat industry can offer the following: manufacturing of packaging that provides the highest level of the barrier for microbes; reduction of components of meat which has a detrimental effect on the health of consumers such as salts, fats, and nitrites; delivery of ingredients having antibacterial or antioxidant properties along with the nutrient quality of meat [34]. Noori et al. developed sodium caseinate and ginger essential oil-based nanoemulsion for edible coating onto chicken breast fillet to enhance its self-life. The coating of nanoemulsion with 6% of ginger essential oil was found to significantly decrease the total aerobic psychrophilic bacteria of refrigerated chicken fillets during 12 days [35].

5.6.3 Dairy Products

Dairy products include milk and its derived products, including cheese, cream, fermented milk, etc. These dairy products have high nutritional quality and play a key constituent in the human diet. However, dairy products are prone to spoilage due to external factors such as oxygen, moisture, light, and microorganism, leading to undesirable changes, including oxidation, discoloration, change in flavors [36]. Among various options available, edible coating of dairy products seems to be most effective in improving the self-life of the dairy products. The preservation of low-fat cut cheese has been a challenge due to its low salt and higher water content leading to its spoilage by bacteria and fungi. Therefore, a multifunctional nanoemulsion comprised of sodium alginate, organic essential oil, tween 80, and mandarin fibers was developed to create an edible coating around low-fat cut cheese. The presence of organic essential oil reduces the growth of bacteria, yeast, and molds in cheese during refrigerated storage. The incorporation of mandarin fiber, a prebiotic, enhanced the nutritional value of the coated cheese [37].

5.6.4 Bakery

Nanoemulsion-based edible films are also used to preserve bakery products. In one such study, nanoemulsions composed of methylcellulose and clove or oregano essential oil was developed and successfully formed edible coat onto packaged bread slides. Mentioned edible coat extended the self-life of the sliced bread for 15 days due to the presence of essential oils [38].

5.6.5 Preservation of Food

Food preservation is an important aspect in terms of protecting its nutritional quality and providing freshness for ready-to-use foodstuffs. In this perspective, antibacterial agents or antifungal agents in the nanoemulsion can offer such preservation. Essential oils are mainly used in nanoemulsion for preservation.

5.6.6 Antibacterial Agent in Nanoemulsion

Thyme essential oils showed exceptional antibacterial properties. Therefore, it has been widely used to preserve foodstuffs against various pathogens. In one attempt, thyme essential oil was loaded into nanoemulsion to preserve ultrafiltered Labneh Cheese. The resulting nanoemulsion eliminated pathogenic count in a span of 1 h and kept the quality of cheese intact up to 6 weeks [39]. In another study, nanoemulsion comprised of gelatine and lecithin with thymol essential oil in order to produce fortified milk. The bacteriostatic effect of thymol showed a gradual reduction in the colony of *Listeria monocytogenes* bacteria, which resulted in the preservation of fortified milk [40]. Moghimi et al. developed nanoemulsion by incorporating essential oil (obtained from *Thymus diagenesis*) and then examined its antibacterial activity against *E. coli*. The minimum inhibitory concentration (MIC) and minimum bactericidal concentration (MBC) against *E. coli* demonstrated higher antibacterial activity compared to the pure form of essential oil, which was attributed to the intimate contact between the cell wall of the bacteria and essential oil [41].

5.6.7 Antifungal Agent in Nanoemulsion

Similar to that of antibacterial activity, few essential oils showed antifungal activity, thereby preserving foodstuffs from fungal pathogens. Nanoemulsion encapsulated with oregano essential showed control over fungal spoilage of Minas Padrao cheese. The resulting nanoemulsion demonstrated higher antifungal activity on *Fusarium* spp. [42]. Wan et al. developed nanoemulsions are having various essential oils such as thyme, lemongrass, cinnamon, clove, or peppermint oil and studied antifungal and mycotoxin inhibition effects against *Fusarium graminearum*. The result showed that the mycotoxin inhibitory activity of the nanoemulsion was enhanced due to the greater solubility of essential oils [43].

Curcumin is a natural ingredient obtained from Turmeric, which showed strong antimicrobial efficacy against various food pathogens. Recently, nanoemulsions containing curcumin were developed with essential oils such as cinnamon and garlic or sunflower oil, high ester pectin, and Tween 80. The resulting emulsion was found to inhibit the growth of fungi and bacteria. Nanoemulsion with incorporated cinnamon oil demonstrated the highest efficacy in controlling microbial population throughout the period of storage compared to other oil [44].

5.6.8 Anti-oxidant in Nanoemulsion

Antioxidant showed their activity basically through two ways: (i) free radical scavenging and (ii) inhibition of enzymes that generate free radicals. Curcumin is a natural antioxidant that eliminates peroxyl radicals from the system [45]. In one investigation, Joung et al. developed a nanoemulsion loaded with curcumin and added into milk to evaluate lipid oxidation and physical Stability during storage. The resulting nanoemulsion demonstrated good physical Stability during storage at room temperature, and lipid oxidation was found to be significantly lower in milk samples fortified with nano curcumin compared to sample without curcumin [46]. In another study, nanoemulsion developed with corn oil showed significant radical scavenging activity [47].

5.6.9 Delivery of Food Components

Various food components such as healthy lipids, vitamins, and excipient foods can be better delivered with nanoemulsion.

5.6.9.1 Healthy Lipids Delivery

Healthy lipids such as omega-3 fatty acids, omega-3 fatty acids rich in algal oil, alpha-linolenic acids, bioactive oil rich in tocopherols are investigated to deliver with nanoemulsion vehicle. Nanoemulsion incorporated with flaxseed oil (a rich source of alpha-linolenic acid) and stabilized with whey protein and sodium alginate was developed. Alpha-linoleic acid is an important source of omega-3 fatty acids that has many health benefits. The resulted in nanoemulsion when feed to broiler chicken showed a higher level of omega-3 fatty acids in the flesh compared to bulk flaxseed oil [48]. In another work, a nanoemulsion was developed to deliver bioactive rich in tocopherols obtained from kenaf seed oil, and β-cyclodextrin, Tween 20, and sodium caseinate were added to stabilize it. After administration of the above nanoemulsion to rats, it was observed that the tocopherol level in the bloodstream was increased rapidly compared to the emulsion and bulk oil. Both the results were attributed to the faster digestion and absorption of small droplets (more surface area) present in nanoemulsion [39]. The same authors investigated the ability of kenaf seed oil in reducing cholesterol levels in hypercholesterolemia rats. The results showed that nanoemulsion-based vehicles demonstrated the strongest cholesterol-lowering ability and also reduced liver fat levels [49].

5.6.9.2 Vitamins Delivery

Various oil-soluble vitamins such as Vitamin A, Vitamin D, and Vitamin E were successfully encapsulated within nanoemulsion to improve their Stability, bioavailability, bioaccessibility, and bioactivity. For instance, a nanoemulsion of Vitamin A (retinol) employing different types of oils and emulsifiers was developed. The above nanoemulsion prevented the degradation of Vitamin A and improved the Stability during the storage period, which was based on the appropriate section of emulsifier and oils [50]. In another work, the bioavailability of Vitamin D3 (cholecalciferol) was found to be enhanced when incorporated in nanoemulsion [51]. In the same way, the bioavailability of Vitamin D2 (ergocalciferol) was investigated by incorporating it into the nanoemulsion [52]. In vitro and in vivo studies demonstrated that bioavailability of both Vitamin D2 and D3 were enhanced due to an increase in bioaccessibility with the decrease in droplets size, which was resulted from the faster rate of digestion and micellar solubilization. Nanoemulsion of α-tocopherol succinate (a vitamin E analog) was studied for its increase in bioactivity against cancer. The result showed that the above nanoemulsion demonstrated enhanced anticancer activity against model breast cancer cell line and human oral epithelial cancer cell line compared to the bulk form of the vitamin [53].

5.6.10 Excipient Nanoemulsion Delivery

Excipient nanoemulsions are basically O/W systems whose structure and composition are prudently controlled to generate an environment inside the gut that improves the bioaccessibility, Stability, and/or absorption of co-ingested bioactive substances [54]. Thus, excipient nanoemulsion enhances the bioavailability of hydrophobic substances present in co-ingested foods, including vitamins and nutraceuticals [55]. The suitability of the nanoemulsion is due to small oil droplets that are quickly digested within the human gut lumen, leading to the rapid formation of mixed micelles, which can solubilize and transport the hydrophobic substances originated from foods or supplements taken with it [56]. On the basis to form a variety of products intended for use in the food or supplement industries, excipient nanoemulsion could be used in various ways such as (i) salad dressings that are added onto salads, (ii) rich creams which are added onto fruits, (iii) cooking sauces which are consumed with cooked vegetables, etc. The excipient nanoemulsion was found to enhance the bioaccessibility of carotenoids present in tomatoes, carrots, spinach, and dietary supplements [57].

5.6.11 Delivery of Nutraceuticals

Nutraceuticals are being encapsulated, protected, and delivered with the help of nanoemulsion. Hydrophobic nutraceuticals are typically dissolved in the oil phase prior to the formation of nutraceuticals. The bioaccessibility of nutraceuticals depends on various factors such as oil droplet size, oil droplet content, oil digestibility, and the type of emulsifier [58]. For instance, beta-carotene showed higher bioaccessibility in nanoemulsions formulated with long-chain triglycerides [59]. Nanoemulsion of quercetin showed 3.4-fold higher absorption and 34-fold higher bioavailability compared to free quercetin when taken orally [60]. In another study, the oral bioavailability of coenzyme Q10 was found to be higher compared to the bulk oily formulation when administered orally to rats [61]. Other nutraceuticals such as lycopene, lutein, astaxanthin [57]. Various applications of nanoemulsion in the food industry with bioactive and results are mentioned in Table 5.1.

Table 5.1 Applications of nanoemulsion in the food industry

Types of application	Bioactive	Results	References
1. Enhancing physical properties of food			
Coloring activity	Curcumin	Good physiochemical properties and improved stability	[15]
2. Increasing food efficacy			
Enhancing the digestibility of food	Curcumin	Higher digestibility	[17]
	Vitamin D3	Fortified whole-fat milk showed stability	[18]
	Vitamin D3	Promises for the fortification of beverages	[19]
Increasing bioavailability	Lipid-free nano CoQ10	Improved solubility and bioavailability	[20]
	Green tea catechin	Enhance permeation	[21]
3. Increasing food safety			
By edible film/coating	Cinnamaldehyde and papaya puree	Effective in inactivating different bacteria	[25]
	Thyme, lemongrass or sage oil	Provides different functional as well as physical properties	[26]
	Lemongrass oil	Preserved fresh-cut Fuji apples for 2 weeks from the microbes	[28]
	Lemongrass oil	Provides exceptional protection against *Salmonella* species	[29]
	Lemon oil	Extend self-life of the rucola leaves for 7 days	[30]
	Carvacrol	Inhibit inoculated *E. coli* growth on fresh green beans for 11 days	[31]
	Ginger oil	Significantly decrease the total aerobic psychrophilic bacteria of refrigerated chicken fillets during 12 days	[35]
	Oregano essential oil and mandarin fibers	Oregano oil reduces the growth of bacteria, yeast, and molds in cheese during refrigerated storage and mandarin fiber enhances the nutritional value of the coated cheese	[37]
	Clove or oregano essential oil	Extended the self-life of the sliced bread for 15 days	[38]

(continued)

Table 5.1 (continued)

Types of application	Bioactive	Results	References
4. Preservation of food			
With antibacterial agent	Thyme essential oil	Eliminated pathogenic count in a span of 1 h and kept the quality of ultrafiltered Labneh cheese intact for up to 6 weeks	[39]
	Thymol essential oil	Preservation of fortified milk	[40]
With antifungal agent	Oregano essential	Showed control over fungal spoilage of Minas Padrao cheese	[42]
	Curcumin, cinnamon, and garlic, or sunflower oil	Inhibit the growth of fungi and bacteria	[44]
With anti-oxidant agent	Curcumin	Provides good physical stability during storage at room temperature, and lipid oxidation was found to be significantly lower in milk samples	[46]
5. Delivery of food components			
Healthy lipids delivery	Alpha-linoleic acid (omega-3 fatty acids)	Broiler chicken showed a higher level of omega-3 fatty acid	[48]
	Kenaf seed oil (tocopherols)	Tocopherol level in the bloodstream of the rat was increased rapidly and reducing cholesterol levels in hypercholesterolemia rats	[49]
Vitamins delivery	Different oils and emulsifiers	Improved the stability of vitamin A (retinol) during the storage period	[50]
		Increase in bioavailability of vitamin D3 (cholecalciferol)	[51]
		Increase the bioavailability of vitamin D2 (ergocalciferol)	[52]
		Increase in bioactivity against cancer α-tocopherol succinate	[53]
Excipient nanoemulsion delivery	Excipient nanoemulsion	Enhance bioaccessibility of carotenoids present in tomatoes, carrots, spinach, and dietary supplements	[57]
Delivery of nutraceuticals	Long-chain triglycerides	Higher bioaccessibility of beta-carotene	[59]
		Quercetin showed 3.4-fold higher absorption and 34-fold higher bioavailability	[60]
		Increase the oral bioavailability of coenzyme Q10	[61]

5.7 Conclusion

Nanoemulsions are extensively used in food science for enhancing the physical properties of food, increasing food efficacy, increasing food safety, preservation of food, and delivery of food components and nutraceuticals.

References

1. Espitia PJP, Fuenmayor CA, Otoni CG. Nanoemulsions: synthesis, characterization, and application in bio-based active food packaging. Compr Rev Food Sci Food Saf. 2019;18(1):264–85.
2. Azrini N, Azmi N, Elgharbawy AAM, Motlagh SR, Samsudin N. Nanoemulsions: factory for food, pharmaceutical, and cosmetics. PRO. 2019;7(617):1–34.
3. Silva HD, Cerqueira MÂ, Vicente AA. Nanoemulsions for food applications: development and characterization. Food Bioprocess Technol. 2012;5(3):854–67.
4. Singh Y, Meher JG, Raval K, Khan FA, Chaurasia M, Jain NK, et al. Nanoemulsion: concepts, development, and applications in drug delivery. J Control Release [Internet]. 2017;252:28–49. https://doi.org/10.1016/j.jconrel.2017.03.008.
5. Chircov C, Grumezescu AM. Nanoemulsion preparation, characterization, and application in the field of biomedicine [Internet]. Nanoarchitectonics in biomedicine. Elsevier Inc.; 2019. p. 169–88. https://doi.org/10.1016/B978-0-12-816200-2.00019-0
6. Abbas S, Hayat K, Karangwa E, Bashari M, Zhang X. An overview of ultrasound-assisted food-grade nanoemulsions. Food Eng Rev. 2013;5(3):139–57.
7. Borthakur P, Boruah PK, Sharma B, Das MR. 5-Nanoemulsion: preparation and its application in food industry A2—Grumezescu, Alexandru Mihai BT—emulsions [Internet]. Nanotechnology in the agri-food industry. Elsevier Inc.; 2016. p. 153–91. https://www.sciencedirect.com/science/article/pii/B9780128043066000052
8. Dasgupta N, Ranjan S. Food nanoemulsions: stability, benefits, and Applications. Singapore: Springer; 2018. p. 19–48.
9. Ashaolu TJ. Nanoemulsions for health, food, and cosmetics: a review. Environ Chem Lett [Internet]. 2021;19(4):3381–95. https://doi.org/10.1007/s10311-021-01216-9.
10. Majeed A, Bashir R, Farooq S, Maqbool M. Preparation, characterization and applications of nanoemulsions: an insight. J Drug Deliv Ther. 2019;9(2):520–7.
11. Gupta A, Eral HB, Hatton TA, et al. Nanoemulsions: formation, properties, and applications. Soft Matter. 2016;12:2826.
12. Chime SA, Kenechukwu FC, Attama AA. Nanoemulsions-advances in formulation, characterization, and applications in drug delivery. In: Application of nanotechnology in drug delivery. Rijeka: INTECH; 2014. p. 77–126.
13. Gurpreet K, Singh SK. Review of nanoemulsion formulation and characterization techniques. Ind J Pharm Sci. 2018;80(5):781–9.
14. Kumar KS, Tan KX, Loo SCJ. Developing nano-delivery systems for agriculture and food applications with nature-derived polymers. iScience. 2020;23:101055.
15. Kumar DD, Mann B, Pothuraju R, Sharma R, Bajaj RM. Formulation and characterization of nano encapsulated curcumin using sodium caseinate and its incorporation in ice cream. Food Funct. 2016;7:417–24.
16. Riquelme N, Zúñiga RN, Arancibia C. Physical stability of nanoemulsions with emulsifier mixtures: replacement of tween 80 with quillaja saponin. LWT. 2019;111:760–6.
17. Rutvij JP, Gunjant JP, Bharadia PD, et al. Nanoemulsion: an advanced concept of the dosage form. Int J Pharm Cosmetol. 2011;1:122–33.
18. Golfomitsou I, Mitsou E, Xenakis A, et al. Development of food-grade O/W nanoemulsions as carriers of vitamin D for the fortification of emulsion-based food matrices: a structural and activity study. J Mol Liq. 2018;268:734–42.
19. Maurya VK, Aggarwal M. A phase inversion-based nanoemulsion fabrication process to encapsulate vitamin D3 for food applications. J Steroid Biochem Mol Biol. 2019;190:88–98.
20. Zhou H, Liu G, Zhang J. Novel lipid-free nanoformulation for improving oral bioavailability of coenzyme Q10. Biomed Res Int. 2014;2014:793879.
21. Bhushani JA, Karthik P, Anandharamakrishnan C. Nanoemulsion-based delivery system for improved bioaccessibility and Caco-2 cell monolayer permeability of green tea catechins. Food Hydrocoll. 2016;56:372e82.

22. Wahab A, Rahim AA, Hassan S, Egbuna C, Manzoor MF, Okere KJ, Walag AMP. Application of nanotechnology in the packaging of edible materials. In: Egbuna C, Mishra AP, Goyal MR, editors. Preparation of phytopharmaceuticals for the management of disorders. Academic; 2021. p. 215–25.
23. Hygreeva D, Pandey MC. Novel approaches in improving the quality and safety aspects of processed meat products through high-pressure processing technology-a review. Trends Food Sci Technol. 2016;54:175–85.
24. Acevedo-Fani A, Soliva-Fortuny R, Martın-Belloso O. Nanoemulsions as edible coatings. Cur Opin Food Sci. 2017;15:43–9.
25. Otoni CG, de Moura MR, Aouada FA, et al. Antimicrobial and physical-mechanical properties of pectin/papaya puree/cinnamaldehyde nanoemulsion edible composite films. Food Hydrosol. 2014;41:188–94.
26. Acevedo-Fani A, Salvia-Trujillo L, Rojas-Graü MA, et al. Edible films from essential-oil-loaded nanoemulsions: physicochemical characterization and antimicrobial properties. Food Hydrosol. 2015;47:168–77.
27. Rekha Chawla R, Sivakumar S, Kaur H. Antimicrobial edible films in food packaging: current scenario and recent nanotechnological advancements—a review. Carbohydr Polym Technol Appl. 2021;2:100024.
28. Salvia-Trujillo L, Rojas-Grau MA, Soliva-Fortuny RC, et al. Use of antimicrobial nanoemulsions as edible coatings: impact on safety and quality attributes of fresh-cut Fuji apples. Postharvest Biol Technol. 2015;105:8–16.
29. Oh YA, Oh YJ, Song AY, et al. Comparison of the effectiveness of edible coatings using emulsions containing lemongrass oil of different size droplets on grape berry safety and preservation. LWT Food Sci Technol. 2017;75:742–50.
30. Sessa M, Ferrari G, Donsı F. The novel edible coating contains essential oil nanoemulsions to prolong the shelf life of vegetable products. Chem Eng Trans. 2015;43:55–60.
31. Severino R, Ferrari G, Vu KD, et al. Antimicrobial effects of modified chitosan-based coating containing nanoemulsion of essential oils, modified atmosphere packaging, and gamma irradiation against *Escherichia coli* O157:H7 and *Salmonella typhimurium* on green beans. Food Control. 2015;50:215–22.
32. Acevedo-Fani A, Soliva-Fortuny R, Martín-Belloso O. Nanostructured emulsions and nanolaminates for delivery of active ingredients: improving food safety and functionality. Trends Food Sci Technol. 2017;60:12–22.
33. Valdés A, Ramos M, Beltrán A, et al. State of the art of edible antimicrobial coatings for food packaging applications. Coatings. 2017;7(4):56.
34. Gomez B, Barba FJ, Dominguez R, et al. Microencapsulation of antioxidant compounds through innovative technologies and their specific application in meat processing. Trends Food Sci Technol. 2018;82:135–47.
35. Noori S, Zeynali F, Almasi H. Antimicrobial and antioxidant efficiency of nanoemulsion-based edible coating containing ginger (*Zingiber officinale*) essential oil and its effect on safety and quality attributes of chicken breast fillets. Food Control. 2018;84:312–20.
36. Chen H, Wang J, Cheng Y, et al. Application of protein-based films and coatings for food packaging: a review. Polymers (Basel). 2019;11(12):2039.
37. Artiga-Artigas M, Acevedo-Fani A, Martín-Belloso O. They are improving the shelf life of low-fat cut cheese using nanoemulsion-based edible coatings containing oregano essential oil and mandarin fiber. Food Control. 2017;76:1–12.
38. Otoni CG, Pontes SFO, Medeiros EAA, et al. Edible films from methylcellulose and nanoemulsions of clove bud (*Syzygium aromaticum*) and oregano (*Origanum vulgare*) essential oils as shelf-life extenders for sliced bread. J Agric Food Chem. 2014;62(22):5214–9.
39. Bedoya-Serna CM, Dacanal GC, Fernandes AM, et al. Antifungal activity of nanoemulsions encapsulating oregano (*Origanum vulgare*) essential oil: in vitro study and application in Minas Padrão cheese. Braz J Microbiol. 2018;49(4):929–35.
40. Xue J, Davidson PM, Zhong Q. Inhibition of Escherichia coli O157: H7 and Listeria monocytes growth in milk and cantaloupe juice by thymol nanoemulsions prepared with gelatin and lecithin. Food Control. 2017;73:1499–506.

41. Moghimi R, Ghaderi L, Rafati H, Aliahmadi A, et al. Superior antibacterial activity of nanoemulsion of *Thymus diagenesis* essential oil against *E. coli*. Food Chem. 2016;194:410–5.
42. El-Sayed SM, El-Sayed HS. Antimicrobial nanoemulsion formulation based on thyme (*Thymus vulgaris*) essential oil for UF Labneh preservation. J Mater Res Technol. 2021;10:1029–41.
43. Wan J, Zhong S, Schwarz P, et al. Enhancement of antifungal and mycotoxin inhibitory activities of food-grade thyme oil nanoemulsions with natural emulsifiers. Food Control. 2019;106:106709.
44. Abdou ES, Galhoum GF, Mohamed EN. Curcumin-loaded nanoemulsions/pectin coatings for refrigerated chicken fillets. Food Hydrocoll. 2018;83:445–53.
45. Hewlings S, Kalman D. Curcumin: a review of its effects on human health. Foods. 2017;6:92.
46. Joung HJ, Choi MJ, Kim JT, et al. Development of food-grade curcumin nanoemulsion and its potential application to food beverage system: antioxidant property and in vitro digestion. J Food Sci. 2016;81:N745–53.
47. Shah BR, Zhang C, Li Y, et al. Bioaccessibility and antioxidant activity of curcumin after encapsulated by nano and Pickering emulsion based on chitosan-tripolyphosphate nanoparticles. Food Res Int. 2016;89:399–407.
48. Abbasi F, Samadi F, Jafari SM. Ultrasound-assisted preparation of flaxseed oil nanoemulsions coated with alginate-whey protein for targeted delivery of omega-3 fatty acids into the lower sections of the gastrointestinal tract to enrich broiler meat. Ultrason Sonochem. 2019;50:208–17.
49. Cheong AM, Tan CP, Nyam KL. Effect of emulsification method and particle size on the rate of in vivo oral bioavailability of kenaf (*Hibiscus cannabinus* L.) seed oil. J Food Sci. 2018;83(7):1964–9.
50. Park H, Min S, Kim YR. UV and storage stability of retinol contained in oil-in-water nanoemulsions. Food Chem. 2019;272:404–10.
51. Kadappan AS, Guo C, Gumus CE, et al. The efficacy of nanoemulsion-based delivery to improve vitamin D absorption: comparison of in vitro and in vivo studies. Mol Nutr Food Res. 2018;62(4):836–860.
52. Salvia-Trujillo L, Fumiaki B, Park Y. The influence of lipid droplet size on the oral bioavailability of vitamin D-2 encapsulated in emulsions: an in vitro and in vivo study. Food Funct. 2017;8(2):767–77.
53. Gao YA, Qi XJ, Zheng YP, et al. Nanoemulsion enhances alpha-tocopherol succinate bioavailability in rats. Int J Pharm. 2016;515(1–2):506–14.
54. McClements DJ, Xiao H. Excipient foods: designing food matrices that improve the oral bioavailability of pharmaceuticals and nutraceuticals. Food Funct. 2014;5(7):1320–33.
55. Kopec RE, Failla ML. Recent advances in the bioaccessibility and bioavailability of carotenoids and effects of other dietary lipophiles. J Food Compos Anal. 2018;68:16–30.
56. Yuan X, Xiao J, Liu X, et al. The gastrointestinal behavior of emulsifiers used to formulate excipient emulsions impacts the bioavailability of beta-carotene from spinach. Food Chem. 2019;278:811819.
57. McClements DJ. Advances in edible nanoemulsions: digestion, bioavailability, and potential toxicity. Prog Lipid Res. 2021;81:101081.
58. Zhang RJ, Zhang ZP, McClements DJ. Nanoemulsions: an emerging platform for increasing the efficacy of nutraceuticals in foods. Colloids Surf B Biointerfaces. 2020;194:111202.
59. Salvia-Trujillo L, Qian C, Martín-Belloso O, et al. Modulating β-carotene bioaccessibility by controlling oil composition and concentration in edible nanoemulsions. Food Chem. 2013;139:878–84.
60. Pangeni R, Kang SW, Oak M, et al. Oral delivery of quercetin in oil-in-water nanoemulsion: in vitro characterization and in vivo anti-obesity efficacy in mice. J Funct Foods. 2017;38:571–81.
61. Belhaj N, Dupuis F, Arab-Tehrany E, et al. Formulation, characterization, and pharmacokinetic studies of coenzyme Q(10) PUFA's nanoemulsions. Eur J Pharm Sci. 2012;47(2):305–12.

Chapter 6
Food System Application of Nanomaterials in the Food Industry

Syeda Konain Mizba, Tafadzwa Justin Chiome, Chukwuebuka Egbuna, Asha Srinivasan, and Raghu Ram Achar

6.1 Introduction

Nanotechnology was conceptualized by Nobel Laureate Richard P. Feynman in his 1959 lecture "There's Plenty of Room at the Bottom", in which he spoke about manipulating matter at molecular and atomic scales. The word "nano" is derived from the Greek word "nanos" which literally means "dwarf". Nanomaterials can fall under different dimensional classes while having a size range of 1 to 100 nm which gives rise to its unique phenomenon and novel applications. When compared to their macroscale counterparts, these nanomaterials have interesting physical, chemical, biological properties, including a large surface to volume ratio, altered solubility, and toxicity, due to which they are gaining momentum in various industries like agriculture, food, biomedical sciences, healthcare systems, water treatments and other industries [1, 2]. Nanotechnology is gaining popularity and opening up new opportunities for the food industry. To maintain market leadership in the

S. K. Mizba · T. J. Chiome · A. Srinivasan (✉)
Division of Nanoscience & Technology, School of Life Sciences, JSS Academy of Higher Education & Research, Mysuru, Karnataka, India
e-mail: justinchiome@jssuni.edu.in; asha.srinivasan@jssuni.edu.in

C. Egbuna
Department of Biochemistry, Faculty of Natural Sciences, Chukwuemeka Odumegwu Ojukwu University, Igbariam, Anambra State, Nigeria

Nutritional Biochemistry and Toxicology Unit, World Bank Africa Centre of Excellence, Centre for Public Health and Toxicological Research (ACE-PUTOR), University of Port-Harcourt, Port-Harcourt, Rivers State, Nigeria

R. R. Achar (✉)
Division of Biochemistry, School of Life Sciences, JSS Academy of Higher Education & Research, Mysuru, Karnataka, India
e-mail: rracharya@jssuni.edu.in

C. Egbuna et al. (eds.), *Application of Nanotechnology in Food Science, Processing and Packaging*, https://doi.org/10.1007/978-3-030-98820-3_6

food processing business, new technologies are required to produce fresh, authentic, convenient, and flavorful food products while also prolonging shelf life and freshness through better food quality [3, 4]. Concerns about the quality of food and health benefits from consumers are the driving force for research aimed at improving food quality while preserving the product's nutritional value [1]. Because of its sub-microscopic nature, nanotechnology has great potential in ensuring color, flavor, nutritional value modification, increasing the shelf life of food, and monitoring the integrity of food via barcodes such as cold chain, i.e. whenever there is a slight change in food storage conditions [5].

6.2 Conventional Food Practices

Food storage has come a long way. Cave-men stored their fresh kill in caves which provided a damp environment that in turn prevented decay. Since the time of the cave-men, food storage has seen incredible innovations to the point where shelf-life has increased by leaps and bounds. Even so, the drying and fermentation processes that have been around since millennia are still used today with slight modifications [6]. Chilling, fermenting, salting, sun drying, roasting, oven baking, steaming, curing, pickling, canning, bottling, jellying, irradiation, carbonation of food, vacuum packing, and use of chemical or artificial preservatives are a few methods of preservation that are commonly used on a day-to-day basis. These traditional preservation methods have one simple goal, retarding the growth of disease-causing organisms. However, these techniques were applied without full knowledge and understanding of the mechanisms that governed them [7].

According to archaeological evidence, these preservation techniques were used in ancient civilizations of Greece, Rome, and Egypt. The Egyptians sun-dried their food, Romans introduced the idea of pickling whereas, the Greeks were responsible for jellying of food by adding additives like honey and sugar [8]. In an article by National Centre for Home Food Preservation, French revolutionary Nicolas Appert presented the observation that the application of heat to food in sealed glass jars increased the shelf life of the food. Unbeknownst to him, he had discovered the process of canning without fully understanding its mechanism. In the year 1784, William Cullen made the first technologically breakthrough in food preservation techniques by inventing a crude method of artificial refrigeration. In 1809, Nicolas Appert invented the technique of vacuum bottling to facilitate the supply of food to the French troops, a method which gradually evolved into tinning and canningr. Pasteurization was yet another breakthrough that helped increase the shelf-life of milk, wine, and beer [7]. The concentration, bioavailability, and bioactivity of phytochemicals in food can be significantly altered by traditional food preservation methods [8]. However, these techniques were primitive and were incapable of preserving food for an extended period of time which led to the dire need for an innovative, permanent, and reliable solution for the preservation of food [7].

6.3 Nanomaterials in Food Systems

All organisms are made up of a collection of nanoscale-sized objects. The earth is an ecosystem formed by the interaction of various nanostructures. Food molecules like carbohydrates, proteins, and fats consist of sugars, amino acids, and fatty acids merged at the nanoscale. Food consumed is broken down into organic constituents, which range in size from large polymers to smaller molecules in the nano-range. Nanoparticles interact easily with these food products, and at the cellular level of human cells, making them more appealing to the food industry [9, 10].

Nanotechnology impacts on all stages of production in the food industry beginning with primary production (farming), due to advancements in pesticide efficacy and delivery (novel formulations and improved crop adherence) [11], processing where emulsification and encapsulation have progressed to the nanoscale [12], and packaging where barrier improvement has been enhanced through the use of nanoscale fillers [13]. Nanotechnological innovations in the food industry could lead to macroscale innovations in sensory food characteristics such as texture, taste, as well as coloring strength, processability, and shelf-life stability, resulting in a variety of new products. Furthermore, nanotechnology has the potential to improve solubility, thermal stability, and oral bioavailability of bioactive compounds [14, 15].

The application of nanotechnology in the food sector is summarized in two main groups (Fig. 6.1).

1. Food Nanostructured Ingredients
2. Food Nano sensing.

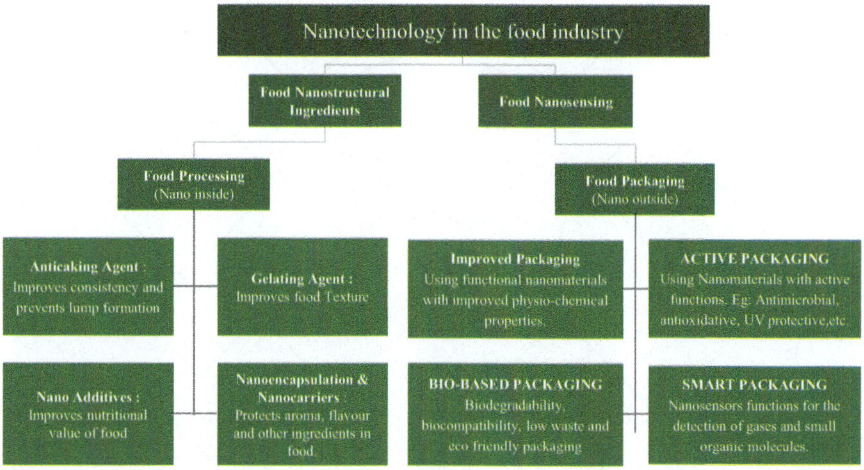

Fig. 6.1 Application of nanotechnology in food industry

6.3.1 Food Processing

Food processing is the transformation of agro-allied raw materials into marketable food products. It consists of procuring raw materials, sorting and grading, primary processing, packaging, transportation, and storage [16]. Toxin removal, pathogen prevention, preservation, and improving consistency of foods, fortification, and enrichment with micronutrients are all part of food processing. Nanotechnology has been incorporated into food processing to ensure longer shelf life of while maintainng its freshness and suitable for long-distance transit [17]. The application of nanotechnology in food processing is broadly classified into,

1. Direct usage: Mixing fragrances, coloring agents, nano preservatives, antioxidants, and bioactive compounds such as fatty acids, vitamins, and polyphenols are some of the most common direct applications.
2. Indirect Usage: It includes the use of nanosized substances in packaging, nanosensors, and catalysts in fat hydrogenation [18].

In functional food products, various types of food additives are used to perform a specific function (e.g., to preserve food or increase the nutritional value of the food). The most common food additives used are antimicrobial agents, antioxidants, flavoring agents, anti-browning, anti-caking, and nutrients. These additives have specific properties depending on their chemistry. Nano additives are combined with nanocarriers (nanocapsules, nanoemulsions) which makes their application more efficient and valuable [19]. The various nano techniques used for the processing of these delivery systems (nano additive + nanocarrier) were discribed (Fig. 6.2).

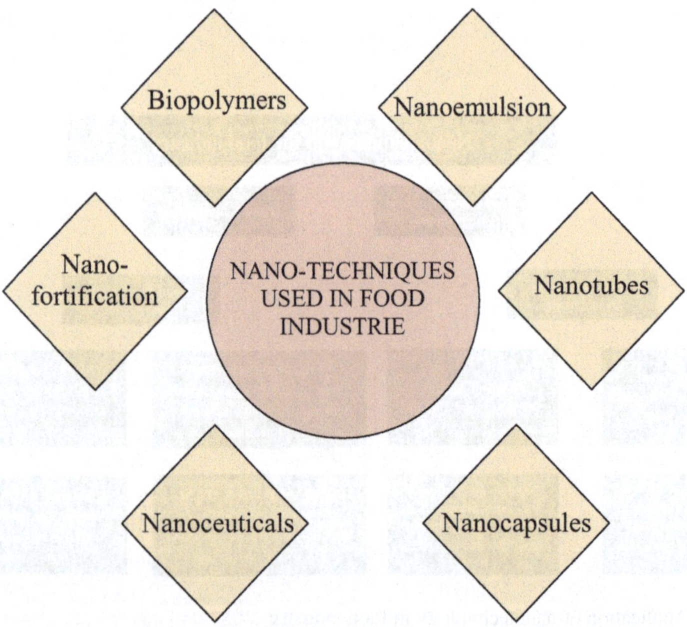

Fig. 6.2 Nano techniques used in food processing

6.3.1.1 Biopolymers

Biopolymers are made up of a long chain of covalently bound monomeric units that have a higher affinity for specific ions. To perform better in their specific environment, all biochemical materials that perform functions in living organisms, such as proteins, carbohydrates, enzymes, and so on, are in the form of polymers [9, 10]. In food science, they are used to bind specific nutrients to increase their availability and stability within the food product itself. They are biodegradable and derived from living organisms, plant oils, fats, resins, and proteins [20]. Proteins and Polysaccharides are used as biopolymers. They have nanoscale dimensions and self-assemble in aqueous media via polymer to polymer or polymer to solute attraction. Furthermore, they have the ability to alter their physical binding by modifying the factors in the food (i.e., temperature, pH, and addition of salts). Weak intermolecular forces, such as H-bonding and electrostatic forces, cause selected biopolymers to form fibers, aggregates, or complex structures such as food gels or dispersions [21]. Some nanoparticles are used to improve the nutritional quality of certain food products by adding nutrients the foods are deficient in.

Polylactic Acid (PLA) is a biodegradable, biopolymeric nanoparticle and organic acid. It is used for the enrichment of iron, calcium, and vitamin in food products. A study showed that iron absorption is enhanced by organic acids. i.e. on treatment with organic acids, these nutrients release metal micronutrients which enhances their bioavailability [22]. The required material is encapsulated within PLA with an associative compound such as polyethylene glycol to perform its function effectively [23].

6.3.1.2 Nanoemulsions

Micro dispersion or film encapsulation at nanometre scales is described by the term "nanoencapsulation". Nano-Structured Encapsulation Layer (NSEL) protects the food/flavor molecules/ingredients from the environment. The active ingredient is often within the molecule in nanoscale form [5]. The nanoencapsulation system provides numerous advantages, including improved food product stability and integrity, oxidation and rancidity prevention, volatile ingredient retention, taste masking, pH and moisture triggered controlled release, consecutive multiple active ingredient delivery, flavor character change, long-lasting organoleptic properties, and enhanced bioavailability [24]. These nanostructured functional compounds are encapsulated in carriers made of oil (liposomes) or proteins (micelles). Nanocarriers protect food additives from thermal degradation while also masking their flavors. Lycopene, citric acid, benzoic acid, ascorbic acid, omega-3 omega-6 fatty acids, fat-soluble vitamins A and E, isoflavones, β-carotene, and lutein are some of the commercially available nano encapsulated food additives [25, 26]. Nanocapsules were reportedly used by Australian company George Weston Foods to mask the taste and odor of tuna fish oil (an omega-3 fatty acid source) that was baked into bread. Because the nanocapsules only rupture when they reach the stomach, the unpleasant fish oil flavor is avoided. To promote healthy gut function, nanocapsules have been used to protect and release beneficial live probiotic species in a controlled manner [27, 28].

6.3.1.3 Nanotubes

Nanotubes are unique one-dimensional wire-like structures that can augment the mechanical properties of materials by their tensile strength and elastic properties. The most commonly available nanotubes are boron nitride nanotubes, Carbon-Based Nanotubes (CNT), alumina nanotubes, and rutile nanotubes, among which CNTs are widely used. In the food industry, CNTs are used to remove toxins from foods and in microfluidic devices. CNTs are coated with specific antibody adsorbents (anti-SEB IgG) and then spread across the food surface. CNTs' large surface area binds to toxins produced by harmful microbes, which immunosensors can detect [29, 30].

6.3.1.4 Nanocapsules

Nanocapsules are nanoshells that are made from a non-toxic polymer. It is a colloidal drug carrier system composed of an aqueous or oily core surrounded by a thin polymer membrane. A drug, gene, protein, or any other bioactive substance can be incorporated into the interior cavity (core) which is surrounded by a distinctive polymer membrane [30]. In food systems, they are used as nanocarriers and formed by using bioabsorbable polymers like alginate, chitosan, collagen, albumin, gelatin, nanoliposomes, archaeosomes. These nanocarriers are used in food as flavoring agents, stabilizing agents, and pathogen control agents. Food flavors are mostly volatile, and their aromatic compounds change as a result of oxidation, chemical interactions, and heating. Encapsulation prevents aromatic compounds from volatilizing, thereby entrapping their flavors. Zein, the main protein in corn, has attracted much interest in food nanotechnology. Zein nanomaterials possess the potential to form a microorganism-resistant tubular network. Zein nanoparticles or nanobeads can be employed as edible carriers for flavor compounds or nutraceutical encapsulation, as well as to enhance the strength of plastic and bioactive packaging [10, 31]. One of the most commonly used flavoring aromatic compounds used is Citral which is an α,β-unsaturated aldehyde with an additional double bond. However, its degradation produces off-flavor compounds. To prevent this, the nanoemulsion of Citral was paired with natural anti-oxidants like β-carotene, tanshinone, and black tea extract. These compounds had high chemical stability during storage and also led to decreased formation of off-flavour compounds upon its degradation [32].

6.3.1.5 Nanoceuticals

Nutraceuticals are substances that are either food or components of food that provides medical or health benefits, including disease prevention and treatment. Functional food contains vitamins, fats, proteins, and carbohydrates that are essential for an individual's health [30]. Few of the nanotechnology-based available products include Nanoparticles of carotenoids that are dispersed in water and added to the juice product to improve the bioavailability of Carotenoids. Another example is the use of canola oil nano-sized micelles to provide vitamins and minerals to the body.

6.3.1.6 Nano Fortification

Adding one or more essential nutrients to food to prevent or correct a demonstrated deficiency of the respective nutrients in specific population groups is called Food Fortification (FF). It generally makes use of staple foods as carriers to deliver micronutrients that are deficient in the diet of a population [33]. Nanofood fortification has a wide variety of advantages such as the protection of phytochemicals through an encapsulation technique. Micronutrients that are rapidly degraded or are not properly absorbed by the body can also be aided by food fortification on the nanoscale [34]. The bioavailability of iron and iodine was considerably increased before salt fortification by encapsulating it with modified starches, gelatin, sodium hexa-meta-phosphate, and purified sodium chloride [35].

6.4 Food Packaging

Food is perishable and could be contaminated or degraded at any point across the supply chain, from processing to consumer delivery. Food packaging can help maintain and/or improve food quality and safety by preventing product deterioration, preserving the benefits of processing, and extending shelf life. It provides essential food information and simplifies food processing from distribution to the consumer's table. Food packaging's goal is to contain food in a cost-effective manner that meets industry and consumer demands, ensures food safety, and reduces environmental impact. The four basic functions of conventional packaging are containment (ease of transportation and handling), convenience (consumer-friendly), protection, and preservation (longer shelf-life by protecting from external factors to avoid contamination and avoids leakage) [35, 36].

Traditionally leaves, vegetable fibers, textiles (muslin cloth) were used for food preservation and local sales as well as for domestic consumption. Food packaging is limited in its ability to meet consumer demands for high-quality, mildly preserved, convenient, and long-lasting food products [37]. The methods have been replaced by materials such as glass, metals (aluminum, laminated, tinplate, and steel), paper, and cardboard. But these materials do not meet the consumer demands and also have low mechanical barrier properties. Hence, plastics and polymers have revolutionized the packaging industry because of their economical, functional, versatile properties. Plastics can be rigid (bottles, jars, boxes, cases), thermoformed (food trays), or flexible (woven mesh, multi-layer, films). But, extensive use of plastics has caused serious environmental issues worldwide, because they are petroleum-based and non-biodegradable. These materials pollute the environment during the manufacturing and disposal stage. Therefore, the development of new eco-friendly packaging along with innovative packaging concepts represents a great alternative to protect the environment and are more efficient options for food packaging and also increase the shelf life of these packaged foods [38].

6.4.1 *Improved Packaging*

The incorporation of functionalized nanomaterials into polymer matrices to improve physicochemical properties such as gas barrier properties, temperature resistance, humidity, mechanical strength, and flexibility is the basis of improved packaging [38]. Polymers are preferred for food packaging because it is lightweight, low cost, easy to process, and are known for their formability and diversity. The most commonly used polymers are polyolefins such as polypropylene (PP) and various grades of polyethylene (HDPE, LDPE, etc.), polyethylene terephthalate (PET), polystyrene (PS), and polyvinyl chloride (PVC). Polymers have revolutionized the packaging industry but they have several drawbacks like inherent permeability to gases and other small molecules [39]. To overcome these limitations, usually nano clay and silica nanoparticles are incorporated into these polymer matrices in a 5% (w/w) ratio. These polymer nanocomposites are created by dispersing an inert, nanoscale filler throughout a polymer matrix. The filler material could be clay, silica nanoparticles, cellulose, carbon nanotubes, etc. [39]. The various nanotechnology-based techniques used for improved packaging are discussed in the following sections.

6.4.1.1 Nano-Coatings

A coating is a coherent layer generated by applying a coating substance to a substrate once or several times. Nanocoatings are tiny films of 1–100 nm thickness that are grown on surfaces. Nanocoatings do not change the topography of the surface, do not fill in imperfections or smoothen the surface like paint does, and do not withstand abrasion and wear. As gas-barrier coatings, hydrophilic/hydrophobic or oleophobic qualities, improved corrosion resistance, and enhanced insulating or conductive capabilities, are utilized to impart a specific chemical or physical function(s) to a surface [40]. Nanocoatings can be single molecular layers thick or multi-molecular layered. Nanocoatings are used on a variety of substrates, including metals, glass, ceramics, polymers, and other materials. Smart coatings are multifunctional coatings having additional functionalities such as thermal insulation, controlled release of active substances, or self-healing [2]. Edible nano-coatings (5 nm thin coatings) are used in meat processing, agriculture (for fruit and vegetable protection), cheese and bread industry, as well as other industries to add flavor, color, enzymes, antioxidants, and anti-browning ingredients to products [41]. Materials that are used for edible coating films include Hydrocolloids—Polysaccharides with water-soluble properties. Derivatives of cellulose, starch, pectin, chitosan, alginate, gums, etc. Lipids—fatty acids, acylglycerols, waxes, essential oils, and extracts, etc. Proteins—Milk and soy proteins, gelatin, myofibrillar protein, zein, etc.

6.4.1.2 Nano-Laminates

Nanolaminates consist of two or more layers of material of nanometre dimensions that are physically or chemically bonded to each other. One powerful method in nanolamination is the layer-by-layer deposition approach, in which the charged surfaces are covered with interfacial films possessing numerous nanolayers of different materials [42]. Natural polyelectrolytes (proteins, polysaccharides), charged lipids (phospholipids, surfactants), and colloidal particles (micelles, vesicles, droplets) are some of the strongest absorbing candidates that can be used for layering [43]. Incorporation of active functional agents such as antimicrobials, anti-browning agents, antioxidants, flavors, enzymes, and colors into the films could increase the shelf life, aesthetic value, and quality of packaged food. The functional properties of laminated films depend upon the characteristics and properties of the materials used for their preparation.

6.4.1.3 Nano Clays

Nano clays are naturally occurring fine-grained minerals. These when incorporated with nanofillers for polymer-clay nanocomposites enhance the packaging material properties like mechanical strength, thermal stability, gas barrier properties. These packages also retard the migration of potentially harmful additives into packaged food. Clay platelets are impermeable in general, therefore their integration into matrix polymers creates a convoluted pathway for a permeate to diffuse the nanocomposite. The inclusion of clay fillers causes the diffusion of permeate to take a longer mean path rather than a straight line [44].

6.4.2 *Active Packaging*

Active nano packaging is the incorporation of bioactive nanoparticles into traditionally used polymer matrices. These nanoparticles interact directly with food or the environment, usually endowed with antimicrobial and/or antioxidant properties to extend the shelf life of the packaged food. Examples of nanoparticles that are used for antimicrobial properties are silver nanoparticles, titanium dioxide nanoparticles, carbon nanotubes. Active packaging is mainly used as antimicrobial films, oxygen scavenging films, and UV Absorbing films [44].

6.4.3 *Antimicrobial Films*

The main goal of antimicrobial films is the incorporation of an active antimicrobial agent into the polymer matrix or applying a coating layer within the packaging material which inhibits the growth of microorganisms to extend the shelf life of the packaged food. There are two mechanisms of antimicrobial action: Inhibition of the essential metabolic pathways of the organism and destruction/rupture of cell wall/membrane. Most commonly used nanoparticles in antimicrobial films include inorganic nanoparticles like AgNP, TiO_2, ZnO, MgO, carbon nanotubes, peptides (nisin), etc. Research has shown that Titanium dioxide (TiO_2) nanoparticles are non-toxic to the human body and hence are approved in food processing and packaging systems. Nisin acts as a depolarizing agent on bacterial cell membranes and creates pores in the lipid bilayer [45]. A study by Qi et al. showed that nanoscale chitosan particles also exhibited antimicrobial properties by the interactions between positively charged chitosan and negatively charged cell membranes, increasing membrane permeability and eventually causing rupture and leakage of intracellular material [46].

1. **Oxygen Scavenging Films:** Oxygen permeability into packaged food can turn fats rancid, cause flavor deterioration, deplete nutritional value, supplement aerobic bacterial growth. Therefore, incorporation of Oxygen scavengers into food packaging is essential to maintain low levels of Oxygen as they extend the shelf life of food from 3 to 4 days to 14 days or more [47]. The use of oxygen scavenging packing materials reduces the amount of oxygen dissolved in the food or present in the headspace of the packaged food to levels significantly lower than those achieved by modified atmosphere packaging [48]. Materials like activated iron, zinc, ascorbic acid, catechol sulfite, TiO_2, Ag nanomaterials incorporated into polymer matrices via coatings or immobilization have been effective. Ideal oxygen scavengers must meet certain criteria, including biocompatibility, absorbing oxygen in the package at an appropriate rate, not producing toxic substances or undesirable gas or odor, and providing consistent quality and performance [49].

2. **UV absorption films:** UV radiation triggers the formation of free radicals which causes oxidative stress in the packaged food hence pigments like chlorophyll, riboflavin, carotenoids, etc. are physically available food. Protein and fat-rich foods are more susceptible to light damage i.e. photosensitizer, induced photo-oxidation [50]. Titanium dioxide has been used as a photocatalyst and UV-blocking agent. Akihiko Hashimoto in their study prepared a UV-Blocking film using Titanium dioxide nanoparticles. The coating had antibacterial properties which were induced by TiO_2 catalysis [51].

6.4.4 Smart Packaging

The use of nanoparticles in smart packaging allows for the monitoring of biochemical and microbiological changes within the food as well as in the environment around the product. The nanomaterials used in smart packaging have the ability to communicate the condition of the product but do not interact with the product directly in any way whereas these products interact with the environment surrounding the packaged food. The main objective of Smart packaging is to monitor the packed product and transmit the information to the consumers. The various nanosensors used in smart packaging for food applications include [52] (a) Time and Temperature Indicators, (b) Integrity Indicators, (c) Freshness Indicators, and (d) Radio Frequency Identification. The various types of packaging methods and their examples are mentioned in Table 6.1.

6.5 Conclusion and Future of Nanomaterials in the Food Industry

This chapter has highlighted the various use of nanoscale materials, their innovative applications, and scientific advancements in food industry. The R & D for nanotechnology in the food industry is progressively important at every level. The utilization of nanomaterials by the food industries has transformed the entire food sector and innovations in food nanotechnologies are expected to play a major role to evade global food scarcity, food safety, and food storage. This has been possible by the scientific advancements with the use of nanotechnologies that have influenced the essential aspects of agri-food, manufacturing from food protection to the molecular synthesis of new food products and ingredients. The new and future innovations in R & D nanotechnology will be exceptionally focused on increased food production, through precision farming techniques with geo-spatial adaptions, use in nanomaterials for plant gene delivery for enhancing plant micro-nutrient bioavailability, enhancing food texture and packaging for extended freshness and shelf-life.

Table 6.1 Various types of packaging methods and examples

Advanced food packaging type	Subclass	Nano-material used	Functions	References
Improved packaging	Nano-coatings	Carboxymethyl cellulose/sodium montmorillonite clay/titanium dioxide (TiO_2)	The addition of NPs decremented water vapor permeability, while moisture content, density, and glass transition temperature were incremented slightly.	[1]
		Whey protein isolate/cellulose nanofibers/TiO_2/ rosemary essential oil	Improved physicomechanical, antibacterial, and antioxidant properties.	[2]
	Nano-laminates	Alginate/chitosan/ folic acid	Improved stability under ultraviolet light exposure after folic acid encapsulation.	[3]
		Polyethylene terephthalate/ aluminum oxide (Al_2O_3)/Zinc oxide (ZnO)	Good barrier properties	[4]
	Nano-clay	Chitosan-clay nanocomposites	Addition of clay significantly increased the strength and stiffness of neat chitosan nanocomposite.	[5]
		Corn starch/natural montmorillonite/ anthocyanin	Active and pH-sensitive bionanocomposites with improved mechanical and thermal properties.	[6]

(continued)

Table 6.1 (continued)

Advanced food packaging type	Subclass	Nano-material used	Functions	References
Active Packaging	Antimicrobials	Nisin-loaded Pectin Nanoparticles.	Antimicrobial activity against Gram-positive (Arthrobacter sp. and *Bacillus subtilis*) and Gram-negative (*E. coli* and *Klebsiella* sp.) bacteria.	[7]
		Titanium dioxide incorporated Polyethylene films	Showed effectiveness against *S. aureus* (Gram positive) and *E. coli* (Gram negative) bacteria	[8]
	Oxygen Scavenging	Metallic (Iron powder, Activated iron, Zn.)	Prevention of fat oxidation	[9]
		Organic (Ascorbic acid, tocopherol, catechol.)		
		Inorganic (Sulfite, thiosulfate, ZnO.)		
	UV absorbing Films	TiO_2 and Ag Nanoparticles embedded in Polyvinyl chloride	Increased Photocatalytic activity under UV irradiation with enhanced antibacterial properties	[10]

(continued)

Table 6.1 (continued)

Advanced food packaging type	Subclass	Nano-material used	Functions	References
Smart Packaging	Freshness indicators	Ag/CuNPs, poly(thiophene), polyaniline, protein-based NPs, carbon nanotubes	Provides accurate information about the freshness of the product.	[11–16]
	Spoilage and Pathogens Indicators	CdSe/ZnS QDs, UCNPs	Detection of the spoilage associated changes, detection of pathogens and chemical contaminants detection of CO_2, proteins, nucleic acids, NH_3, enzymes	[17–19]
	Time–temperature indicators	Au nanorods, Ag nanoplates, Au/Ag nanorods	Indications of time–temperature variations (e.g., freeze–thaw–refreeze)	[20, 21]
	Toxins indicators	Detection of aflatoxin B1 aflatoxin B2, ochratoxin A	Dendrimer-based electrochemical sensors	[22, 23]
	Contaminants indicators	Dendrimer-based voltammetric immunosensor	Detection of atrazine	[5]
		AgNPs/GNRs-based electrochemical sensors	Detection of methyl parathion	[24]

Acknowledgment All the authors thank JSS Academy of Higher Education & Research, Mysuru for the support towards research activities.

References

1. Singh T, Shukla S, Kumar P, Wahla V, Bajpai VK, Rather IA. Application of nanotechnology in food science: perception and overview. Front Microbiol. 2017;8:1501.
2. Nile SH, Baskar V, Selvaraj D, Nile A, Xiao J, Kai G. Nanotechnologies in food science: applications, recent trends, and future perspectives. Nano-Micro Lett. 2020;12(1):45.
3. Ameta SK, Rai AK, Hiran D, Ameta R, Ameta SC. Use of nanomaterials in food science. In: Ghorbanpour M, Bhargava P, Varma A, Choudhary DK, editors. Biogenic nano-particles and their use in agro-ecosystems. Singapore: Springer; 2020. p. 457–88. http://link.springer.com/10.1007/978-981-15-2985-6_24. Accessed 27 Oct 2021.
4. Alfadul SM, Elneshwy AA. Use of nanotechnology in food processing, packaging and safety – review. Afr J Food Agric Nutr Dev. 2010;10(6):1. http://www.ajol.info/index.php/ajfand/article/view/58068. Accessed 27 Oct 2021
5. Aigbogun I, Mohammed S, Orukotan A, Tanko J. The role of nanotechnology in food industries- a review. J Adv Microbiol. 2018;7(4):1–9.

6. Hamad AF, Han J-H, Kim B-C, Rather IA. The intertwine of nanotechnology with the food industry. Saudi J Biol Sci. 2018;25(1):27–30.
7. Pradhan N, Singh S, Ojha N, Shrivastava A, Barla A, Rai V, et al. Facets of nanotechnology as seen in food processing, packaging, and preservation industry. Biomed Res Int. 2015;2015:1–17.
8. Grzegorzewski F. Influence of non-thermal plasma species on the structure and functionality of isolated and plant-based 1,4-benzopyrone derivatives and phenolic acids. 2011. https://depositonce.tu-berlin.de/handle/11303/3075. Accessed 27 Oct 2021.
9. Ravichandran R. Nanotechnology applications in food and food processing: innovative green approaches, opportunities and uncertainties for global market. Int J Green Nanotechnol Phys Chem. 2010;1(2):P72–96.
10. Cushen M, Kerry J, Morris M, Cruz-Romero M, Cummins E. Nanotechnologies in the food industry – recent developments, risks and regulation. Trends Food Sci Technol. 2012;24(1):30–46.
11. Silva MS, Cocenza DS, Grillo R, de Melo NFS, Tonello PS, de Oliveira LC, et al. Paraquat-loaded alginate/chitosan nanoparticles: preparation, characterization and soil sorption studies. J Hazard Mater. 2011;190(1–3):366–74.
12. Rao J, McClements DJ. Food-grade microemulsions, nanoemulsions and emulsions: fabrication from sucrose monopalmitate & lemon oil. Food Hydrocoll. 2011;25(6):1413–23.
13. Neethirajan S, Jayas DS. Nanotechnology for the Food and Bioprocessing Industries. Food Bioprocess Technol. 2011;4(1):39–47.
14. Ezhilarasi PN, Karthik P, Chhanwal N, Anandharamakrishnan C. Nanoencapsulation techniques for food bioactive components: a review. Food Bioprocess Technol. 2013;6(3):628–47.
15. Huang Q, Yu H, Ru Q. Bioavailability and delivery of nutraceuticals using nanotechnology. J Food Sci. 2010;75(1):R50–7.
16. Siddiqui S, Alrumman SA. Influence of nanoparticles on food: an analytical assessment. J King Saud Univ Sci. 2021;33(6):101530.
17. Wahab A, Rahim AA, Hassan S, Egbuna C, Manzoor MF, Okere KJ, Walag AMP. Application of nanotechnology in the packaging of edible materials. In: Egbuna C, Mishra AP, Goyal MR, editors. Preparation of phytopharmaceuticals for the management of disorders. London: Academic Press; 2021. p. 215–25.
18. Stanković M, Gabrovska M, Krstić J, Tzvetkov P, Shopska M, Tsacheva T, et al. Effect of silver modification on structure and catalytic performance of Ni-Mg/diatomite catalysts for edible oil hydrogenation. J Mol Catal A Chem. 2009;297(1):54–62.
19. Adeyeye SAO, Fayemi OE. Nanotechnology and food processing: between innovations and consumer safety. J Culin Sci Technol. 2019;17(5):435–52.
20. Mohan S, Oluwafemi OS, Kalarikkal N, Thomas S, Songca SP. Biopolymers – application in nanoscience and nanotechnology. In: Parveen FK, editor. Recent advances in biopolymers. Rijeka: InTech; 2016. http://www.intechopen.com/books/recent-advances-in-biopolymers/biopolymers-application-in-nanoscience-and-nanotechnology. Accessed 27 Oct 2021.
21. Semenova MG. Biopolymers in food colloids: thermodynamics and molecular interactions. London: CRC Press; 2012. p. 384.
22. Riley T, Govender T, Stolnik S, Xiong CD, Garnett MC, Illum L, et al. Colloidal stability and drug incorporation aspects of micellar-like PLA-PEG nanoparticles. Colloids Surf B: Biointerfaces. 1999;1–4(16):147–59.
23. Khare AR, Vasisht N. Chapter 14-Nanoencapsulation in the food industry: technology of the future. In: Microencapsulation in the food industry. London: Academic Press; 2014. p. 151–5.
24. Chaudhry Q, Scotter M, Blackburn J, Ross B, Boxall A, Castle L, et al. Applications and implications of nanotechnologies for the food sector. Food Addit Contam Part A. 2008;25(3):241–58.
25. Sahoo M, Vishwakarma S, Panigrahi C, Kumar J. Nanotechnology: current applications and future scope in food. Food Front. 2021;2(1):3–22.

26. European Commission. CORDIS. Final report summary - NANOFOODS (Development of foods containing nanoencapsulated ingredient). FP7. 2010. https://cordis.europa.eu/project/id/222006/reporting. Accessed 27 Oct 2021.
27. Yang M, Kostov Y, Rasooly A. Carbon nanotubes based optical immunodetection of Staphylococcal Enterotoxin B (SEB) in food. Int J Food Microbiol. 2008;127(1–2):78–83.
28. Lugani Y, Sooch BS, Singh P, Kumar S. Nanobiotechnology applications in food sector and future innovations. In: Microbial biotechnology in food and health. Amsterdam: Elsevier; 2021. p. 197–225. https://linkinghub.elsevier.com/retrieve/pii/B9780128198131000086. Accessed 27 Oct 2021.
29. Couvreur P, Barratt G, Fattal E, Vauthier C. Nanocapsule technology: a review. Crit Rev Ther Drug Carrier Syst. 2002;19(2):99–134.
30. Sozer N, Kokini JL. Nanotechnology and its applications in the food sector. Trends Biotechnol. 2009;27(2):82–9.
31. Dudeja P, Gupta RK. Nutraceuticals. In: Food safety in the 21st century. Amsterdam: Elsevier; 2017. p. 491–6. https://linkinghub.elsevier.com/retrieve/pii/B9780128017739000406. Accessed 27 Oct 2021.
32. Yang X, Tian H, Ho C-T, Huang Q. Inhibition of citral degradation by oil-in-water nanoemulsions combined with antioxidants. J Agric Food Chem. 2011;59(11):6113–9.
33. Clinton T. Nanofood fortification benefits within the protection of phytochemicals by victimization. Afr J Food Agric Nutr Dev. 2021;12(7):1.
34. Diosady LL, Alberti JO, Venkatesh Mannar MG. Microencapsulation for iodine stability in salt fortified with ferrous fumarate and potassium iodide. Food Res Int. 2002;35(7):635–42.
35. Jagus RJ, Gerschenson LN, Ollé Resa CP. Combinational approaches for antimicrobial packaging. In: Antimicrobial food packaging. Amsterdam: Elsevier; 2016. p. 599–608. https://linkinghub.elsevier.com/retrieve/pii/B9780128007235000498. Accessed 27 Oct 2021.
36. Sharma C, Dhiman R, Rokana N, Panwar H. Nanotechnology: an untapped resource for food packaging. Front Microbiol. 2017;8:1735.
37. Han JH. A review of food packaging technologies and innovations. In: Innovations in food packaging. Amsterdam: Elsevier; 2014. p. 3–12. https://linkinghub.elsevier.com/retrieve/pii/B9780123946010000011. Accessed 27 Oct 2021.
38. Salgado PR, Di Giorgio L, Musso YS, Mauri AN. Recent developments in smart food packaging focused on biobased and biodegradable polymers. Front Sustain Food Syst. 2021;5:630393.
39. Vasile C. Polymeric nanocomposites and nanocoatings for food packaging: a review. Materials. 2018;11(10):1834.
40. Müller K, Bugnicourt E, Latorre M, Jorda M, Echegoyen Sanz Y, Lagaron J, et al. Review on the processing and properties of polymer nanocomposites and nanocoatings and their applications in the packaging, automotive and solar energy fields. Nanomaterials. 2017;7(4):74.
41. Decher G, Schlenoff JB, editors. Multilayer thin films: sequential assembly of nanocomposite materials. Weinheim: Wiley-VCH; 2012. p. 2. 2nd, compl. rev. and enl. ed ed.
42. Ingale AG, Chaudhari AN. Nanotechnology in the food industry. In: Gothandam KM, Ranjan S, Dasgupta N, Ramalingam C, Lichtfouse E, editors. Nanotechnology, food security and water treatment, Environmental chemistry for a sustainable world. Cham: Springer International Publishing; 2018. p. 87–128. http://link.springer.com/10.1007/978-3-319-70166-0_3. Accessed 27 Oct 2021.
43. Majeed K, Jawaid M, Hassan A, Abu Bakar A, Abdul Khalil HPS, Salema AA, et al. Potential materials for food packaging from nanoclay/natural fibres filled hybrid composites. Mater Des. 2013;46:391–410.
44. Adeyeye SAO. Food packaging and nanotechnology: safeguarding consumer health and safety. NFS. 2019;49(6):1164–79.
45. Sahl H-G, Kordel M, Benz R. Voltage-dependent depolarization of bacterial membranes and artificial lipid bilayers by the peptide antibiotic nisin. Arch Microbiol. 1987;149(2):120–4.
46. Qi L, Xu Z, Jiang X, Hu C, Zou X. Preparation and antibacterial activity of chitosan nanoparticles. Carbohydr Res. 2004;339(16):2693–700.

47. Gaikwad KK, Singh S, Lee YS. Oxygen scavenging films in food packaging. Environ Chem Lett. 2018;16(2):523–38.
48. Zerdin K, Rooney ML, Vermuë J. The vitamin C content of orange juice packed in an oxygen scavenger material. Food Chem. 2003;82(3):387–95.
49. Souza R, Peruch G, dos Santos Pires AC. Oxygen scavengers: an approach on food preservation. In: Amer Eissa A, editor. Structure and function of food engineering. Rijeka: InTech; 2012. http://www.intechopen.com/books/structure-and-function-of-food-engineering/oxygen-scavengers-an-approach-on-food-preservation. Accessed 27 Oct 2021.
50. Kwon S, Orsuwan A, Bumbudsanpharoke N, Yoon C, Choi J, Ko S. A short review of light barrier materials for food and beverage packaging. Korean J Packag Sci Technol. 2018;24(3):141–8.
51. Hashimoto A, Sakamoto K. UV-blocking film for food storage using titanium dioxide. Food Sci Technol Res. 2011;17(3):199–202.
52. Primožič M, Knez Ž, Leitgeb M. (Bio)nanotechnology in food science—food packaging. Nanomaterials. 2021;11(2):292.

Chapter 7
Shelf-Life Improvement of Foodstuffs through Nanotechnology Engineered Application

Saira Sattar, Amna Javed, Muhammad Faisal Nisar, Uzma Javaid, Muhammad Saad Hashmi, and Obinna Chukwuemeka Uchenna Adumanya

7.1 Introduction

The food industry is inventing techniques that will help minimize food quality deterioration in order to ensure food safety limits in terms of shelf life. The shelf life of food is known as the period in which food is declared safe for human health with respect to the nutritional value and sensory attributes and profile [1–3]. Because of the high rate of population growth, there is a strong need for food, which is causing food sectors and industries to work overtime to meet the demand. However, it has been noted that such high levels of food production may not be able to meet sustainability goals [4, 5].

Food losses occur majorly at two stages, uppermost during the period of harvesting, post-harvesting, and final stages of production while contrarily through distribution and food consumption levels. There is always a need to address these losses by using advanced approaches to food processing [6, 7].

The food losses can be reduced by implementing multiple approaches including (a) product shelf-life improvement by articulate monitoring (b) increasing the awareness among retailers, stakeholders, and end-users about advanced monitoring labels and systems depicting food item shelf life (c) approaching real-time and

S. Sattar (✉) · A. Javed · M. F. Nisar
Department of Food Sciences, Government College University Faisalabad, Faisalabad, Pakistan

U. Javaid
Department of Anatomy, Sargodha Medical College, Sargodha, Pakistan

M. S. Hashmi
Institute of Food Science and Nutrition, Bahauddin Zakariya University, Multan, Pakistan

O. C. U. Adumanya
Department of Biochemistry, University of Agriculture and Environmental Science, Umuagwo, Nigeria

C. Egbuna et al. (eds.), *Application of Nanotechnology in Food Science, Processing and Packaging*, https://doi.org/10.1007/978-3-030-98820-3_7

accurate demand of food products [8, 9]. Another trend that has emerged in recent years is "ready to eat products," which refers to sophisticated food packaging for half-cooked meals. There is a high demand for these products by consumers that now became the uttermost need of the manufacturer [10]. Many analytical indicators, advanced tools enhancing the resulting credibility of modern technologies with computational approaches estimate the shelf life of the food product. Many previous researchers is in consensus with the benefits of advanced food packaging material on the improvement of shelf life [11–13].

Food-borne diseases in addition to food-borne crud are the major concerns in considering packaging materials as they influence human health. To address this issue, anti-microbial packaging materials are considered in order to retain the quality of food [14, 15]. The general mode of action to retain and preserve for best quality of food items is by protecting them from moisture, light, oxygen, physical and physiochemical stress [16].

To reach all attributes and characteristics of good food packaging, nanotechnology-based food packaging material (NBFPM) has found diverse applications in the food sectors. Previous studies documented the high demand for NBFPM due to its multipurpose functionality [17, 18]. Figure 7.1 demonstrates the role of nanotechnology in different sectors of the food industry.

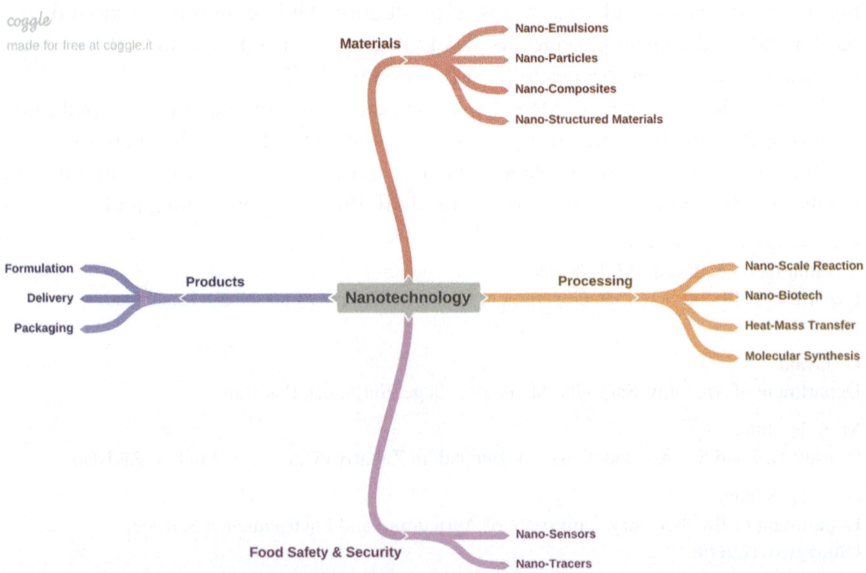

Fig. 7.1 Nanotechnology in the food industry

7.2 Nano-Packaging Materials: Boon for the Food Sector

Packaging materials provide a shield against biochemical, physical, chemical, and microbiological stress to the shelf life of a product. Nano-based food packaging material (NBFPM) has a diverse range that is being used in the food sector all over the world, which enhances the shelf life of the product. However, there are two main types of NBFPM, (a) Improved Packaging (IP) and (b) Active Packaging (AP). IP intensifies the gas barrier characteristics for instance clay nano-composites by adding nano-materials into the matrix of the polymer. There is a direct mode of action of AP with the food products and their surroundings that enhance the protection capability. Different factors such as temperature and pH could affect the AP performance with the addition of anti-fungal compounds [19]. The IP labeling indicates the physicochemical attributes of the food item and environment within the package [18]. Both types of packaging comprise the NBFPM approach that accounted for the fastest-growing segment of packaging in upcoming years. Other types of packaging material include edible coatings, biodegradable film, smart packaging material, and bioactive packaging, etc. [18, 20].

7.2.1 *Edible Coating Material*

Edible coatings (EC) have a wide range of applications to many food items by retaining their quality characteristics. A very thin layer protects the food item from mass transfer by hindering the exchange of gaseous compounds, moisture, and oxygen, etc., made by the edible components which are easily digestible. The incorporation of Nanoparticles into EC enhances the physicochemical such as mechanical strength and sensory attributes like the texture of the food. The composition of EC includes proteins, polysaccharides, and lipids. Lipids-based coatings showed less absorption of moisture whereas proteins and polysaccharides were poor for gaseous exchange [20–23]. Low cost and excessive sources of starch, chitosan, and galactomannans made them the most widely use polysaccharides in food packaging [10].

7.2.2 *Biodegradable Film Material*

The environmentally friendly attributes with good film-forming capacity, made biodegradable film (BF) packaging one of the most attentions gaining packaging. BF is produced from the class of biopolymers including proteins, lipids, and their variants; it is easily decomposed by microbes and bacteria. Bionanocomposites are known as Nano-composite films made from biopolymers. The network of organic and inorganic material made the polymer matrix (dimensions >100 nm). An important factor of BF polymers is their recovery from the fermentation process of microbes such as "pullulan" [24, 25].

7.2.3 *Smart Packaging Material*

Smart packaging (SP) materials are used to monitor the packed food and its outer condition. The cheap and low-cost stickers and labels on the surfaces of bottles, cans, pouches, and containers, etc. are examples of SP which facilitate supply chain communications [26]. The nanoparticle examples and uses are shown in Table 7.1.

7.3 Anti-Microbial Transformation of Nano-Packaging

Many factors influence the shelf life of food a product such as enzymes, pH, water activity, light, humidity, physical handling and exchange of gases, etc. NBFPM changed their properties to make them antimicrobial. Antimicrobial packaging demonstrated an efficient positive response in the presence of antimicrobial compounds incorporated into the food, which are most frequently found in perishable foods [32].

The nanoparticles showing antimicrobial behavior include Fe, ZnO, Ag, MgO with the inclusion of Nano-emulsions or encapsulations. These particles are utilized by using different bondings such as hydrogen and covalent or applying electrostatic charges to provide anti-microbial attributes. One of the most common examples is silver nano-particle surface attachments to plastic materials to smoothen the lessened release of ions (silver) to slow down their addition to food [33].

Table 7.1 Applications of nano-units in the food sector

Nanoparticles	Records	References
Platinum (engineered)	pH of food packaging material, used in the detection of moisture and oxygen during smart packaging application	[27]
Copper (carbon-coated tensile film)	Moisture content sensor, Humidity detection	[28]
Iron oxide incorporated with Composite film	Humidity sensors and records humidity efficiently	[29]
Titanium dioxide, Silica in a combined form with reactive dye[a]	Detects oxygen, efficiently used in NBFPM[b]	[30]
Tungsten oxide	Detection of ethylene gas to control the ripening process in fruits	[31]
Titanium dioxide		

[a] Methylene blue
[b] Nano-based food packaging material

The organic and inorganic NBFPM played important role in food packaging [21, 34]. Silver–polyamide Nano-composites showed excellent anti-microbial behavior towards *Staphylococcus aureus* and *E. coli* [35]. Nanoparticles such as Nano-silver, TiO_2, ZnO, and UV-C ultraviolet light-activated TiO_2 showed anti-microbial properties in the food sector during surface cleaning, food processing, and food transportation to prevent bio-contamination [36].

7.4 Health Concerns

The technology used in nanotechnology uses nanonmaterial manipulation for diverse applications in fields of agriculture, food, human health, and many more [17]. With the development of food packaging by applying nanotechnology, there are a few facts and concerns regarding human health, safety, and interlinked toxicity. It has been reported that nanoparticles showed a tendency to migrate in food from packaging causing toxic effects in end consumers [37].

NBFPM showed direct exposure to the gastrointestinal functions of humans as reported by previous researchers [38]. The nanoparticles have a noticeable impact on gastric functions by being trapped by stomach enzymes and acids. These particles are delivered to the human body from NBFPM to food items being used by the end consumers. However, they might be translocated to the stomach through inhalation followed by trapping of the respiratory system [39].

The ingested nanoparticles showed severe toxic effects by damaging the epithelial linings and posing risk on protection roles in the small intestine. The key function of epithelial lining is to provide a shielding effect by keeping away the lumen of the intestine from circulating blood. Once they crossed the shield, they can damage other sensitive organs of the human body such as the kidney, brain, and liver [40–43]. There is still an essential need to enlighten in detail about the specificity of Nanotechnology regarding the food industry to address the uncertainties of health safety. According to a survey, it was recorded that consumers are reluctant to use NBFMP items due to health concerns [44, 45] whereas, positive public acceptance has also been published [46].

7.5 Nanotechnology Advance Applications and Techniques

Food packaging applications have been classified into three types of polymers (a) Starch-based (b) Cellulose-based and (c) Protein-based. Nanolaminates, a term used for multi-layered material in packaging comprises of one or more than one layer of materials. The dimension of layers ranges in nanometer. The excellent technique used in NBFPM is the layer-by-layer deposition technique which works by charged surfaces coatings on the food items. Due to the highly refined thin layer, it is acclaimed to be the best coating for food items in addition to barrier properties [47, 48].

Nanolaminates' function can be enhanced by adding functional agents such as enzymes, antimicrobials, flavoring agents, and antioxidants. These functional agents could be applied directly through the dipping method or spray technology [49, 50]. Carbon nanotubes have been used as nanofiller in the formulation of gelatin film. The film can enhance mechanical properties by strengthening tensile strength [51]. The potential plane charge created in engineered water nanostructures showed antimicrobial behavior. The deactivation process can be applied on stainless steel surfaces for *Listeria innocua*, *E. coli*, and *Salmonella enteric* effectively while it's efficiently used on tomatoes [52].

Nanomaterials are widely used in the food sector to enhance the shelf life of products. Silicon dioxide, titanium dioxide, zinc oxide, silver nanoparticles, polymeric-nanoparticles showed diversified properties such as hygroscopic, anticaking, antimicrobial, and bactericidal with an efficient delivery mechanism to enhance the shelf life of food products [53–56]. Chitosan provides anti-fungal properties and improved the shelf life of mandarin and fresh fruits [57]. The meat and dairy industry mostly used gold nanoparticles AuNPs while Al_2O_3 efficiently purifies the water [58]. Vacuum proof food packaging can impart shelf life through the application of by carbon nanotubes [59].

7.6 Nanotechnology Shelf-Life Extension of Perishable Food: Fruits and Vegetables

The nutritional content of food items is the key feature to access the quality characteristics and sensory profile. The perishable food used to be discarded once they started to lose their sensory profile including of-taints and off-colors with wrinkled skin [60]. Due to good vitamin content and caloric ratio, fruits and vegetables are considered necessary for human health. They contain never-ending biological interactions even after the harvesting such as the transpiration process, respiration, etc. which ultimately lead to quality defects, loss of sensory and nutritional characteristics making them unfit for human consumption. These biological limitations ended up with almost 20–40% deterioration worldwide followed by poor methods for extending shelf life [61–63].

With the advancement of food processing and preservation techniques, Nanotechnology is known as one of the fastest-growing technology to overcome the traditional shelf-life extending methods by using diverse approaches such as crushing (Top-down) and re-building (Bottom-up) techniques [64]. The basics of every preservation method follow three important points to control in order to preserve the food item (a) Respiration process (b) Microorganism (c) Water evaporation [65].

Nanotechnology used the approach of quantum mechanics which enhances the shielding and barrier effect of nanoparticles [66]. Nanoparticles exhibit a greater surface-to-volume ratio and form an excellent interfacial interaction with polymer providing an effective shelf-life extension effect by nano-composite technique [67]. Due to excellent mechanical properties, fruits and vegetables showed more stability towards the shelf life as suitable chemical interactions and reinforcing effect of stress transfer [68, 69]. The thermal property of Nano-composites facilitates post-processing treatments of fruits and vegetables such as pasteurization [70]. A wide range of vegetables such as carrot, cucumber, and Chinese yam was preserved by nano-units zinc oxide NP, chitosan and $CaCO_3$ NP respectively [71–73]. The applications of nano-units in the enhancement of fruits shelf life are shown in Table 7.2.

7.7 Future Trend and Perspective

Nanotechnology has a great role in the food industry in extending the shelf life of foodstuffs, but these preservation strategies are still at laboratory investigations with a lot of challenges that need to be overcome. Nanoparticles when added as an ingredient in foodstuff may pass through their membrane via diffusion, absorption, and dissolution which could be necessary for the safety of food items and their use for

Table 7.2 Application of nanotechnology to enhance the shelf life of fruit

Tested fruit	Nano-units	Applications in the food industry	References
Melon	Silver NP[a]	Microflora related spoilage control	[74]
		Retarded the process of senescence in freshly-cut melons	
Apple	Zinc oxide NP	Prolonged shelf life by 6 days in freshly-cut apples in comparison to samples packed with traditional packaging	[75]
	Silver NP	Showed anti-microbial behavior in food packaging	[76]
	Titanium dioxide NP		
Pomegranate	Zinc oxide NP	Nano-coatings showed antimicrobial behavior, preserves weight loss, vitamin C, and antioxidant capacity	[77]
Longan fruit	Nano-silica	Inhibits quality deterioration and nutrient loss, enhances shelf life, and preserves anthocyanin content	[78]
Strawberry	Zinc oxide NP	Showed anti-microbial activity in addition to preserving weight loss and functional nutrients	[79]
	Silver NP	Preserved sensory attributes, prevented quality deterioration (physicochemical, and physiological) during storage	[80]
	Titanium dioxide NP		
Barberry	Silver NP	Showed anti-microbial activity, maintained appearance, taste, and aroma of dried samples	[81]
Kiwifruit	Zinc oxide NP	Retained good texture with reduced water loss and lowered ethylene production in fresh-cut kiwifruit	[82]

[a] *NP* nanoparticle

human health. Hence, the utilization of nanoparticles in food packaging could be less harmful as compared to their use as food ingredients but there is a need to understand the proper safety aspects of nanomaterials and standard policies and guidelines should be made for their use in the food industry and to determine their efficacy in food systems. Moreover, for consumer acceptability and marketing, the appropriate labeling of nano-foods is required on food packaging. Likewise, for the production of complex nanoparticles highly efficient processing techniques are required so the appropriate optimization of processing procedures and improvement in existing technologies is necessary for better output.

7.8 Conclusion

According to consumer demands, the delivery of high-quality food with an extended shelf life is the foremost task of food industrialists nowadays. Hence, the role of many innovative food processing technologies is under consideration to fulfill this task and to overcome conventional methods which along with shelf life extension may deteriorate the food quality and its sensory attributes. Nanotechnology is one of the advanced technologies which are keenly used for the extension of the shelf life of food products due to its unique properties like the production of high-quality food products by preventing contamination in a shorter time. A nanotechnology-based delivery system also enhances the bioavailability of phytochemicals in food. Moreover, nanoparticles due to their efficiency are used in different divisions in the food industry like food packaging, food processing, food safety, food quality, etc.

References

1. Van Boekel MA. Kinetic modeling of food quality: a critical review. Compr Rev Food Sci Food Saf. 2008;7(1):144–58.
2. Van Boekel MA. Kinetic modeling of reactions in foods. Boca Raton, FL: CRC Press; 2008.
3. Jedermann R, Nicometo M, Uysal I, Lang W. Reducing food losses by intelligent food logistics. Philos Trans A Math Phys Eng Sci. 2014;372:20130302.
4. DeFries R, Fanzo J, Remans R, Palm C, Wood S, Anderman TL. Metrics for land-scarce agriculture. Science. 2015;349(6245):238–40.
5. Augustin MA, Riley M, Stockmann R, Bennett L, Kahl A, Lockett T, Osmond M, Sanguansri P, Stonehouse W, Zajac I, Cobiac L. Role of food processing in food and nutrition security. Trends Food Sci Technol. 2016;56:115–25.
6. van Boekel M, Fogliano V, Pellegrini N, Stanton C, Scholz G, Lalljie S, Somoza V, Knorr D, Jasti PR, Eisenbrand G. A review on the beneficial aspects of food processing. Mol Nutr Food Res. 2010;54(9):1215–47.
7. Beccaria M, Oteri M, Micalizzi G, Bonaccorsi IL, Purcaro G, Dugo P, Mondello L. Reuse of dairy product: evaluation of the lipid profile evolution during and after their shelf-life. Food Anal Methods. 2016;9(11):3143–54.
8. Manzocco L, Alongi M, Sillani S, Nicoli MC. Technological and consumer strategies to tackle food wasting. Food Eng Rev. 2016;8(4):457–67.

9. Bibi F, Guillaume C, Gontard N, Sorli B. A review: RFID technology having sensing aptitudes for food industry and their contribution to tracking and monitoring of food products. Trends Food Sci Technol. 2017;62:91–103.
10. Rai M, Ingle AP, Gupta I, Pandit R, Paralikar P, Gade A, Chaud MV, dos Santos CA. Smart nanopackaging for the enhancement of food shelf life. Environ Chem Lett. 2019;17(1):277–90.
11. Carbone M, Donia DT, Sabbatella G, Antiochia R. Silver nanoparticles in polymeric matrices for fresh food packaging. J King Saud Univ Sci. 2016;28(4):273–9.
12. Xiao Y, Yang S. The retail chain design for perishable food: the case of price strategy and shelf space allocation. Sustainability. 2017;9(1):12.
13. Shi C, Wu Y, Fang D, Pei F, Mariga AM, Yang W, Hu Q. Effect of nanocomposite packaging on postharvest senescence of Flammulina velutipes. Food Chem. 2018;246:414–21.
14. Pal M. Nanotechnology: a new approach in food packaging. J Food Microbiol Saf Hyg. 2017;2(02):8–9.
15. Appendini P, Hotchkiss JH. Review of antimicrobial food packaging. Innovative Food Sci Emerg Technol. 2002;3(2):113–26.
16. Wesley SJ, Raja P, Raj AA, Tiroutchelvamae D. Review on-nanotechnology applications in food packaging and safety. Int J Eng Res. 2014;3(11):645–51.
17. Duncan TV. Applications of nanotechnology in food packaging and food safety: barrier materials, antimicrobials and sensors. J Colloid Interface Sci. 2011;363(1):1–24.
18. Echegoyen Y. Nano-developments for food packaging and labeling applications. In: Nanotechnologies in food and agriculture. Cham: Springer; 2015. p. 141–66.
19. Van Long NN, Joly C, Dantigny P. Active packaging with antifungal activities. Int J Food Microbiol. 2016;220:73–90.
20. Wahab A, Rahim AA, Hassan S, Egbuna C, Manzoor MF, Okere KJ, Walag AMP. Application of nanotechnology in the packaging of edible materials. In: Egbuna C, Mishra AP, Goyal MR, editors. Preparation of phytopharmaceuticals for the management of disorders. Academic Press; 2021. p. 215–25.
21. Saeed F, Javaid A, Ahmed N, Nadeem MT, Arshad MS, Imran A, Sohaib M, Khan AU. Influence of edible coating techniques on quality characteristics of eggs. J Food Process Preserv. 2017;41(2):e12815.
22. Weiss J, Takhistov P, McClements DJ. Functional materials in food nanotechnology. J Food Sci. 2006;71(9):R107.
23. Valdés A, Ramos M, Beltrán A, Jiménez A, Garrigós MC. State of the art of antimicrobial edible coatings for food packaging applications. Coatings. 2017;7(4):56.
24. Cabedo L, Luis Feijoo J, Pilar Villanueva M, Lagarón JM, Giménez E. Optimization of biodegradable nanocomposites based on aPLA/PCL blends for food packaging applications. Macromol Symp. 2006;233(1):191–7.
25. Garavand F, Rouhi M, Razavi SH, Cacciotti I, Mohammadi R. Improving the integrity of natural biopolymer films used in food packaging by crosslinking approach: a review. Int J Biol Macromol. 2017;104:687–707.
26. Yam KL. The Wiley encyclopedia of packaging technology. 3rd ed. New York, NY: Wiley; 2009.
27. Martins AJ, Benelmekki M, Teixeira V, Coutinho PJ. Platinum nanoparticles as pH sensor for intelligent packaging. J Nano Res. 2012;18:97–104.
28. Luechinger NA, Loher S, Athanassiou EK, Grass RN, Stark WJ. Highly sensitive optical detection of humidity on polymer/metal nanoparticle hybrid films. Langmuir. 2007;23(6):3473–7.
29. Taccola S, Greco F, Zucca A, Innocenti C, de Julián Fernández C, Campo G, Sangregorio C, Mazzolai B, Mattoli V. Characterization of free-standing PEDOT: PSS/iron oxide nanoparticle composite thin films and application as conformable humidity sensors. ACS Appl Mater Interfaces. 2013;5(13):6324–32.
30. Pradhan N, Singh S, Ojha N, Shrivastava A, Barla A, Rai V, Bose S. Facets of nanotechnology as seen in food processing, packaging, and preservation industry. Biomed Res Int. 2015;2015:365672.

31. Pimtong-Ngam Y, Jiemsirilers S, Supothina S. Preparation of tungsten oxide–tin oxide nanocomposites and their ethylene sensing characteristics. Sensors Actuators A Phys. 2007;139(1–2):7–11.
32. Emblem A. Plastics properties for packaging materials. In: Packaging technology. New Delhi: Woodhead Publishing; 2012. p. 287–309.
33. Morris MA, Padmanabhan SC, Cruz-Romero MC, Cummins E, Kerry JP. Development of active, nanoparticle, antimicrobial technologies for muscle-based packaging applications. Meat Sci. 2017;132:163–78.
34. Nile SH, Baskar V, Selvaraj D, Nile A, Xiao J, Kai G. Nanotechnologies in food science: applications, recent trends, and future perspectives. Nano-Micro Lett. 2020;12(1):1–34.
35. Fedotova AV, Snezhko AG, Sdobnikova OA, Samoilova LG, Smurova TA, Revina AA, Khailova EB. Packaging materials manufactured from natural polymers modified with silver nanoparticles. Int Polym Sci Technol. 2010;37(10):59–64.
36. Khan ST, Al-Khedhairy AA, Musarrat J. ZnO and TiO 2 nanoparticles as novel antimicrobial agents for oral hygiene: a review. J Nanopart Res. 2015;17(6):276.
37. Golja V, Dražić G, Lorenzetti M, Vidmar J, Ščančar J, Zalaznik M, Kalin M, Novak S. Characterisation of food contact non-stick coatings containing TiO2 nanoparticles and study of their possible release into food. Food Addit Contam Part A. 2017;34(3):421–33.
38. Gupta I, Duran N, Rai M. Nano-silver toxicity: emerging concerns and consequences in human health. In: Nano-antimicrobials. Berlin: Springer; 2012. p. 525–48.
39. Wang L, Nagesha DK, Selvarasah S, Dokmeci MR, Carrier RL. Toxicity of CdSe nanoparticles in Caco-2 cell cultures. J Nanobiotechnol. 2008;6(1):1–5.
40. Egbuna C, Parmar VK, Jeevanandam J, Ezzat SM, Patrick-Iwuanyanwu KC, Adetunji CO, Khan J, et al. Toxicity of nanoparticles in biomedical application: nanotoxicology. J Toxicol. 2021;2021:9954443., 21 pages. https://doi.org/10.1155/2021/9954443.
41. Choi K, Riviere JE, Monteiro-Riviere NA. Protein corona modulation of hepatocyte uptake and molecular mechanisms of gold nanoparticle toxicity. Nanotoxicology. 2017;11(1):64–75.
42. Kononenko V, Repar N, Marušič N, Drašler B, Romih T, Hočevar S, Drobne D. Comparative in vitro genotoxicity study of ZnO nanoparticles, ZnO macroparticles and ZnCl2 to MDCK kidney cells: size matters. Toxicol in Vitro. 2017;40:256–63.
43. Yin Y, Hu Z, Du W, Ai F, Ji R, Gardea-Torresdey JL, Guo H. Elevated CO2 levels increase the toxicity of ZnO nanoparticles to goldfish (Carassius auratus) in a water-sediment ecosystem. J Hazard Mater. 2017;327:64–70.
44. Siegrist M, Cousin ME, Kastenholz H, Wiek A. Public acceptance of nanotechnology foods and food packaging: the influence of affect and trust. Appetite. 2007;49(2):459–66.
45. Bieberstein A, Roosen J, Marette S, Blanchemanche S, Vandermoere F. Consumer choices for nano-food and nano-packaging in France and Germany. Eur Rev Agric Econ. 2013;40(1):73–94.
46. Cobb MD, Macoubrie J. Public perceptions about nanotechnology: risks, benefits and trust. J Nanopart Res. 2004;4(6):395–405.
47. Kour H, Malik AA, Ahmad N, Wani TA, Kaul RK, Bhat A. Nanotechnology-new lifeline for food industry. Crit Rev Food Sci Nutr. 2015;60:1.
48. Momin JK, Joshi BH. Nanotechnology in foods. In: Nanotechnologies in food and agriculture. Cham: Springer; 2015. p. 3–24.
49. GuhanNath S, Aaron SI, Raj AA, Ranganathan TV. Recent innovations in nanotechnology in food processing and its various applications—a review. Int J Pharm Sci Rev Res. 2014;29(2):116–24.
50. Ramachandraiah K, Han SG, Chin KB. Nanotechnology in meat processing and packaging: potential applications—a review. Asian Australas J Anim Sci. 2015;28(2):290.
51. Ortiz-Zarama MA, Jiménez-Aparicio A, Perea-Flores MJ, Solorza-Feria J. Barrier, mechanical and morpho-structural properties of gelatin films with carbon nanotubes addition. J Food Eng. 2014;120:223–32.

52. Pyrgiotakis G, Vasanthakumar A, Gao Y, Eleftheriadou M, Toledo E, DeAraujo A, McDevitt J, Han T, Mainelis G, Mitchell R, Demokritou P. Inactivation of foodborne microorganisms using engineered water nanostructures (EWNS). Environ Sci Technol. 2015;49(6):3737–45.
53. Jones N, Ray B, Ranjit KT, Manna AC. Antibacterial activity of ZnO nanoparticle suspensions on a broad spectrum of microorganisms. FEMS Microbiol Lett. 2008;279(1):71–6.
54. Zhao R, Torley P, Halley PJ. Emerging biodegradable materials: starch-and protein-based bio-nanocomposites. J Mater Sci. 2008;43(9):3058–71.
55. Acosta E. Bioavailability of nanoparticles in nutrient and nutraceutical delivery. Curr Opin Colloid Interface Sci. 2009;14(1):3–15.
56. Arshak K, Adley C, Moore E, Cunniffe C, Campion M, Harris J. Characterisation of polymer nanocomposite sensors for quantification of bacterial cultures. Sensors Actuators B Chem. 2007;126(1):226–31.
57. Xing Y, Xu Q, Li X, Chen C, Ma L, Li S, Che Z, Lin H. Chitosan-based coating with antimicrobial agents: preparation, property, mechanism, and application effectiveness on fruits and vegetables. Int J Polym Sci. 2016;2016:1–24.
58. Ozdemir C, Yeni F, Odaci D, Timur S. Electrochemical glucose biosensing by pyranose oxidase immobilized in gold nanoparticle-polyaniline/AgCl/gelatin nanocomposite matrix. Food Chem. 2010;119(1):380–5.
59. Thangavel G, Thiruvengadam S. Nanotechnology in food industry—a review. Int J Chem Tech Res. 2014;6(9):4096–101.
60. Aiello G, La Scalia G, Micale R. Simulation analysis of cold chain performance based on time–temperature data. Prod Plan Control. 2012;23(6):468–76.
61. El-Ramady HR, Domokos-Szabolcsy É, Abdalla NA, Taha HS, Fári M. Postharvest management of fruits and vegetables storage. In: Sustainable agriculture reviews. Cham: Springer; 2015. p. 65–152.
62. Sanzani SM, Reverberi M, Geisen R. Mycotoxins in harvested fruits and vegetables: insights in producing fungi, biological role, conducive conditions, and tools to manage postharvest contamination. Postharvest Biol Technol. 2016;122:95–105.
63. Caprioli G, Iannarelli R, Cianfaglione K, Fiorini D, Giuliani C, Lucarini D, Papa F, Sagratini G, Vittori S, Maggi F. Volatile profile, nutritional value and secretory structures of the berry-like fruits of Hypericum androsaemum L. Food Res Int. 2016;79:1.
64. Peters RJ, Bouwmeester H, Gottardo S, Amenta V, Arena M, Brandhoff P, Marvin HJ, Mech A, Moniz FB, Pesudo LQ, Rauscher H. Nanomaterials for products and application in agriculture, feed and food. Trends Food Sci Technol. 2016;54:155–64.
65. Yan WQ, Zhang M, Huang LL, Tang J, Mujumdar AS, Sun JC. Studies on different combined microwave drying of carrot pieces. Int J Food Sci Technol. 2010;45(10):2141–8.
66. Sanchez-Garcia MD, Lopez-Rubio A, Lagaron JM. Natural micro and nanobiocomposites with enhanced barrier properties and novel functionalities for food biopackaging applications. Trends Food Sci Technol. 2010;21(11):528–36.
67. Khan A, Khan RA, Salmieri S, Le Tien C, Riedl B, Bouchard J, Chauve G, Tan V, Kamal MR, Lacroix M. Mechanical and barrier properties of nanocrystalline cellulose reinforced chitosan based nanocomposite films. Carbohydr Polym. 2012;90(4):1601–8.
68. Sanchez-Garcia MD, Lagaron JM, Hoa SV. Effect of addition of carbon nanofibers and carbon nanotubes on properties of thermoplastic biopolymers. Compos Sci Technol. 2010;70(7):1095–105.
69. Abdollahi M, Alboofetileh M, Rezaei M, Behrooz R. Comparing physico-mechanical and thermal properties of alginate nanocomposite films reinforced with organic and/or inorganic nanofillers. Food Hydrocoll. 2013;32(2):416–24.
70. Cheng S, Zhang Y, Cha R, Yang J, Jiang X. Water-soluble nanocrystalline cellulose films with highly transparent and oxygen barrier properties. Nanoscale. 2016;8(2):973–8.
71. Luo Z, Wang Y, Jiang L, Xu X. Effect of nano-CaCO3-LDPE packaging on quality and browning of fresh-cut yam. LWT - Food Sci Technol. 2015;60(2):1155–61.

72. Mohammadi A, Hashemi M, Hosseini SM. Postharvest treatment of nanochitosan-based coating loaded with Zataria multiflora essential oil improves antioxidant activity and extends shelf-life of cucumber. Innovative Food Sci Emerg Technol. 2016;33:580–8.
73. Xu J, Zhang M, Bhandari B, Kachele R. ZnO nanoparticles combined radio frequency heating: a novel method to control microorganism and improve product quality of prepared carrots. Innovative Food Sci Emerg Technol. 2017;44:46–53.
74. Fernandez A, Picouet P, Lloret E. Reduction of the spoilage-related microflora in absorbent pads by silver nanotechnology during modified atmosphere packaging of beef meat. J Food Prot. 2010;73(12):2263–9.
75. Li X, Li W, Jiang Y, Ding Y, Yun J, Tang Y, Zhang P. Effect of nano-ZnO-coated active packaging on quality of fresh-cut 'Fuji' apple. Int J Food Sci Technol. 2011;46(9):1947–55.
76. Metak AM, Ajaal TT. Investigation on polymer based nano-silver as food packaging materials. Int J Chem Mol Eng. 2013;7(12):1103–9.
77. Saba MK, Amini R. Nano-ZnO/carboxymethyl cellulose-based active coating impact on ready-to-use pomegranate during cold storage. Food Chem. 2017;232:721–6.
78. Shi S, Wang W, Liu L, Wu S, Wei Y, Li W. Effect of chitosan/nano-silica coating on the physicochemical characteristics of longan fruit under ambient temperature. J Food Eng. 2013;118(1):125–31.
79. Sogvar OB, Saba MK, Emamifar A, Hallaj R. Influence of nano-ZnO on microbial growth, bioactive content and postharvest quality of strawberries during storage. Innovative Food Sci Emerg Technol. 2016;35:168–76.
80. Yang FM, Li HM, Li F, Xin ZH, Zhao LY, Zheng YH, Hu QH. Effect of nano-packing on preservation quality of fresh strawberry (Fragaria ananassa Duch. cv Fengxiang) during storage at 4 C. J Food Sci. 2010;75(3):C236–40.
81. Valipoor Motlagh N, Hamed Mosavian MT, Mortazavi SA. Effect of polyethylene packaging modified with silver particles on the microbial, sensory and appearance of dried barberry. Packag Technol Sci. 2013;26(1):39–49.
82. Meng X, Zhang M, Adhikari B. The effects of ultrasound treatment and nano-zinc oxide coating on the physiological activities of fresh-cut kiwifruit. Food Bioprocess Technol. 2014;7(1):126–32.

Chapter 8
Roles of Nanotechnology for Efficient Nutrient Delivery of Foods

Shahira M. Ezzat, Maha Salama, Nehal El Mahdi, and Mohamed Salem

8.1 Introduction

The word "nano" is a Greek word that means 'dwarf'. A nanomaterial is one whose size ranges from 1 to 100 nm which can be structured in various forms [1]. This material can then be "structured" to the micro level by different means to have varieties of applications. Nanomaterials could be in the form of nanowires, nanorods, nanoparticles (NPs), thin films with nanoscale thickness, or bulk materials with nanoscale building blocks which should have at least one dimension in nanometer [2].

Food nanotechnology is a growing and interesting area in food industry which can be applied in all phases of food production cycle. The basic application in food

S. M. Ezzat (✉)
Department of Pharmacognosy, Faculty of Pharmacy, Cairo University, Cairo, Egypt

Department of Pharmacognosy, Faculty of Pharmacy, October University for Modern Science and Arts (MSA), Giza, Egypt
e-mail: shahira.ezzat@pharma.cu.edu.eg; smelkomy@msa.eun.eg

M. Salama
Department of Pharmacognosy, Faculty of Pharmacy, Cairo University, Cairo, Egypt

Department of Pharmacognosy, Faculty of Pharmacy, The British University in Egypt, Cairo, Egypt
e-mail: maha.salama@bue.edu.eg; maha.salama@pharma.cu.edu.eg

N. El Mahdi
Department of Pharmaceutics and Industrial Pharmacy, Faculty of Pharmacy, October University for Modern Sciences and Arts (MSA), Giza, Egypt
e-mail: nmahdy@msa.eun.eg

M. Salem
Department of Pharmacognosy, Faculty of Pharmacy, Menoufia University, Menoufia, Egypt
e-mail: mohamed.salem@phrm.menofia.edu.eg

C. Egbuna et al. (eds.), *Application of Nanotechnology in Food Science, Processing and Packaging*, https://doi.org/10.1007/978-3-030-98820-3_8

nanotechnology includes the development of food production, processing and packaging with the aim of the extension of shelf life and to improve food safety [3]. The application of nanotechnology in food is mainly applied to improve the functional behavior of food. One of the main targets is the preparation of nanomaterials of the original food that will have different functions and activity [4].

For successful application, it is necessary to study and understand the properties of food and food processing at the nanoscale. In the food industry, still the number of food products produced nanotechnology is small although the technique is highly promising in detecting bacteria in packaging, or the production of stronger natural coloring matters or safer flavors. Production of new food products using nanotechnology may be achieved in the near future, but at the present the main benefits of nanotechnology is the improvement of processability, stability, food texture and taste [5].

One of the main applications of nanotechnology in food industry is the improvement of delivery and bioavailability of bioactive nutrients via the use of nano-derived assemblies [6, 7]. Nanotechnology applications also involve the preparation of nanoscale edible coatings in order to act as a barrier for protection against moisture and gas exchange [8]. Despite the advances in food nanotechnology, little is reported about the toxicity of such products [3, 9]. A lot of people do not agree on using engineered food, for this reason many rules and guiding principles are used to control the use of nano-materials in food. In this chapter we presented an overview on the use and applications of nanotechnology for efficient nutrient delivery in food industry.

8.2 Nanotechnology in Nutraceuticals Delivery Applications

The use of nanoparticles in food-based and nutraceutical delivery application have gained wide attention recently due to its vast advantages and applications. The use of nanomaterials provides a large surface area leading to increased bioavailability due to faster absorption and penetration through the epithelial cells as well as faster digestion in GIT [10]. Moreover, the appearance of nanoparticle dispersion is usually transparent or translucent leading to a more elegant appearance. Most importantly, they shield the encapsulated materials from degradation and prevent their interaction with other food ingredients or mask their undesirable flavor. Food nanotechnology has been applied to many aspects, such as food packaging, additives, and food preservation, and safety [11]. Much of these nanoparticles are derived from food-based materials such as polysaccharides, proteins, lipids or minerals, or non-food-grade polymers or metal nanoparticles [12]. It is worth noting that the selection criteria for a particular nanoparticle depends on multiple factors such as nutraceutical or food material type, the safety, biocompatibility, particle size, stability, release mechanisms, delivery method, ease of scalability and production costs [13]. There are several examples of nanoparticles that have been used in literature

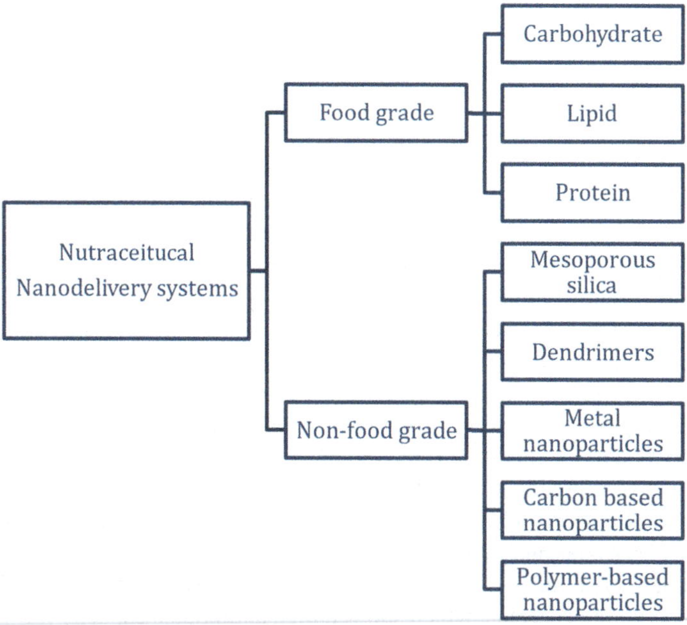

Fig. 8.1 Classification of nanoparticle systems used for nutraceuticals delivery

for nutraceutical delivery summarized in Fig. 8.1 and classified into food-grade and non-food grade nanoparticles.

8.2.1 Food Grade Nanoparticles Delivery Systems

Food-grade biopolymers like lipids, polysaccharides or proteins have been used for the preparation of nanoparticles especially in nutraceutical loading owing to their safety, biocompatibility, established non toxicity as well as abundancy which make them ideal for use in safe nutraceutical formulations.

8.2.1.1 Lipid-Based Nanoparticles

Lipid-based nano delivery systems are considered the most promising and versatile nano-delivery systems applied in the food industry as they are easily produced and upscaled. Moreover, their ability to incorporate hydrophobic bio-moieties is beneficial for food materials like triglycerides, carotenoids, tocopherols, flavonoids, polyphenols, phytosterols, oil-soluble vitamins which thereby increase their solubilization and bioavailability in vivo [14].

Liposomes

Liposomes are a common choice for the delivery of nutrients, as they are spherical shaped and nano-range in size, which are constructed through assembling phospholipids molecules with an inner aqueous core. Liposomes have been extensively exploited for food delivery [15]. Owing to their nano size, they offer a large surface area thereby enhancing their absorption, at the same time they can possess hydrophobic molecules in the phospholipid bilayers with high efficiency [15]. Liu et al. [16] co-loaded Vitamin C and *β*-carotene (*β*C) in liposomes using hydrophilic and hydrophobic cavities simultaneously which showed higher stability as well as better release properties. Rovoli et al. [17] developed a stable vitamin A loaded liposome/ β-lactoglobulin formulation which showed sufficient entrapment and delivery of vitamin A in vivo. Kapoor et al. developed transdermal folate formulation for systemic delivery using liposomes stabilized within 3D matrix of Fuller's earth and henna which showed enhanced room temperature stability which showed high stability as well as a significantly higher folate plasma concentration compared to conventional oral delivery [15]. Niosomes are also lipid nanoparticles composed of non-ionic surfactants and lipids and characterized by high stability and loading efficiency. Talebi et al. utilized niosomes containing cholesterol and polyethylene glycol to load vitamin E-acetate which showed high antioxidant capacity (56%) as well as high encapsulation efficiency (92.9%) [18].

Nano-Emulsions

Nano-emulsions comprise heterogeneous dispersions composed of two immiscible liquids, in which one liquid is dispersed as droplets in the nano range into the other liquid forming continuous phase with and emulsifier as stabilizer. They are classified as either an O/W (oil in water) or W/O (water in oil) emulsion [13, 19] with oils such as olive oil, castor oil, corn oil, or sesame oil to form the nanoemulsion [13]. They are widely exploited for the delivery of nutraceuticals as they have high kinetic and thermodynamic stability and high penetration capacity [19, 20]. Zou et al. examined the influence of lipid droplet size of nanoemulsions on the solubility and bioavailability of curcumin which showed a direct relationship between droplet size and curcumin encapsulation [21] β-Carotene nanoemulsions with different oil compositions demonstrated enhanced β-carotene bioavailability as the long-chain triglyceride (LCT) ratio in the lipid phase increased for low-fat nanoemulsion attributed to the increased solubilization capacity of mixed micelles formed by LCT. However, this effect did not hold for high fat nanoemulsions, due to the opposing effects of lipid digestion and micelle solubilization [22]. β-carotene was also loaded in a tea polyphenols- (TP-BC) oil-in-water (O/W) nanoemulsion which demonstrated increased stability and retention rate of β-carotene compared to conventional nanoemulsion as well as improved oral bioavailability [23]. The same team also utilized TP nanoemulsion to enhance the bioavailability of epigallocatechin gallate (EGCG) compared to that in aqueous solution as well as enhanced bioabsorption with a sustained release profile [24]. Yazdi et al. formulated a Zataria multiflora essential oil nano-emulsion which showed a higher protoscolicidal effect than its conventional emulsion [25].

Solid Lipid Nanoparticles (SLNs) and Nano-Structured Lipid Carriers

Solid lipid nanoparticles (SLNs) are lipid-based delivery systems containing a lipid core matrix such as waxes, fatty acids, and acylglycerol and are stabilized by surfactants such as sphingomyelins, phospholipids, bile salts, and sterols [26]. They have the ability to solubilize lipophilic bioactive materials as well as provide protection against degradation in the GI tract, and increase stability by offering an in vitro controlled release profile [10, 13]. SLNs were used for curcumin loading through novel cold dilution of oil-in-water microemulsion consisting of trilaurin in a partially water-miscible solvent which was able to encapsulate almost 90% of the curcumin, improved stability, and inhibited pancreatic carcinoma cell growth in vitro in a concentration-dependent manner [27]. SLNs were also utilized for EGCG loading for protection during storage against degradation as well as shielding against digestion under simulated gastrointestinal pH conditions using a lipid matrix consisting of pure cocoa butter as well as sodium stearoyl-2-lactylate (SSL) and mono- and diglycerides (MDG) as surfactants. The prepared SLNs stabilized the encapsulated EGCG during the entire storage period at neutral pH values [28].

Similarly, nanostructured lipid carriers (NLC) are produced through mixing either solid lipids or liquid lipids or both [13, 26]. NLCs offer the same advantages of SLNs but overcome the shortcomings of SLNs as it provides higher encapsulation of bioactive molecules, decreased aggregation and material leakage from the NLC via crystal transformation either during preparation, processing or storage [26]. NLCs were utilized for loading Oleuropein (OLE), a naturally occurring polyphenol from olive leaves (*Olea europaea* L.), to increase its stability and improve its antioxidant efficacy. The cellular in vitro assays confirmed the biocompatibility of the NLCs. The formula also enhanced OLE antioxidant efficacy in the A549 and CuFi-1 lung epithelial cell lines, respectively as well as maintained its stability [29]. NLC's were also utilized by Kamel et al. for encapsulation of curcumin which improved its cell penetration as well as cytotoxic anticancer properties [30].

8.2.1.2 Carbohydrate-Based

Carbohydrates are mainly composed of polysaccharides, oligosaccharides, and monosaccharides which offer an advantage of being a natural biocompatible, bioavailable inexpensive material capable of interaction with several bioactive compounds through their functional groups regardless of whether they are hydrophilic or hydrophobic [31]. Moreover, they have high stability in higher temperatures compared to lipid or protein-based nanoparticles due to the likelihood of either melting or denaturation making them ideal in nano-delivery systems of nutraceuticals [13]. Examples for carbohydrate-based delivery are summarized below.

Chitosan

Chitosan is a biopolymer derived from chitin and composed of *N*-acetyl-D-glucosamine units linked by β(1,4)-glycosidic linkages [32]. Owing to its biocompatibility and biodegradability, bioactivity, chitosan was used in various applications ranging from wastewater treatment, biosensor, tissue engineering as well as drug and gene deliveries [33]. Chitosan has been used in food applications as an emulsifier, encapsulation nanoparticles, edible coating, preservative, and stabilizer [33].

In nanotechnology food delivery, several attempts to harness the advantages of chitosan were applied in food delivery applications. Wang et al. developed Chitosan red blood cells (RBC)-hitchhiking nanoparticles loading vitamin K (VK-CSNPs) using ionotropic gelation method. VK-CSNPs adsorbed onto RBCs (RBC-VK-CSNPs) via electrostatic interactions rapidly to release the drug, then desorbed to be eliminated in vivo offering prolonged circulation [34]. Yilmaz et al. fabricated *Origanum vulgare* essential oil loaded-chitosan nanoparticles using electrospraying which was thermostable nanoparticles with increased loading efficiency and particle stability [35]. ε-Polylysine-loaded chitosan-sodium alginate nanoparticles (ε-PL NPs) were prepared by Liu et al. which showed a threefold increase in antibacterial activity of ε-PL NPs than free ε-PL in terms of inhibition zone diameter [36]. Du et al. investigated the mechanism of entrapment of egg white derived peptides (EWDP) in chitosan-tripolyphosphate nanoparticles (CS-TPP NPs). The peptides interacted with CS-TPP NPs through strong hydrogen bonds and electrostatic interactions leading to enhanced entrapment [37]. Ng et developed Curcumin-encapsulated chitosan nanoparticles which showed enhanced bioavailability and enhanced antiviral effects of curcumin against Feline infectious peritonitis viral infection [38].

Curdlan

Curdlan is a β-1,3-glucan linear polysaccharide, with no branching produced by the microorganism *Agrobacterium biovar*. It is currently in use in food applications due to its thermal gelling properties [39]. Yu et al. developed a biopolymer–polyphenol conjugate-stabilized oil-in-water emulsion system was formulated by preparing a β-carotene (BC) ferulic acid-grafted curdlan conjugate (Cur-D-*g*-FA) which was able to enhance the chemical stability of BC in different environmental stress conditions and led to increased bioavailability of BC in vitro offering great potential to protect lipid-soluble [40]. The same team also studied curdlans with different carboxylate contents, molecular weights (MWs), and chain conformations while utilizing them as hydrophilic coatings for hydrophobic zein nanoparticles (ZNPs) to load curcumin. The anionic carboxylic curdlan coating on the surface of zein enhanced curcumin water dispersibility, photostability, and thermostability, as well as offering a sustained release behaviour and stronger in vitro antioxidant activity [41].

Starch

Starch is an abundant polysaccharide in plants, formed of glucose units linked to α-d-(1 → 4) and/or α-d-(1 → 6) linkages. Starch is digestible, biodegradable, and biocompatible. However it is hydrophilic, which restricts its use only to hydrophilic nutraceuticals, another limitation is its degradation in GIT acid media and amylase hydrolysis which limits its applicability in nutraceutical loading [13]. Nevertheless, Ahmad et al. formulated starch-based nanoparticles from horse chestnut (HSC), water chestnut (WSC), and lotus stem (LSC) for loading of catechin which showed controlled release of catechin in the intestine and higher levels of encapsulation [42]. Ramanan utilized starch nanoparticles prepared from quinoa and maize starch (QR and MR) for rutin encapsulation. The prepared nanoparticles showed increased rutin bioavailability during simulated *in vitro* digestion with a significant increase in ($p < 0.05$) *in vitro* antioxidant activities [43].

Cellulose

Cellulose is also a common polysaccharide in plants formed from glucose linked with β-d (1 → 4) linkages, It suffers from poor solubility and large dimensions, it is also not digested by the colon or microbial enzymes which considered to account their absorption and potential toxicity [12]. Nevertheless, cellulose nanocarriers have been studied for the delivery of nutraceuticals. One such example is nanogels (NGs) formed from macromolecular self-assembly of lysozyme (Ly) and carboxymethyl cellulose (CMC) was formulated to encapsulate and protect tea polyphenols (TPs) (Ly–CMC NGs) and formed based on molecules driven by electrostatic interaction and hydrophobic forces. TP-loaded NGs successfully encapsulated TP and proved to have a sustained release profile as well as inducing apoptosis in hepatoma cells and cell cycle arrest [44].

Pectin

Pectin is a linear anionic polysaccharide consist of linear α-d-(1 → 4) galacturonic acid residues. It has the ability to resist enzymatic degradation in the upper digestive system but it is digestible by the microflora in the colon making it a suitable choice for oral delivery of nutraceuticals that are degraded in stomach acidic media [12]. Pectin was successfully incorporated as an emulsifier in water-in-oil-in-water ($W_1/O/W_2$) double emulsions which resisted sedimentation and creaming cementing its use as an appropriate emulsifier for stabilizing double emulsions [20].

Carrageenan

Carrageenan is a linear hydrophilic polysaccharide extracted from marine algae with 15–40% ester sulphate content and an average molecular weight above 100 kDa which is commonly used for bioactive molecules delivery in the form of nanoparticles due to its gelling properties, solubility, and electronegativity [45]. Yew et al. developed a pH-dependent κ-carrageenan-chitosan nanoparticle which showed high encapsulation of hydrophilic compounds such as ascorbic acid as well as slow controlled release [46]. Sun et al. combined *k*-carrageenan (Kc), with tween 80 for zein nanoparticles stabilization which showed high encapsulation for curcumin as well as adequate particle size and stability [47].

8.2.1.3 Protein-Based Delivery

Protein is abundant and acquired from bacterial, plant, animal, and fungal sources. Protein nanocarriers can be applied in the delivery of both hydrophilic and hydrophobic bioactive molecules [13]. Many proteins are natural nanovesicles for nutraceuticals as they can bind to bioactive molecules forming nano complexes transforming hydrophobic nutraceuticals into a dispersible and water-soluble form [48, 49]. Thus, application of nutraceutical-based proteins has been widely attempted to improve the dispersibility and solubility of poorly soluble drugs such as curcumin (CUR), resveratrol, thymol etc. [14].

Liu et al. formulated a novel protein-lipid composite nanoparticle with a three-layered structure (a barley protein layer, α-tocopherol layer, and phospholipid layer) with an inner aqueous core for encapsulation of hydrophilic nutraceuticals vitamin B12 as well as providing controlled release profile and increased serum levels upon oral delivery [50].

β-lactoglobulin (BLG)

β-lactoglobulin (BLG) has been also widely used owing to its good colloidal stability, the ability to be engineered for specific ligand binding, and its resistance to acidic media while degradation in alkaline condition making it a good candidate for oral delivery of acid-liable molecules [14]. Thus, BLG was used by Pujara et al. for loading of resveratrol which showed increased aqueous solubility, in vitro dissolution, and in vivo as well as a higher anti-inflammatory effect against ulcerative colitis in mice [51].

Casein

Caseins are the main proteins in milk accounting for 78% of total milk protein. Caseins tend to interact with one another and form supramolecular aggregate in milk, known as 'casein micelle' or 'casein supramolecule' [52] Casein micelles show pH-dependent behavior as they tighten at lower pH and swell at higher pH values which can be favorable for the controlled release of bioactive components taken orally in addition to their safety, biocompatibility, and ability to load both hydrophilic and hydrophobic drugs makes them excellent tools for delivery of nutraceuticals [53]. Casein micelles were utilized by Xu et al. by formulating curcumin, casein, and soy soluble polysaccharide ternary complex nanoparticles with a loading efficiency of ~97%, curcumin and enhanced the oral bioavailability of curcumin by 3.4-fold compared with the curcumin/Tween 20 [54]. Du et al. developed oral self-assembled L-arginine (Arg)- or L-lysine (Lys)-functionalized chitosan–casein nanoparticles delivery systems for the loading, stabilizing, and achieving a controlled release profile of hydrophobic curcumin and hydrophilic egg white-derived peptides (EWDP). The developed formula was spherical in shape, has pH-responsive properties, and showed outstanding stability during for 28 days. The *in vitro* controlled release profile of the core–shell NPs for curcumin and EWDP verified that curcumin and EWDP showed better bioavailability in Arg/Lys-CS–CA NPs [55]. Ghatak et al. encapsulated quercetin with casein particles for improvement of its water solubility as well as bioavailability and succeeded to achieve an encapsulation yield of 96% at pH 7.09 [56].

Albumin

Albumin has been widely applied in drug delivery and disease diagnosis due to its accessibility and ability to bind to both hydrophilic and hydrophobic drugs to form nanoparticles. Albumin can be extracted from different sources such as Ovalbumin (OVA), Bovine Serum Albumin (BSA), Human Serum Albumin (HSA), and Rat Serum Albumin (RSA) [57]. Albumin nanoparticles were utilized for co-delivery of curcumin and doxorubicin to MCF-7 resistant breast cancer cells to cause efficient cell killing through increased intracellular internalization of doxorubicin and cell killing compared to the sequential drug co-administration. Nevertheless, the simultaneous drug co-administration led to decreased intracellular uptake of curcumin and lysosomotropism which is the aggregation and entrapment of curcumin in the lysosomes immediately after its release from the albumin nanoparticles [58].

Gelatin and Collagen

Animal proteins including gelatin and collagen have been used for loading of nutraceuticals due to their stability in high temperatures, safety biocompatibility, and biodegradability [13]. Gelatin is made from the controlled hydrolysis of collagen from different animal sources and has various applications in the food, pharmaceutical, and cosmetic industries [59]. It shows an amphoteric behavior in solution depending on the solution's pH, can act as an emulsifier, and enhances the stability of bioactive materials [60]. Gelatin nanoparticles were used for loading of two flavonoids, low-molecular-weight Genistein (GEN) and high-molecular-weight Icariin (ICA) (FLA@GNPs), results showed higher drug encapsulation of FLA@GNPs than that of the single drug-loaded (GEN or ICA) with enhanced stability under acidic conditions at room temperature thus preserving the bioactivity for at least 180 days [61]. Wang et al. utilized electrospinning to formulate a multilayer film with ethyl cellulose nanofibers as the outer layer and curcumin-loaded gelatin nanofibers as the inner layer. The formed nanofibers showed adequate stability, a sustained release profile of curcumin for 96 h, compared to the burst release within 30 min from the gelatin with retaining the antioxidant activities of curcumin [62]. Castro et al. also utilized gelatin nanoparticles loaded with buriti oil to increase its bioavailability and stability. The resultant formula showed adequate stability and enhanced antimicrobial activity against *Pseudomonas aeruginosa*, *Klebsiella pneumonia*, and *Staphylococcus aureus* [60]. Selected Examples for different food-grade nanoparticles for delivery of nutraceuticals are presented in Table 8.1.

8.2.2 Non-food Grade Nano-Delivery Systems for Food Application

Synthetic polymers or non-food grade polymers such as metal or carbon based nanoparticles have also been utilized in food and nutraceutical delivery owing to their versatility, ease of modification and formulation. However, toxicity of some of these materials are yet to be established to achieve their full potential.

8.2.2.1 Polymeric Systems

Nanocarriers based on synthetic polymers have been widely used in nutraceutical delivery owing to their low toxicity, stability, particle size in the nano range, biocompatibility, ease of production and scaling up, and the possibility of sustaining the release of nutraceuticals. They can also be easily modified to achieve specific characteristics of bioactive molecule release or specific targeting site [12]. Polymers such as poly(D,L-lactide), poly-lactic acid (PLA), poly(D,L-glycolide) (PLG), poly(lactide-co-glycolide) (PLGA), and poly(cyanoacrylate) (PCA) were attempted to encapsulate both hydrophilic and hydrophobic nutraceutical agents. However, the selection of the polymer is critical and is based on biocompatibility, biodegradability, mechanical strength, and safety [13]. Haggag et al. developed Poly(lactic-co-glycolic acid) nanocapsules comprising amphiphilic biosurfactant sophorolipids using a dispersion-based procedure. The prepared nanoparticles displayed increased

Table 8.1 Selected examples for different food-grade nanoparticles for delivery of nutraceuticals

Type of food-grade nanoparticle	Nanodelivery system	Nutraceutical compound	Reference
Lipid-based	Liposome	Vitamin C and B carotene	[16]
		Vitamin A and B lactoglobulin	[17]
		Folate	[15]
	Noisome	Vitamin E	[18]
	Nano-emulsion	Curcumin	[11]
		β-carotene	[13, 22]
		Epigallocatechin gallate	[24]
		Zataria multiflora essential oil	[25]
	Solid lipid nanoparticles	Curcumin	[27]
		Epigallocatechin gallate	[17]
	Nanostructured lipid carriers	Oleuropein	[29]
		Curcumin	[30]
Carbohydrate-based	Chitosan	Vitamin-K	[34]
		Origanum vulgare	[35]
		Curcumin	[38]
	Chitosan and sodium alginate	ε-Polylysine	[36]
	Chitosan-tripolyphosphate	Egg white derived peptides	[37]
	Curdlan	Bovine serum albumin	[39]
		A β-carotene, ferulic acid	[40]
	Curdlan-Zein	Curcumin	[41]
	Starch	Catechin	[42]
		Rutin	[43]
	Carboxymethyl cellulose	Tea polyphenols	[44]
	Carrageenan	Ascorbic acid	[46]
		Curcumin	[47]
Protein-based delivery	Barley protein, α-tocopherol layer, and phospholipid layer	Vitamin B12	[50]
	β-lactoglobulin	Resveratrol	[51]
	Casein	Curcumin	[45]
		Quercetin	[56]
	L-arginine (Arg)- or L-lysine (Lys)-functionalized chitosan–casein nanoparticles	Curcumin and egg white-derived peptides	[55]
	Albumin nanoparticles	Curcumin	[58]
	Gelatin nanoparticles	Genistein Icariin	[61]
		Buriti oil	[60]
	Ethyl cellulose-gelatin nanofibres	Curcumin	[62]

uptake by tumors and a longer circulation in blood. Studies in animals showed increased tumor growth inhibition of 57% compared to controls [63]. Mousa et al. encapsulated anticancer natural products, such as 3,3′-diindolylmethane (DIM) and ellagic acid (EA) with confirmed anticancer activity against various types of cancers in poly(lactic-co-glycolic acid) (PLGA) and poly(ethylene glycol) (PEG) nanoparticles (PLGA-PEG NPs) nanoparticles which resulted in fast suppression of pancreatic cancer cell viability/proliferation within 24 h ($P < 0.01$) compared to free drugs [64]. PLGA was also used by Ahmed et al. to load phenolic-rich extracts from *Callistemon citrinus* and berberine which enhanced their bioactivity against the three breast cancer cell lines by approximately twofold as well as linear decreased growth of cells compared to phenolic concentration, with a nearly 100% maximum inhibition at 0.1 mg/mL compared to control [65].

8.2.2.2 Metal-Based Nanomaterials

Metal and metal oxide nanoparticles such as aluminum (Al), iron (Fe), lead (Pb), cadmium (Cd), copper (Cu), silver (Ag), cobalt (Co), and zinc (Zn) are often utilized as nanoparticles as their size can be easily manipulated, they are reactive and able to load several bioactive compounds, with oxides usually more reactive than their metal counterparts [14]. Moreover, the encapsulated nutraceutical in NPs possesses the cumulative biological properties of both the nutraceutical and the NP [66]. A wound dressing material was developed by caging plumbagin on silver nanoparticles which showed enhanced microbial activity against *E. coli* [67]. Gold nanoparticles loaded with gallic acid showed an enhanced ability to inhibit the proliferation of cholangiocarcinoma cells compared to free gallic acid alone [68]. Shah et al. also explored gold nanoparticle-loaded gallic acid after surface-functionalization of magnetite as antimicrobial agents which showed enhanced anti-microbial ability and stability.

8.2.2.3 Carbon-Based Nanoparticles

Carbon nanoparticles such as carbon nanotubes, quantum dots, and fullerenes, graphene, and nanodiamonds are made entirely of carbon with their flexible structures, tunable surface chemistry, and easily engineered size. Modification of surface functional groups of carbon nanomaterials with hydroxyl and carboxyl groups as well as polymers allows them to encapsulate a wide range of materials [13, 14]. However, they are limited by their un-established toxicity in vivo which limits their use [13]. Sadeghi et al. to formulated poly-D-lysine (PDL) and BSA to formulate bionanotubes (BNTs) through layer-by-layer deposition and load it with curcumin as a model drug [69]. Sahoo et al. also utilized curcumin loaded single-walled carbon nanotubes (SWCNT) prepared through layer by layer technique for coating the nanoparticles by 1-palmitoyl-2-oleoyl phosphatidylcholine (POPC) lipid bilayer [70].

8.2.2.4 Mesoporous Silica

Mesoporous silica holds great potential as a nanomaterial owing to its high surface area, porosity, thermal and chemical stability, the capability of surface functionalization as well as high biodegradability and biocompatibility [71]. Elbialy et al. [72] synthesized curcumin loaded smart mesoporous silica nanoparticles which showed greatly enhanced curcumin bioavailability, as well as pH-triggered drug release in an acidic pH which is the tumor microenvironment thereby increasing drug targeting. Curcumin-loaded mesoporous silica nanoparticles were also studied after modification with polyethylenimine by Harini et al. [73] which showed 50% mortality of MCF-7 breast cancer cell line as well as sustained in vitro release in acidic pH. Chen et al. [74] also studied curcumin-loaded mesoporous silica nanoparticles decorated with folic acid for tumor targeting which showed significantly increased the cytotoxicity of curcumin against MCF-7 cell than HEK-293T cells attributed to the enhanced cellular uptake of nanocarriers [74]. Juère et al. formulated resveratrol-loaded silica nanoparticles, which enhanced the solubility of resveratrol as well as significantly increased permeability of resveratrol through the human colon carcinoma cell monolayer (Caco-2) cell line [75].

8.2.2.5 Dendrimers

Dendrimers are three dimensional, hyperbranched and monodispersed, molecules composed of an inner core, branching units, and functional groups which can interact with bioactive molecules and load them [76]. Due to the presence of branches, dendrimers have high solubility, and reactivity and are easily modified. However, studies are still investigating their toxicity in various tissues [77]. Thus, dendrimers have been widely exploited for loading of nutraceuticals, one such example is curcumin-loaded G4 PAMAM dendrimer-Palmitic acid core-shell nanoparticle loaded with curcumin as an antistress therapeutic agent which showed enhanced pharmacokinetic parameters as well as non-toxicity in mice [78]. PAMAM dendrimers were also used for berberine encapsulation which showed sustained release pattern as well as significantly higher anticancer activity against MCF-7 and MDA-MB-468 breast cancer cells and improved bioavailability [79].

Selected Examples for different non-food grade nanoparticles for delivery of nutraceuticals are presented in Table 8.2.

8.3 Conclusion and Future Prospective

Despite significant advances in food nanotechnology, little is known about nanoparticle occurrence, fate, or toxicity. Nanotechnology is still a mystery to the general public. The use and application of nanosystems in food is regulated by several worldwide organizations such as Food and Drug Administration (FDA). The

Table 8.2 Selected Examples for different non-food grade nanoparticles for delivery of nutraceuticals

Type of non-food-grade nanoparticle	Nanodelivery system	Nutraceutical compound	Reference
Polymeric systems	PLGA nanocapsules	Sophorolipids	[54]
	PLGA and PEG nanoparticles	3,3′-diindolylmethane (DIM) and ellagic acid (EA)	[64]
	PLGA	Callistemon citrinus and berberine	[65]
Metal Based nanoparticles	Silver nanoparticle	plumbagin	[67]
	Gold nanoparticle	Gallic acid	[59]
Carbon-based nanoparticles	poly-D-lysine (PDL) and BSA to form bionanotubes (BNTs)	Curcumin	[69]
	Single-walled carbon nanotube	curcumin	[70]
Mesoporous silica nanoparticles		Curcumin	[63, 73, 74]
		Resveratrol	[75]
Dendrimers	G4 PAMAM dendrimer	Curcumin	[78]
	PAMAM dendrimer	Berberine	[79]

European Parliament and Council Legislation are in charge of enacting restrictions on the size of nanoparticles in food additives. Precautionary principle should be taken into consideration so freely engineered nanomaterials in less incorporated in food products. Introducing nanomaterials into food items require proper study and research. EC Food Law Regulation points out that nanomaterials engineered for food application should be free from toxic materials, mycotoxins and heavy metals.

References

1. Singh T, Shukla S, Kumar P, Wahla V, Bajpai VK, Rather IA. Application of nanotechnology in food science: perception and overview. Front Microbiol. 2017;8:1501.
2. Pathakoti K, Manubolu M, Hwang H-M. Nanostructures: current uses and future applications in food science. J Food Drug Anal. 2017;25(2):245–53.
3. Wahab A, Rahim AA, Hassan S, Egbuna C, Manzoor MF, Okere KJ, Walag AMP. Application of nanotechnology in the packaging of edible materials. In: Egbuna C, Mishra AP, Goyal MR, editors. Preparation of phytopharmaceuticals for the management of disorders. Academic Press; 2021. p. 215–25.
4. Ottaway P. Nanotechnology in supplements and foods–EU concerns. Nutraceuticals International; 2009. p. 1.
5. Ennajar M, Bouajila J, Lebrihi A, Mathieu F, Abderraba M, Raies A, et al. Chemical composition and antimicrobial and antioxidant activities of essential oils and various extracts of Juniperus phoenicea L.(Cupressacees). J Food Sci. 2009;74(7):M364–M71.

6. Shimoni E. Nanotechnology for foods: focus on delivering health. In: Mortimer A, Colonna P, Lineback D, Spiess W, Buckle K, Barbosa-Canovas G, editors. Global issues in food science and technology. Amsterdam: Elsevier; 2009. p. 411–24.
7. Chen H, Weiss J, Shahidi F. Nanotechnology in nutraceuticals and functional foods. Food Technol. 2006;60(3):30–6.
8. Azeredo HM, Mattoso LHC, Wood D, Williams TG, Avena-Bustillos RJ, McHugh TH. Nanocomposite edible films from mango puree reinforced with cellulose nanofibers. J Food Sci. 2009;74(5):N31–N5.
9. Kampers FW. Opportunities for bionanotechnology in food and the food industry. Bionanotechnol Glob Prosp. 2008:79–90.
10. Mishra V, Bansal KK, Verma A, Yadav N, Thakur S, Sudhakar K, et al. Solid lipid nanoparticles: emerging colloidal nano drug delivery systems. Pharmaceutics. 2018;10(4):191.
11. He X, Deng H, Hwang H-M. The current application of nanotechnology in food and agriculture. J Food Drug Anal. 2018;27(2019):1–21.
12. Assadpour E, Mahdi JS. A systematic review on nanoencapsulation of food bioactive ingredients and nutraceuticals by various nanocarriers. Crit Rev Food Sci Nutr. 2019;59(19):3129–51.
13. Rashidi L. Different nano-delivery systems for delivery of nutraceuticals. Food Biosci. 2021;43:101258.
14. Singh AR, Desu PK, Nakkala RK, Kondi V, Devi S, Alam MS, et al. Nanotechnology-based approaches applied to nutraceuticals. Drug Deliv Transl Res. 2022;12:485.
15. Kapoor MS, D'Souza A, Aibani N, Nair SS, Sandbhor P, Kumari D, et al. Stable liposome in cosmetic platforms for transdermal folic acid delivery for fortification and treatment of micronutrient deficiencies. Sci Rep. 2018;8(1):16122.
16. Liu X, Wang P, Zou Y-X, Luo Z-G, Tamer TM. Co-encapsulation of vitamin C and β-carotene in liposomes: storage stability, antioxidant activity, and in vitro gastrointestinal digestion. Food Res Int. 2020;136:109587.
17. Rovoli M, Pappas I, Lalas S, Gortzi O, Kontopidis G. In vitro and in vivo assessment of vitamin A encapsulation in a liposome-protein delivery system. J Lipos Res. 2019;29(2):142–52.
18. Talebi V, Ghanbarzadeh B, Hamishehkar H, Pezeshki A, Ostadrahimi A. Effects of different stabilizers on colloidal properties and encapsulation efficiency of vitamin D3 loaded nanoniosomes. J Drug Deliv Sci Technol. 2021;61:101284.
19. Peng S, Zou L, Zhou W, Liu W, Liu C, McClements DJ. Encapsulation of lipophilic polyphenols into nanoliposomes using pH-driven method: advantages and disadvantages. J Agric Food Chem. 2019;67(26):7506–11.
20. Gharehbeglou P, Jafari SM, Hamishekar H, Homayouni A, Mirzaei H. Pectin-whey protein complexes vs. small molecule surfactants for stabilization of double nano-emulsions as novel bioactive delivery systems. J Food Eng. 2019;245:139–48.
21. Zou L, Zheng B, Liu W, Liu C, Xiao H, McClements DJ. Enhancing nutraceutical bioavailability using excipient emulsions: influence of lipid droplet size on solubility and bioaccessibility of powdered curcumin. J Funct Foods. 2015;15:72–83.
22. Salvia-Trujillo L, Qian C, Martín-Belloso O, McClements DJ. Modulating β-carotene bioaccessibility by controlling oil composition and concentration in edible nanoemulsions. Food Chem. 2013;139(1):878–84.
23. Meng Q, Long P, Zhou J, Ho C-T, Zou X, Chen B, et al. Improved absorption of β-carotene by encapsulation in an oil-in-water nanoemulsion containing tea polyphenols in the aqueous phase. Food Res Int. 2019;116:731–6.
24. Peng Y, Meng Q, Zhou J, Chen B, Xi J, Long P, et al. Nanoemulsion delivery system of tea polyphenols enhanced the bioavailability of catechins in rats. Food Chem. 2018;242:527–32.
25. Karimi Yazdi M, Haniloo A, Ghaffari A, Torabi N. Antiparasitic effects of Zataria multiflora essential oil nano-emulsion on larval stages of Echinococcus granulosus. J Parasit Dis. 2020;44(2):429–35.

26. Gordillo-Galeano A, Mora-Huertas CE. Solid lipid nanoparticles and nanostructured lipid carriers: a review emphasizing on particle structure and drug release. Eur J Pharm Biopharm. 2018;133:285–308.
27. Chirio D, Peira E, Dianzani C, Muntoni E, Gigliotti CL, Ferrara B, et al. Development of solid lipid nanoparticles by cold dilution of microemulsions: curcumin loading, preliminary in vitro studies, and biodistribution. Nanomaterials. 2019;9(2):230.
28. Shtay R, Keppler JK, Schrader K, Schwarz K. Encapsulation of (—)-epigallocatechin-3-gallate (EGCG) in solid lipid nanoparticles for food applications. J Food Eng. 2019;244:91–100.
29. Huguet-Casquero A, Moreno-Sastre M, López-Méndez TB, Gainza E, Pedraz JL. Encapsulation of oleuropein in nanostructured lipid carriers: biocompatibility and antioxidant efficacy in lung epithelial cells. Pharmaceutics. 2020;12(5):429.
30. Kamel AE, Fadel M, Louis D. Curcumin-loaded nanostructured lipid carriers prepared using Peceol and olive oil in photodynamic therapy: development and application in breast cancer cell line. Int J Nanomedicine. 2019;14:5073–85.
31. Kim ES, Lee J-S, Lee HG. Improvement of antithrombotic activity of red ginseng extract by nanoencapsulation using chitosan and antithrombotic cross-linkers: polyglutamic acid and fucoidan. J Ginseng Res. 2021;45(2):236–45.
32. Akbari-Alavijeh S, Shaddel R, Jafari SM. Encapsulation of food bioactives and nutraceuticals by various chitosan-based nanocarriers. Food Hydrocoll. 2020;105:105774.
33. Cheba BA. Chitosan: properties, modifications and food nanobiotechnology. Proc Manufact. 2020;46:652–8.
34. Wang Y, Zhou C, Ding Y, Liu M, Tai Z, Jin Q, et al. Red blood cell-hitchhiking chitosan nanoparticles for prolonged blood circulation time of vitamin K1. Int J Pharm. 2021;592:120084.
35. Yilmaz MT, Yilmaz A, Akman PK, Bozkurt F, Dertli E, Basahel A, et al. Electrospraying method for fabrication of essential oil loaded-chitosan nanoparticle delivery systems characterized by molecular, thermal, morphological and antifungal properties. Innovative Food Sci Emerg Technol. 2019;52:166–78.
36. Liu J, Xiao J, Li F, Shi Y, Li D, Huang Q. Chitosan-sodium alginate nanoparticle as a delivery system for ε-polylysine: preparation, characterization and antimicrobial activity. Food Control. 2018;91:302–10.
37. Du Z, Liu J, Zhang T, Yu Y, Zhang Y, Zhai J, et al. A study on the preparation of chitosan-tripolyphosphate nanoparticles and its entrapment mechanism for egg white derived peptides. Food Chem. 2019;286:530–6.
38. Jiao W-H, Xu Q-H, Ge G-B, Shang R-Y, Zhu H-R, Liu H-Y, et al. Flavipesides A–C, PKS-NRPS hybrids as pancreatic lipase inhibitors from a marine sponge symbiotic fungus aspergillus flavipes 164013. Org Lett. 2020;22:1825.
39. Kim BS, Jung ID, Kim JS, Lee J-h, Lee IY, Lee KB. Curdlan gels as protein drug delivery vehicles. Biotechnol Lett. 2000;22(14):1127–30.
40. Yu Y-B, Cai W-D, Wang Z-W, Yan J-K. Emulsifying properties of a ferulic acid-grafted curdlan conjugate and its contribution to the chemical stability of β-carotene. Food Chem. 2021;339:128053.
41. Yu Y-B, Wu M-Y, Wang C, Wang Z-W, Chen T-T, Yan J-K. Constructing biocompatible carboxylic curdlan-coated zein nanoparticles for curcumin encapsulation. Food Hydrocoll. 2020;108:106028.
42. Ahmad M, Mudgil P, Gani A, Hamed F, Masoodi FA, Maqsood S. Nano-encapsulation of catechin in starch nanoparticles: characterization, release behavior and bioactivity retention during simulated in-vitro digestion. Food Chem. 2019;270:95–104.
43. Remanan MK, Zhu F. Encapsulation of rutin using quinoa and maize starch nanoparticles. Food Chem. 2021;353:128534.
44. Liu C, Zhang Z, Kong Q, Zhang R, Yang X. Enhancing the antitumor activity of tea polyphenols encapsulated in biodegradable nanogels by macromolecular self-assembly. RSC Adv. 2019;9(18):10004–16.

45. Dong Y, Wei Z, Xue C. Recent advances in carrageenan-based delivery systems for bioactive ingredients: a review. Trends Food Sci Technol. 2021;112:348–61.
46. Yew H-C, Misran M. Preparation and characterization of pH dependent κ-carrageenan-chitosan nanoparticle as potential slow release delivery carrier. Iran Polym J. 2016;25(12):1037–46.
47. Sun X, Pan C, Ying Z, Yu D, Duan X, Huang F, et al. Stabilization of zein nanoparticles with k-carrageenan and tween 80 for encapsulation of curcumin. Int J Biol Macromol. 2020;146:549–59.
48. Livney YD. Milk proteins as vehicles for bioactives. Curr Opin Colloid Interface Sci. 2010;15(1–2):73–83.
49. Tang C-H. Assembly of food proteins for nano-encapsulation and delivery of nutraceuticals (a mini-review). Food Hydrocoll. 2021;117:106710.
50. Liu G, Huang W, Babii O, Gong X, Tian Z, Yang J, et al. Novel protein-lipid composite nanoparticles with an inner aqueous compartment as delivery systems of hydrophilic nutraceutical compounds. Nanoscale. 2018;10(22):10629–40.
51. Pujara N, Wong KY, Qu Z, Wang R, Moniruzzaman M, Rewatkar P, et al. Oral delivery of β-lactoglobulin-nanosphere-encapsulated resveratrol alleviates inflammation in winnie mice with spontaneous ulcerative colitis. Mol Pharm. 2021;18(2):627–40.
52. Tang C-h. Strategies to utilize naturally occurring protein architectures as nanovehicles for hydrophobic nutraceuticals. Food Hydrocoll. 2021;112:106344.
53. Sadiq U, Gill H, Chandrapala J. Casein micelles as an emerging delivery system for bioactive food components. Foods. 2021;10(8):1965.
54. Xu G, Li L, Bao X, Yao P. Curcumin, casein and soy polysaccharide ternary complex nanoparticles for enhanced dispersibility, stability and oral bioavailability of curcumin. Food Biosci. 2020;35:100569.
55. Du Z, Liu J, Zhang H, Chen Y, Wu X, Zhang Y, et al. L-Arginine/l-lysine functionalized chitosan-casein core-shell and pH-responsive nanoparticles: fabrication, characterization and bioavailability enhancement of hydrophobic and hydrophilic bioactive compounds. Food Funct. 2020;11(5):4638–47.
56. Ghatak D, Iyyaswami R. Selective encapsulation of quercetin from dry onion peel crude extract in reassembled casein particles. Food Bioprod Process. 2019;115:100–9.
57. Karami E, Behdani M, Kazemi-Lomedasht F. Albumin nanoparticles as nanocarriers for drug delivery: focusing on antibody and nanobody delivery and albumin-based drugs. J Drug Deliv Sci Technol. 2020;55:101471.
58. Motevalli SM, Eltahan AS, Liu L, Magrini A, Rosato N, Guo W, et al. Co-encapsulation of curcumin and doxorubicin in albumin nanoparticles blocks the adaptive treatment tolerance of cancer cells. Biophys Rep. 2019;5(1):19–30.
59. Kumar S, Shukla A, Baul PP, Mitra A, Halder D. Biodegradable hybrid nanocomposites of chitosan/gelatin and silver nanoparticles for active food packaging applications. Food Packag Shelf Life. 2018;16:178–84.
60. Castro GMMA, Passos TS, Nascimento SSC, Medeiros I, Araújo NK, Maciel BLL, et al. Gelatin nanoparticles enable water dispersibility and potentialize the antimicrobial activity of Buriti (Mauritia flexuosa) oil. BMC Biotechnol. 2020;20(1):55.
61. Song X, Gan K, Qin S, Chen L, Liu X, Chen T, et al. Preparation and characterization of general-purpose gelatin-based co-loading flavonoids nano-core structure. Sci Rep. 2019;9(1):6365.
62. Wang P, Li Y, Zhang C, Feng F, Zhang H. Sequential electrospinning of multilayer ethylcellulose/gelatin/ethylcellulose nanofibrous film for sustained release of curcumin. Food Chem. 2020;308:125599.
63. Haggag Y, Elshikh M, El-Tanani M, Bannat IM, McCarron P, Tambuwala MM. Nanoencapsulation of sophorolipids in PEGylated poly(lactide-co-glycolide) as a novel approach to target colon carcinoma in the murine model. Drug Deliv Transl Res. 2020;10(5):1353–66.
64. Mousa DS, El-Far AH, Saddiq AA, Sudha T, Mousa SA. Nanoformulated bioactive compounds derived from different natural products combat pancreatic cancer cell proliferation. Int J Nanomedicine. 2020;15:2259–68.

65. Ahmed R, Tariq M, Ahmad IS, Fouly H, Fakhar IA, Hasan A, et al. Poly(lactic-*co*-glycolic acid) nanoparticles loaded with *Callistemon citrinus* phenolics exhibited anticancer properties against three breast cancer cell lines. J Food Qual. 2019;2019:2638481.
66. Ali MA, Mosa KA. Encapsulation of metal and metal oxide nanoparticles by nutraceuticals: implications for biological activities. Curr Nutraceut. 2021;2(2):159–65.
67. Duraipandy N, Lakra R, Vinjimur Srivatsan K, Ramamoorthy U, Korrapati PS, Kiran MS. Plumbagin caged silver nanoparticle stabilized collagen scaffold for wound dressing. J Mater Chem B. 2015;3(7):1415–25.
68. Rattanata N, Daduang S, Wongwattanakul M, Leelayuwat C, Limpaiboon T, Lekphrom R, et al. Gold nanoparticles enhance the anticancer activity of gallic acid against cholangiocarcinoma cell lines. Asian Pac J Cancer Prev. 2015;16(16):7143–7.
69. Sadeghi R, Kalbasi A, Emam-jomeh Z, Razavi SH, Kokini J, Moosavi-Movahedi AA. Biocompatible nanotubes as potential carrier for curcumin as a model bioactive compound. J Nanopart Res. 2013;15(11):1931.
70. Sahoo AK, Kanchi S, Mandal T, Dasgupta C, Maiti PK. Translocation of bioactive molecules through carbon nanotubes embedded in the lipid membrane. ACS Appl Mater Interfaces. 2018;10(7):6168–79.
71. Manzano M, Vallet-Regí M. Mesoporous silica nanoparticles for drug delivery. Adv Funct Mater. 2020;30(2):1902634.
72. Elbialy NS, Aboushoushah SF, Sofi BF, Noorwali A. Multifunctional curcumin-loaded mesoporous silica nanoparticles for cancer chemoprevention and therapy. Microporous Mesoporous Mater. 2020;291:109540.
73. Harini L, Karthikeyan B, Srivastava S, Suresh SB, Ross C, Gnanakumar G, et al. Polyethylenimine-modified curcumin-loaded mesoporus silica nanoparticle (MCM-41) induces cell death in MCF-7 cell line. IET Nanobiotechnol. 2017;11(1):57–61.
74. Chen C, Sun W, Wang X, Wang Y, Wang P. Rational design of curcumin loaded multifunctional mesoporous silica nanoparticles to enhance the cytotoxicity for targeted and controlled drug release. Mater Sci Eng C. 2018;85:88–96.
75. Juère E, Florek J, Bouchoucha M, Jambhrunkar S, Wong KY, Popat A, et al. In vitro dissolution, cellular membrane permeability, and anti-inflammatory response of resveratrol-encapsulated mesoporous silica nanoparticles. Mol Pharm. 2017;14(12):4431–41.
76. Yousefi M, Narmani A, Jafari SM. Dendrimers as efficient nanocarriers for the protection and delivery of bioactive phytochemicals. Adv Colloid Interf Sci. 2020;278:102125.
77. Mehta P, Kadam S, Pawar A, Bothiraja C. Dendrimers for pulmonary delivery: current perspectives and future challenges. New J Chem. 2019;43(22):8396–409.
78. Tripathi PK, Gupta S, Rai S, Shrivatava A, Tripathi S, Singh S, et al. Curcumin loaded poly (amidoamine) dendrimer-plamitic acid core-shell nanoparticles as anti-stress therapeutics. Drug Dev Ind Pharm. 2020;46(3):412–26.
79. Gupta L, Sharma AK, Gothwal A, Khan MS, Khinchi MP, Qayum A, et al. Dendrimer encapsulated and conjugated delivery of berberine: a novel approach mitigating toxicity and improving in vivo pharmacokinetics. Int J Pharm. 2017;528(1):88–99.

Chapter 9
Anticaking Agents in Food Nanotechnology

Muhammad Afzaal, Farhan Saeed, Aftab Ahmed, Fakhar Islam, Muhammad Armghan Khalid, Muzzamal Hussain, Waqas Anjum, and Atka Afzal

9.1 Introduction

Nanotechnology has gained popularity in recent years as a promising technique that has changed the food industry. It is a nanometric technique that works with atoms, molecules as well as macromolecules having a diameter of 1–100 nm to generate and utilize materials containing new characteristics. The developed nanomaterials have 1 or more exterior dimensions, as well as an interior composition, on a level ranging from 1 to 100 nm, allowing for nanoscale observations but also manipulation [1]. The primary benefit of manufacturing nanoparticles is that it results in materials with increased surface area as well as mechanical properties, and stability. These properties are critical for allowing raw and processed nanoparticles to be used in a variety of scientific domains [2]. Food nano-structured components, food nanosensing both are the two fundamental domains of nanotechnology used in the food industry. Food nano-structured components are utilized in a variety of applications, ranging from food manufacturing to food packing. These nanoparticles can be employed in food manufacturing as additives, vehicles for effective nutrient supply, anticaking agents, antibacterial agents, and fillers to enhance the structural strength and stability of packaging materials [3]. Mechanical isolation, lubrication, competing for adsorbed water, including the elimination of electrostatic ions as well as molecular forces are among the 13 processes through which anti-caking substances work. Anti-caking materials work best when employed to the item's surface, where

M. Afzaal (✉) · F. Saeed · F. Islam · M. A. Khalid · M. Hussain · W. Anjum · A. Afzal
Department of Food Sciences, Government College University Faisalabad, Faisalabad, Pakistan
e-mail: muhammadafzaal@gcuf.edu.pk

A. Ahmed
Department of Nutritional Sciences, Government College University Faisalabad, Faisalabad, Pakistan

C. Egbuna et al. (eds.), *Application of Nanotechnology in Food Science, Processing and Packaging*, https://doi.org/10.1007/978-3-030-98820-3_9

the bulk of particle exchanges occur within anticaking agent particles rather than across item surface particles. Caking is generated by a mixture of manufacturing and storage circumstances such as relative humidity, temperature, tension, structure, and particle diameter. Anticaking substances are frequently used in food processes to enhance physical characteristics and durability, and the choices of anti-caking agents rely on a variety of factors [4]. Anticaking compounds have been demonstrated to combat humidity, function as a vapor barrier, reduce surface resistance, and prevent crystal lattice shape development or alteration. Nanoparticles might be utilized to improve the anticaking characteristics of powdered materials. The basic process in anticaking is that fluids are absorbed or powdery particles are coated to prevent clump generation [5]. Although food items' liquid soaking may be reversed by further manufacturing, the cost rise and quality loss of those goods, also after decaking, need the development of innovative solutions to this detrimental consequence in food items. As a result, nano-structured anti-caking agents are vital for extending the storage life of goods because they enhance the surface area accessible for liquid sorption (anti-caking agents). Amorphous silica, silicon dioxide, for example, are mostly employed as anti-caking, anti-foaming agents, and food additives. SiNPs can be utilized in applications that improve sensory qualities, nutrient administration, and packaging barrier features [6]. SiNPs, for example, have allowed a biopolymer (pullulan) layer to be applied to the top of polypropylene sheets. SiNPs with such a large surface area had a stronger bond to a pullulan covering, resulting in a significant improvement in O_2 and CO_2 barrier characteristics. The presence of nanoparticles also had a significant impact on hydrophilicity, since they increased the hydrophobicity of a surface [7]. In this chapter, we mainly focus on the anti-caking agents used in food nanotechnology. These anticaking agents are categorized on the source of their origin such as natural, artificial and some of them are organic as well as inorganic. These anticaking agents are silicon dioxide, calcium phosphates, magnesium silicates sodium aluminum silicate, corn starch, cellulose powder, etc. are under study.

9.2 Anticaking Agents

Anticaking agents prevent powdered and granular components from aggregating, whereas humectants keep foods wet. Food additives might originate from native resources or be made using chemical or synthetic components. Dried powder milk, flours, baking powders, baked goods, and brown sugar are just a few examples of smooth materials which advantage from anticaking additives, which prevent lump formation and keep the goods flowing smoothly. Vending device coffee granules would not work correctly without these, and items like shredded cheese may agglomerate and turn sticky. The type of the substance determines the caking processes. Crystalline solids are frequently cake due to the development of liquid bridges and subsequent microcrystal combinations. It is manufactured by covering particles with a water-repellent layer [8]. Some anticaking chemicals are water-soluble, while others need the use of liquors or various organic solvents to dissolve.

Calcium silicate, a popular anticaking ingredient found in sodium chloride and other products, is capable of absorbing both oil and water. Anticaking substances, although being food additives, also have a variety of additional uses [9]. Various anticaking chemicals soak excessive humidity or encapsulate particles with a water repulsive layer. The anticaking substance calcium silicate is widely utilized. Anticaking additives, both powder, and granule are frequently employed in a variety of foodstuffs. Powder form egg, powder form soups, *S. cerevisiae* in powder form, confectionery items, vending device powders (dairy, cocoa, cream powders), shredded cheese, powder form favors, salt as well as spices, powder form sauces, are among these [10]. Anticaking additives are often used in powder processes to defer or avoid caking, although little is known about their impact on the powdered chemical and structural durability [11]. For analyzing the effectiveness of anticaking chemicals, there is no specialized analytical technique. The major cause of this issue is that numerous independent variables influence the caking ability of powders (moisture, temperature, characteristics of the host powder, and so on).

9.3 Commonly Used Anticaking Agents in Food

9.3.1 Aluminum Silicates

Precipitation of dissolvable aluminum salts with a suitable metal is a typical way to make aluminum silicates. They're frequently characterized as white, unstructured powders. Aluminum silicate minerals are smooth powders employed in the food sector as free-flowing ingredients [12]. It is found in sweet powder including beverage powder. When contrast to certain other anticaking compounds that influence flow behavior, aluminum silicates are very cheap. It is frequently used because of its inexpensive cost and ability to improve powder flow, which is then precipitated [13]. A smooth, white, free-flowing powdery material is characterized as sodium aluminum silicate. The sodium aluminum silicate should have a silicon dioxide concentration of not below 66% and not greater than 88%. It should have an aluminum oxide concentration of not below 5% and not higher than 15%. The amount of sodium oxide should be no lower than 5% and no over than 8.5% [14]. Potassium aluminum silicate is used more frequently than sodium aluminum silicate. As a transporter, potassium aluminum silicate is frequently employed. Furthermore, potassium aluminum silicate is employed to decrease the salt level of products when necessary.

9.3.2 Calcium Carbonate

Calcium carbonate is a legal food ingredient that improves a variety of food characteristics. It can be used in foodstuffs as an acidity stabilizer, food coloring, and anticaking agent, among other things. Carbonic acid calcium salt, calcite, and chalk are some of the other names for it [15]. Calcium carbonate is an odorless, colorless inorganic salt with a molar mass of 100.1 g/mol and having E-170. Microcrystalline (anhydrous crystallized; limestone, calcite crystals, the mineral vaterite, and moistened crystals; crystalline monohydrocalcite but also ikaite) or unstructured powder $CaCO_3$ [16]. Amorphous spherical calcium carbonate nanoparticles have a particle size ranges from 40 to 120 nm, whereas crystallized calcium carbonate molecules have a particle size ranges from 1 to 10 μm. $CaCO_3$ with nanoparticle scale is unsuitable as a food ingredient. The typical particle dimension of food-grade $CaCO_3$ is around 5 μm and higher boundary. Calcium hydrogen carbonate, commonly called calcium bicarbonate and having (E-170), is a kind of $CaCO_3$. $Ca(HCO_3)_2$ has a molecular mass of 162.1 g/mol and a chemical formula of $Ca(HCO_3)_2$. $Ca(HCO_3)_2$ is a granular white powder that dissolves in liquid at a rate of 16.6 g/100 mL (20 °C). An interaction between $CaCO_3$ and carbonic acid can create calcium bicarbonate. As calcium bicarbonate is heated, it disintegrates into $CaCO_3$, CO_2, and water. It is utilized in foodstuff as a color stabilizer and anticaking agent. Calcium bicarbonate has no detailed material [17].

9.3.3 Phosphate of Calcium E341 and Magnesium E340

Phosphoric acid in the form of tri-calcium phosphate (E-341) is a calcium salt of phosphoric acid. Phosphoric acid, which is derived from a phosphate mine, is used to make it commercially. The daily consumption limit was set at 70 mg/kg of body weight. Phosphoric acid as well as phosphates have no nutritional requirements and are usually appropriate for vegetarian populations to consume. Tri-magnesium phosphate (E-340) is a crystalline powder that is white and odorless.

9.4 Organic Anticaking Agents

From the design of the packaging to desirable baking efficiency, consumers demand excellent standard shredded cheese. In the ability of manufacturers to offer elevated quality shredded Mozzarella cheese, manufacturing aids including anticaking ingredients have become necessary for improved flow and separation of cheese fragments. Decreased product movement efficiency can be caused by particles association and the humidity and fat level of the foodstuff [18]. By minimizing the cohesiveness and shear that occurs during the packing and transportation of shredded mozzarella items, anticaking additives can improve product mobility and decrease agglomeration. Anticaking chemicals' impact on the operational characteristics of warmed chopped Mozzarella cheese has received little scientific attention [19]. The cheese packing manufacturer employs the maximum cellulose-based anticaking chemicals, especially powdery cellulose [18]. The capability to stabilize both oil-water state components and dispersion uniformly on cheese chopped surfaces are two advantages of this manufacturing equipment. Anticaking additives made of cellulose tend to generate white slight discoloration over the top of the mozzarella shreds because of the high particle diameter of the food item. The dry, powdery nature of powdered cellulose might cause issues in a production environment. As a consequence, starch-based additives have been created to replace cellulose-based additives. Rice flour was shown to be effective in removing the white staining observed on the cheese surface. The addition of a flour-based anticaking ingredient has been shown to improve the manufacturing aid's effectiveness, particularly in terms of softness and melt [20]. Potato starch, like rice flour, has been noted as having a 'clean' taste when incorporated into food ingredients. Potato starch's increased humidity concentration can be employed to avoid dusting throughout the manufacturing process. Conventional cellulose-based anticaking additives may be replaced with starch-based anticaking agents in current cellulose-based items, providing a functionally better and cost-effective alternative [21]. When potato starch is added to an existing cellulose product, the moisture content increases, enhancing the characteristics of both components and increasing the anticaking agent effectiveness [22]. The effect of changing the anticaking agent components on cheese functioning characteristics is the key theme of this study. Melt ability, stretchability (appearance viscous), as well as free oil generation are the three primary functional characteristics of chopped Mozzarella cheese that will be studied. Two factors influence the total fusibility of Mozzarella: fat concentration as well as protein-protein associations with water retained inside the protein framework.

9.5 Anticaking Agents in Food Nanotechnology

9.5.1 Polydimethylsiloxane

The widespread industrial use of the polydimethylsiloxane polymer has prompted much research into composites enhanced with nanoscale inorganic fillers. The addition of a tiny quantity of transitions and/or rare earth metals nano-oxides to polymers is recognized to increase certain features like thermal durability, conductance, appearance, hydrophobicity, interface behavior, etc. PDMS/silica is indeed an example of a mixed nanocomposite polymer that is already produced and used in massive magnitudes. A significant portion of PDMS/silica composites is used in sectors where heat endurance, as well as oxidation tolerance, are required. For instance, PDMS/silica is largely employed as an anti-caking agent in confectionery and flour items, and also an antifoaming ingredient in consumable oils, in food-related operations [23]. For all of these projects, PDMS, the most commonly used organosilicon polymer, is used approximately 7% of the time. PDMS is approved to be used in a broad range of consumables at the maximum permitted concentrations of 10–100 mg/kg food in the latest edition of the general standard for food additives (GSFA) manuscript, including veggie oils including fats, which are typically used for cooking as well as frying. Understanding that the insertion of highly distributed fillers might change the decomposed chemistry and heat resistance of the polymer [24]. Applying a hybrid of temperature program depletion mass spectrometry and thermo-gravimetric analyses, researchers have investigated the thermally generated mechanisms in PDMS/silica as well as PDMS/silica/ceria nano-composites. These findings revealed that PDMS adsorption on SiO_2 as well as CeO_2/SiO_2 nano oxides undergo three phases of heat degeneration: (1) covalently bonding of PDMS with silica by electrophilic replacement of a surfaces silanol from a polymers terminals trimethylsilyl position; (2) creation of the cyclic oligomer hexamethyl-cyclo-tri-siloxane (HMCTS); (3) high thermal disintegration of the polymers with the generation of methane as well as ethylene. Overall, the studied PDMS depolymerization, as well as radical disintegration mechanisms, were determined to be effective with those previously described. Moreover, there is some debate in the literary works about the use of PDMS at high temperatures. On the one hand, numerous studies, particularly those focusing on polymer uses in industry, indicate the polydimethylsiloxane-based polymer are well appropriate for foaming stabilization in food manufacturing because of their heat stability as well as chemical inertness. We explore dimethylsilanone production by PDMS or its composite materials with nanoscale silica and ceria/silica throughout this research to throw more light on these fascinating topics. Detachable silanones have been unidentified under ambient settings due to their high instability earlier to this innovation in the synthesis of stabilized compounds containing Si=O linkages. They could only be identified and investigated through spectroscopy in the solid inert gas matrix at cold temperatures previously. Valerini et al. [25] In the gaseous state, the presence of silanone molecules, as well as ions, was also detected. Several experiments have shown that silanones occur as precursors in the thermal interactions of low molecular organo-silicon compounds including silenes, hydridosilylketenes, allyloxysilanes, alkoxyvinylsilanes, and polysilylated. Dimethylsilanone has been mentioned to be produced using straight and cyclic PDMS. Lewicki et al. [26] the majority of these assertions about dimethylsilanone synthesis were dependent on kinetics data or chemical capturing studies. Matrix separation Infrared spectroscopy was utilized as one of the direct physiological approaches to detect these chemicals as precursors in the temperature dispersion of several low molecular cyclic as well as straight siloxanes. Interestingly, there have been only single publications on mass spectrometric identification of dimethylsilanone as a result of heat decomposition in polydimethylsiloxane/montmorillonite nanocomposites [27]. Purification methods of biogas are also extremely important commercially, as biogas generation is now one of the possible paths that nations are considering to enable them to accomplish current renewable energy objectives. The most effective method presently utilized is the elimination of unstable polysiloxane contaminants from biogas using high-surface-area adsorbents such as silica gels, zeolites, as well as hybrid alumina/silica processes. Rücker et al. [28] But, cost-effective techniques for renewing adsorbents, which are currently accomplished by the thermal dissociation of adsorbed contaminants, have yet to be discovered.

9.5.2 *Titanium Dioxide*

In 1996, the United States Food and drug administration certified TiO_2 Nanoparticles as a food ingredient and also as a food interaction component [29]. TiO_2 nanoparticles are commonly used in food packing as a bleaching ingredient, excellent photocatalytic agent, antimicrobial agent, and anti-caking agent. TiO_2 nanoparticles are used in food packing as a coloring and covering ingredient to extend the shelf life by avoiding cake generation. The unsaturated polyphospholipid portion of a bacterial cell membrane could be oxidized by TiO_2 Nanoparticles, resulting in a bactericidal impact. The method includes the ultraviolet or natural light-producing oxygen ions like superoxide ionic species, hydrogen oxide, as well as hydroxyl radical, that immediately destroy bacterial cell walls, inhibiting microbe development [30]. Bright white TiO_2 nanoparticles disperse visible light efficiently in pasteurized milk *Staphylococcus aureus* in readily available chicken flesh, demonstrating ZnO's antimicrobial action [31]. ZnO NPs are also less expensive but also less hazardous to animal species and people as compared to AgNPs. Valerini et al. [25] showed that aluminum-treated ZnO NPs functionalized using poly lactate were effective versus *Escherichia coli*. ZnO NPs release Carbon dioxide and water from ethylene when subjected to Ultraviolet light, which serves to extend the durability of food products by avoiding caking. Biofilms encapsulated with titanium dioxide NPs are inexpensive, heat-stable, chemically inactive, non-toxic, as well as photo-stable. By oxidizing the lipid surface of the bacterial cellular membrane but also affecting the D-alanine metabolic activity, TiO_2 has bactericidal instead of bacteriostatic action versus foodborne pathogens [32]. Direct integration of NPs in foodstuff or packing components is required to improve antibacterial properties and boost anti-caking capacity. TiO_2 is utilized as the food pigmenting ingredient with no negative side effects. This is worth noting that researchers studied the effects of Titanium dioxide on human tissues and discovered that the quantity of intracellular radical oxygen molecules rose without causing significant damage to DNA as well as in the endoplasmic reticulum.

9.5.3 *Silicon Dioxide*

Silicon dioxide is a foodstuff ingredient that is mainly employed as an anti-clumping agent to inhibit powdery food components from adhering together or clogging. The diameter of agglomerates influences E-551's ability to serve like an anticaking ingredient in foodstuff [33]. The functionality of the E-551 agglomerates as a separator as well as an anticaking ingredient is substantially hampered if its diameter is less than 100 nm. When taken orally, edible SiO_2 is reported to have quite minimum toxic effects, with a NOAEL (no observable adverse effect limit) of over 2000 mg/kg. According to Clarson et al. [34], the acidic mechanism of digestion could be sped up by the development of an accumulation or agglomeration of silicon-dioxide particles. The ratio within the electrostatic repulsive force of silanol molecules existing on its substratum as well as van-der-Waals association among particles determines the size of a silicon dioxide agglomeration [35]. SiO_2 NPs are employed as an anticaking ingredient to improve the overall quality of foodstuffs and thus are classified as food additives across both European as well as in American countries on under number E-551 [36]. Because SiO_2 NPs are non-toxic for humans when combined with Zn, they have shown to have excellent antimicrobial and antifungal properties against (both gram-positive and gram-negative bacteria) [37]. Titanium dioxide is also utilized as a bleaching ingredient in the manufacture of cakes but also in pastries with E171 foodstuff additive numbers. Nano-encapsulation of foodstuff and endorsing ingredients aids in the preservation of sensory characteristics such as (taste as well as smell) in a variety of foodstuff, consequently increasing appetite. Scent and natural extracts in food have been claimed to be preserved by SiO_2 NPs [38]. Consistency, flow characteristics, as well as aroma and essences, have been shown to improve the plumping characteristics of aqueous food ingredients. Chen et al. [39] transformed the paper into a water lily-like aqua-phobic surface covered by R812S silica NPs and polydimethylsiloxane silicon-oil that was highly water repellent. In numerous powdery foods also utilized SiO_2 as anti-caking ingredients. Silicon dioxide and carbon with a specific diameter of a few hundreds of nanometers are useful food additives as well as packing materials [40].

9.5.4 Synthetic Amorphous Silica

For years, synthesized amorphous silica has been utilized as a basic food ingredient, commonly called synthetic amorphous silicon dioxide. Synthetic amorphous silica's huge manufacturing quantities and widespread usage in a range of applications could result in considerable environmental, industrial, and customer exposure [41]. Anticaking ingredients, stabilizers, thickeners, adsorbent materials, dispersion stabilizers, free flow ingredients, and transporters are all employed in a range of industries as well as consumer goods, encompassing control of pests, medicines, cosmetics, but also food and feed items [42]. Representatives of Synthesized Amorphous Silica Manufacturers have compiled data from consumers and food organizations on related food categories including E-551 consumption amounts. The raw-material providers have no other method of getting this data. As reported by the study, all primary applications but also carry over through foods were addressed. With the help of the Extensive Europeans Dietary Exposed Model [43], community average consumption estimations of SiO_2 from its usage like a food ingredient E-551 ranged from 0.28–4.53 mg E-551/kg bw/day. The maximum consumption estimated for kids in Bulgaria was 12.7 mg/kg bw/day, which is probably an overvalue because it implies a 100% incidence of E-551 throughout all food groups [44]. Relying on expert evaluation of intake frequencies and quantities, a daily dosage of 9.4 mg/kg bw/day was previously calculated for the Dutch community [45]. Silica is added to dietary additives equal to 700-mg silicon/day [46].

Table 9.1 Anticaking agents used in foods with their minimum and maximum dosage level

Anticaking agent	Maximum capacity of use	Group of foods	References
Calcium silicate	5400–30,000 mg/kg	13 different food groups	[46]
Magnesium silicate	5400–30,000 mg/kg	13 different food groups	[46]
Calcium carbonate	6000 mg/kg	Baking powder	[17]
Magnesium stearate	20,000 mg/kg	Chewing gums	[14]
Silicon dioxide	2000–30,000 mg/kg	22 different food groups	[46]

There have been no figures on the amount of E-551 consumed in medicinal preparations. If utilized as a glidant in pills, the most typical medical use, extremely low quantities of lower than 2% are needed. Toothpaste contains a high concentration of fluoride that may be ingested.

Table 9.1 is an overview of anticaking agents used in different foods with their minimum and maximum dosage levels.

9.6 Conclusion and Future Perspective

The usage of nanoscale materials within the food industry has grown a reputation in recent years, thus attention and activity in this chapter have shifted significantly. As nanobiotechnology advances, nanobiotechnology-based devices and materials grow smaller and much more sensitive. Its use in the fields of anticaking additives, food packing, and food safety is well known. Furthermore, encouraging findings have been obtained in the use of nanomaterials in food conservation, where they could preserve the food against moisture, lipids, gases, off-flavors, and smells. They provide effective delivery channels for bioactive substances to targeted tissues. Whereas nanotechnology advancements are opening up new roads every day, there are yet numerous obstacles and possibilities to enhance present technology, as well as questions concerning nanotechnology's repercussions which should be resolved in an attempt to ease customer concerns. When addressing the growth of nanotechnology in food processes, clarity of safety problems and environmental effects ought to be a concern, thus mandatory testing of nano foodstuffs is necessary whenever they are allowed to the markets.

References

1. Wahab A, Rahim AA, Hassan S, Egbuna C, Manzoor MF, Okere KJ, Walag AMP. Application of nanotechnology in the packaging of edible materials. In: Egbuna C, Mishra AP, Goyal MR, editors. Preparation of phytopharmaceuticals for the management of disorders. Academic Press; 2021. p. 215–25.
2. Dufresne A. Nano cellulose: a new ageless bionanomaterial. Mater Today. 2013;16(6):220–7.

3. Ezhilarasi PN, Karthik P, Chhanwal N, Anandharamakrishnan C. Nanoencapsulation techniques for food bioactive components: a review. Food Bioprocess Technol. 2013;6(3):628–47.
4. Barbosa-Cánovas GV, Ortega-Rivas E, Juliano P, Yan H. Food powders. New York, NY: Springer; 2005.
5. Cozzolino CA, Castelli G, Trabattoni S, Farris S. Influence of colloidal silica nanoparticles on pullulan-coated BOPP film. Food Packag Shelf Life. 2016;8:50–5.
6. Fitzpatrick JJ, Descamps N, O'Meara K, Jones C, Walsh D, Spitere M. Comparing the caking behaviours of skim milk powder, amorphous maltodextrin and crystalline common salt. Powder Technol. 2010;204(1):131–7.
7. Chen YL, Chou JY. Selection of anti-caking agents through crystallization. Powder Technol. 1993;77(1):1–6.
8. Chen M, Wu S, Xu S, Yu B, Shilbayeh M, Liu Y, Zhu X, Wang J, Gong J. Caking of crystals: characterization, mechanisms and prevention. Powder Technol. 2018;337:51–67.
9. Belton DJ, Hickman GJ, Perry CC. Traditional materials from new sources–conflicts in analytical methods for calcium carbonate. Food Addit Contam Part A. 2019;36(3):366–73.
10. Lück E, von Rymon Lipinski GW. Foods, 3. Food additives. In: Ullmann's encyclopedia of industrial chemistry. New York, NY: Wiley; 2000.
11. McGinley EJ, Thomas WR. Microcrystalline cellulose–an anticaking agent for grated and shredded cheese. In: Proceedings 17th Annual Marschall Italian and Specialty Cheese Seminar. Madison, Wisconsin; 1980.
12. EFSA. Panel on Food Additives and Nutrient Sources Added to Food. Younes M. 2017, p. 520–3.
13. Pandey RM, Upadhyay SK. Food additive. London: IntechOpen; 2012. https://doi.org/10.5772/34455.
14. JECFA-CTA. Chemical and Technical Assessment (CTA). 2015. http://www.fao.org/food/foodsafety-quality/scientifc-advice/jecfa/technical-assessments/en/. Accessed date: 09.01.2019.
15. EFSA Panel on Food Additives and Nutrient Sources Added to Food (ANS). Scientific opinion on re-evaluation of calcium carbonate (E 170) as a food additive. EFSA J. 2011;9(7):2318.
16. NCBI. PubChem Compound Summary for CID 10176262, Calcium bicarbonate. Bethesda, MD: National Center for Biotechnology Information; 2020. https://pubchem.ncbi.nlm.nih.gov/compound/Calcium-bicarbonate
17. Opinion S. Scientific opinion on re-evaluation of calcium carbonate (E 170) as a food. EFSA J. 2011;9(7):1–73.
18. Akins ML. Effects of starch-based anti-caking agents on the functional properties of shredded mozzarella cheese. Doctoral dissertation, Virginia Tech. 2002.
19. Joint FA. Evaluations of the joint FAO/WHO expert committee on food additives. CADMIUM. 2020. https://apps.who.int/food-additives-contaminants-jecfa-database//chemicalaspx?.
20. Jane JL. Starch functionality in food processing. In: Frazier PJ, Richmond P, Donald AM, editors. Starch. Structure and functionality. London: RSC; 1997.
21. Kim J, Choi JY, Kim J, Jeong S, Lee SH, Oh Y, Moon KD. Effect of anticaking agents on caking and quality characteristics of garlic cream powder sauce. Korean J Food Preserv. 2021;28(2):181–9.C.
22. Reddy MS, inventor; Reddy MS, assignee. Method of treating a divided cheese product for anticaking. United States patent US 5,626,893. 1997.
23. Clarson SJ. Chapter 1: Advanced materials containing the siloxane bond. In: Clarson SJ, Owen MJ, Smith SD, Van Dyke ME, editors. Advances in silicones and silicone-modified materials, vol. 1051. Washington, DC: American Chemical Society; 2010. p. 3–10.
24. Kulyk K, Borysenko M, Kulik T, Mikhalovska L, Alexander JD, Palianytsia B. Chemisorption and thermally induced transformations of polydimethylsiloxane on the surface of nanoscale silica and ceria/silica. Polym Degrad Stab. 2015;120:203–11.
25. Valerini D, Tammaro L, Di Benedetto F, Vigliotta G, Capodieci L, Terzi R, Rizzo A. Aluminum-doped zinc oxide coatings on polylactic acid films for antimicrobial food packaging. Thin Solid Films. 2018;645:187–92.

26. Lewicki JP, Liggat JJ, Patel M. The thermal degradation behaviour of polydimethylsiloxane/montmorillonite nanocomposites. Polym Degrad Stab. 2009;94(9):1548–57.
27. Raabe G, Michl J. Multiple bonding to silicon. Chem Rev. 1985;85(5):419–509.
28. Rücker C, Kümmerer K. Environmental chemistry of organosiloxanes. Chem Rev. 2015;115(1):466–524.
29. Cerrada ML, Serrano C, Sánchez-Chaves M, Fernández-García M, Fernández-Martín F, de Andres A, Rioboo RJ, Kubacka A, Ferrer M, Fernández-García M. Self-sterilized EVOH-TiO2 nanocomposites: interface effects on biocidal properties. Adv Funct Mater. 2008;18(13):1949–60.
30. Kamat PV. TiO2 nanostructures: recent physical chemistry advances. J Phys Chem C. 2012;116:11849.
31. Tankhiwale R, Bajpai SK. Preparation, characterization and antibacterial applications of ZnO-nanoparticles coated polyethylene films for food packaging. Colloids Surf B: Biointerfaces. 2012;90:16–20.
32. Kumar PS, Francis AP, Devasena T. Biosynthesized and chemically synthesized titania nanoparticles: comparative analysis of antibacterial activity. J Environ Nanotechnol. 2014;3(3):73–81.
33. Lee JH, You SM, Luo K, Ko JS, Jo AH, Kim YR. Synthetic ligand-coated starch magnetic microbeads for selective extraction of food additive silicon dioxide from commercial processed food. Nanomaterials. 2021;11(2):532.
34. Clarson SJ, Owen MJ, Smith SD, Van Dyke ME. Advances in silicones and silicone-modified materials. Washington, DC: American Chemical Society; 2010.
35. EFSA Panel on Food Additives and Nutrient Sources added to Food (ANS), Younes M, Aggett P, Aguilar F, Crebelli R, Dusemund B, Filipič M, Frutos MJ, Galtier P, Gott D, Gundert-Remy U. Re-evaluation of silicon dioxide (E 551) as a food additive. EFSA J. 2018;16(1):e05088.
36. He X, Hwang HM. Nanotechnology in food science: functionality, applicability, and safety assessment. J Food Drug Anal. 2016;24(4):671–81.
37. Huang Z, Zheng X, Yan D, Yin G, Liao X, Kang Y, Yao Y, Huang D, Hao B. Toxicological effect of ZnO nanoparticles based on bacteria. Langmuir. 2008;24:4140–4.
38. Athinarayanan J, Periasamy VS, Alsaif MA, Al-Warthan AA, Alshatwi AA. Presence of nano-silica (E551) in commercial food products: TNF-mediated oxidative stress and altered cell cycle progression in human lung fibroblast cells. Cell Biol Toxicol. 2014;30(2):89–100.
39. Chen W, Wang X, Tao Q, Wang J, Zheng Z, Wang X. Lotus-like paper/paperboard packaging prepared with nano-modified overprint varnish. Appl Surf Sci. 2013;266:319–25.
40. Zhao J, Niu Y, Ren B, Chen H, Zhang S, Jin J, Zhang Y. Synthesis of Schiff base functionalized superparamagnetic Fe3O4 composites for effective removal of Pb (II) and Cd (II) from aqueous solution. Chem Eng J. 2018;347:574–84.
41. Peters R, Kramer E, Oomen AG, Herrera Rivera ZE, Oegema G, Tromp PC, Fokkink R, Rietveld A, Marvin HJ, Weigel S, Peijnenburg AA. Presence of nano-sized silica during in vitro digestion of foods containing silica as a food additive. ACS Nano. 2012;6(3):2441–51.
42. Fruijtier-Pölloth C. The toxicological mode of action and the safety of synthetic amorphous silica—a nanostructured material. Toxicology. 2012;294(2–3):61–79.
43. Tennant DR. Comprehensive European dietary exposure model (CEDEM) for food additives. Food Addit Contam Part A. 2016;33(5):772–81.
44. Dekkers S, Krystek P, Peters RJ, Lankveld DP, Bokkers BG, van Hoeven-Arentzen PH, Bouwmeester H, Oomen AG. Presence and risks of nanosilica in food products. Nanotoxicology. 2011;5(3):393–405.
45. EFSA Panel on Food Additives and Nutrient Sources added to Food (ANS), Younes M, Aggett P, Aguilar F, Crebelli R, Dusemund B, Filipič M, Frutos MJ, Galtier P, Gott D, Gundert-Remy U. Re-evaluation of calcium silicate (E 552), magnesium silicate (E 553a (i)), magnesium trisilicate (E 553a (ii)) and talc (E 553b) as food additives. EFSA J. 2018;16(8):e05375.
46. Yapıcı E, Karakuzu-İkizler B, Yücel S. Anticaking additives for food powders. In: Food powders properties and characterization. Cham: Springer; 2021. p. 109–23.

Chapter 10
Gelling Agents, Micro and Nanogels in Food System Applications

Neelma Munir, Maria Hasnain, Huma Waqif, Babatunde Oluwafemi Adetuyi, Chukwuebuka Egbuna, Michael C. Olisah, Chukwudi Jude Chikwendu, Chukwuemelie Zedech Uche, Kingsley C. Patrick-Iwuanyanwu, and Abeer Mohamed Ali El Sayed

Abbreviations

DSC	Differential scanning calorimeter
FTIR	Fourier transform infrared spectroscopy
GA	Gelling agent
NMR	Nuclear magnetic resonance
SEM	Scanning electron microscopy
TEM	Transmission electron microscopy
XRD	X-ray diffraction

N. Munir (✉) · M. Hasnain · H. Waqif
Department of Biotechnology, Lahore College for Women University, Lahore, Pakistan

B. O. Adetuyi
Biochemistry Unit, Department of Natural Sciences, Precious Cornerstone University, Ibadan, Nigeria

C. Egbuna
Nutritional Biochemistry and Toxicology Unit, Africa Centre of Excellence in Public Health and Toxicological Research (ACE-PUTOR), University of Port-Harcourt, Choba, Rivers State, Nigeria

Department of Biochemistry, Faculty of Natural Sciences, Chukwuemeka Odumegwu Ojukwu University, Igbariam, Anambra State, Nigeria

M. C. Olisah
Department of Medical Biochemistry, Faculty of Basic Medical Sciences, Chukwuemeka Odumegwu Ojukwu University, Igbariam, Anambra State, Nigeria

C. Egbuna et al. (eds.), *Application of Nanotechnology in Food Science, Processing and Packaging*, https://doi.org/10.1007/978-3-030-98820-3_10

10.1 Introduction

A gel is a material that has both solid and liquid characteristics, as well as elastic and liquid properties. They are made up of polymer molecules that have been covalently bonded to produce a tangled and interlinked molecular network that is submerged in a fluid state, which in the food chain is liquid [1]. When gels develop, a sol to gel transition occurs. Several gel forming components are used carefully to produce food gelling agents of adequate standard, notably the number of textural features that indicate the gelling process. The word "gel" is used by food technologists to characterize meals with a high moisture content that retain their shape after being removed from their vessel [2].

A variety of meals are offered in the form of gels in market, which provides consumers with convenience like jellies, jams, sweets, fruit and vegetable-like products are some examples to obtain the required or targeted parameter, one or more gelling agents are invariably used [1]. Generally, food hydrocolloids are used for this purpose. A wide range of proteins and polysaccharides are generated from natural sources of hydrocolloids. Gelling agents are now used in a wide range of industrial applications to accomplish different tasks such as thickening and stabilizing forms, gelling aqueous dispersions, emulsified suspended particles, limiting or minimizing gel contraction, and rising aqueous dispersions [3].

Gel formation entails the interaction of randomly distributed polymeric chains in suspension to create a 3-dimensional network with solvent in the spaces. Two or more polymer chains can create the related areas known as 'junction zones.' The development of these connection zones is essentially what the gelation process is all about [4]. The clusters of basic inter-chain connections into "junction zones," which forms the basis of a gel's 3-dimensional network which is the most frequent structure involved in hydrocolloid gelation. All these factors like the presence of ions, temperature and the structure of the hydrocolloids all have an effect on the physical configuration of the linking zones in the network. Ionotropic gelation, cold-set gelation and heat-set gelation are three important processes involved in gelation [5]. Alginate, carrageenan, and pectin are examples of such systems [6, 7]. Ionotropic

C. J. Chikwendu
Department of Biochemistry, Faculty of Natural Sciences, Chukwuemeka Odumegwu Ojukwu University, Igbariam, Anambra State, Nigeria

C. Z. Uche
Department of Medical Biochemistry and Molecular Biology, Faculty of Basic Medical Sciences, University of Nigeria, Nsukka, Enugu, Nigeria

K. C. Patrick-Iwuanyanwu
Nutritional Biochemistry and Toxicology Unit, Africa Centre of Excellence in Public Health and Toxicological Research (ACE-PUTOR), University of Port-Harcourt, Choba, Rivers State, Nigeria

A. M. A. El Sayed
Department of Pharmacognosy, Faculty of Pharmacy, Cairo University, Cairo, Egypt

gelation can be accomplished by diffusion setting or internal gelling agents. A scattering medium is formed when hydrocolloid particles are dispersed in warm water, which when cooled provides an enthalpically-stabilized chain helix resulting in traces of multiple chains and a 3-dimensional network. This process results in the formation of gel from agar and gelatin. Heat set gels necessitate the heating process to the gel. It is generally only used in meals when a temperature setting is necessary (e.g. the use of starch in sauces).

The heat setting approach includes native starch/protein unfolding/expansion followed by network rearrangement [8]. A food gel that retains its characteristic structural form can be seen as a high moisture 3-dimensional polymeric network that resists flow under pressure. A rigid or hard structure is formed when 3-dimensional structure of polymer chain are interlinked with water molecule inside it that hardens the structure [9].

10.2 Gelling Agents

Gelling agents (GA) are using in thickening and stabilizing the food additives such as jellies and sweets. Heteropolysaccharides and hydrocolloids make up a huge bulk of gelling polysaccharides. These gelling agents are used in sweets, salad dressings, jellies jams, marmalade, jujubes, yogurts, and different items are among the application. Numerous proteins, are utilized in the production of gels. These include numerous animal proteins and zein from maize such as gelling agent and whey proteins [9].

In general, certain colloidal proteins and polysaccharides of microorganism and Plant-derived medium solidifiers or stabilizers create a continuous 3-dimensional structure and serve as medium solidifiers or stabilizers. GA alters its diffusion properties and stiffen the gel medium. The diffusion rate is determined by the medium's viscosity, which is controlled by the concentration of the gelling agent and its physicochemical properties. Some gelling substances may change among liquid and gel forms based on temperature, which contributes to their attraction [10].

GA has a controlled temperature and pH range, and various gelling compounds may be removed by a different group of micro-organisms, requiring the use of various gelling agents. In recent years, traditional gelling agent supplies have been reduced prompting the development of novel gelling agents [11]. New research on gelation that can tolerate a wide range of temperatures and pressures has aided in the cultivation of newer microorganisms, including some extremophiles that could not otherwise be cultivated [11, 12]. Despite their importance in microbiology, essential gelling agents are not synthesized in a single location. Pectin, agar-agar, natural gums, starches and proteins are all commonly used gelling agents that may be generally divided into proteins and poly-saccharides [13].

10.3 Gel Formation Conditions

A basic polymer dispersion or particle suspension coupled with an externally controlled temperature or solution composition causes gel formation. The process of sol gel conversion generally includes particle or macromolecule aggregation, culminating in the creation of a network that spans the whole container volume [14]. Gelation processes can be generally categorized as physically or chemically induced. In the case of protein gels, to unwind the natural structure of proteins, a driving force is required which will be followed by an agglomeration process that indicates a 3-dimensional structure of clustered molecules connected by covalent or non-covalent bonds. The conditions for gel formation are primarily determined by the many physicochemical variables discussed in the following sections [15].

10.3.1 Temperature

Heat-induced gelling agents is most likely the significant and widely used procedure for producing gels. Gelation is a two-stage process in which molecules unfold or dissociate owing to energy input in the first step, exposing reactive sites. The second phase involves the interaction and aggregates of extended substances to produce greater molecular mass complexes. The initial stage of the Gel might be reversible, and the other stage is particularly irreversible [13].

10.3.2 Presence of Enzyme

The insertion of artificial covalent cross-links into dietary proteins is the basis for enzyme-induced gelation. Protein cross-linking reactions include those mediated by trans-glutaminase, polyphenol oxidase and peroxidase amongst many others [16].

10.3.3 Pressure

Because higher pressure may be used as a different process or in combination with another, significantly greater temperature; it allows for greater flexibility in changing the functional properties of molecules. High pressure encourages reactions that lower the total capacity of the medium and the pH of the medium becomes acidic under high pressure and water disassociate into the medium [17].

10.3.4 *pH*

Changes in pH caused by acid addition or microbial fermentation affect the net charge of the molecule, altering the attractive and repulsive forces between molecules as well as the interactions among molecules and solvent, i.e., hydration characteristics. Furthermore, salt solubility varies with the pH, which lead to formation of the gel. The fractal aggregation hypothesis might explain the mechanism of acid gel production [18].

10.4 Gel Formation Methods

There are two types of gel formation mechanisms: chemical cross-linking and physical cross-linking because of the creation of covalent bonds, the chemical cross-linking process enables persistent connection between chains. Because these inter-linked gelling agents are not treated after synthesis, they are referred to as irreversible gels [19].

Irreversible gels can be created by using two techniques either cross-linking throughout the polymerization stage or cross-linking the polymer chain helix. Cross-linking may be produced throughout polymerization by several polymerization methods such as condensation polymerization, free radical polymerization, photopolymerization, and plasma polymerization [20, 21]. In the process of cross-linking polymer chains, however, gel structures inter-link by the reactivity of side chain connected to molecular chains, which can also be done through radiation, cross-linking or photo and plasma cross-linking [22].

Physical cross-linking results in reversible gel with transient bonding within chains when temperature, pH, and solvent level alters. Transient connections such as hydrogen, ionic, hydrophobic association, coordination bonding hydrophobic association, helix formation and are widely known for creating reversible gels [23].

10.4.1 *Gelatin and Carrageenan*

While heating gelatin, it melts and solidified again after cooling down while with liquid/water addition it makes semi-solid colloidal gel by fractional reorganization of triple helices present in collagen while cooling by two steps orientation and condensation [24]. In order to prompt a reactive site polypeptide chain grosses an orientation that condense more chains to create a triple chain helix near the reaction site [25]. In the presence of potassium ions on cooling, carrageenan forms gels which promote both gelation and helix formation [26].

10.4.2 Whey and Soy Proteins

Whey proteins is basically globular proteins by following events: denaturation of proteins, aggregation of denatured proteins, strand formation by aggregates and finally network formation from strands [27]. Gel is formed via heating soybean flour in the presence of calcium and magnesium which introduces clusters of denatured molecules of protein [28].

10.4.3 Milk and Egg Proteins

Casein protein in milk is bonded together via hydrophobic bonds and salt conduits. Casein is hydrolyzed into gelation in the presence of caseino macro-peptide [29]. While heating egg protein (albumen and yolk), gelation occur by denaturation of egg proteins and after that aggregation of denatured proteins [30]. Table 10.1 shows different gelling agents and their applications.

10.4.4 Alginates and Pectins

Alginate gel formed at low pH (less than 4) by the adding polyvalent cations. Guluronic acid gives active binding site for the attachment of cations. The type of polyvalent cations defines the strength of gel such as Barium ion > Strontium ion > Calcium ion > magnesium ion [36]. The properties of gel from pectin defines by the degree of esterification. High methoxyl pectin gels are formed only in the presence of sugars and polyols at pH from 3.0 to 4.5 while High methoxyl pectin gels are formed only in the presence of Calcium ions [37] (Fig. 10.1 represents the Food applications of proteins).

Table 10.1 Proteins that are utilized as gelling agents

Gelling agents	Gelling agent source	Binding blocks of Gelling agents	Applications of gelling agents	Ref.
Gelatin	Animal skin and bones	Proline and glycine	Gelling agent in jelly and jam	[31]
Whey protein	Casein curd	Lactoglobulin	Gelling agent and thickener in food industry	[32]
Soya proteins	Soybeans	Conglycinin	Heat set gel	[33]
Egg proteins	Egg	Albumen	Gelling and thickening agent	[34]
Zein	Corn	Prolamin	Gel coated candy and nuts	[35]

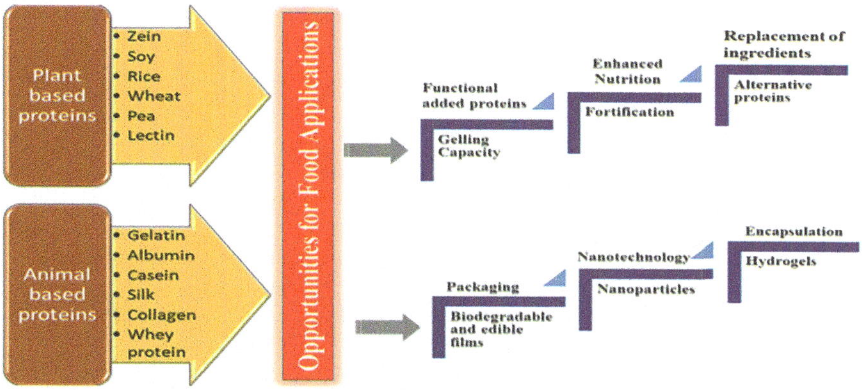

Fig. 10.1 Food applications of proteins

10.4.5 Agar and Starch

Agar gelation is a reversible process because of hydrogen bond. Hot agar on cooling turn into gel by aggregation of helices while starch heated above the limit in the presence of large water cause solubilization of amylose and after cooling this gel is formed [38].

10.4.6 Guar and Gellan Gum

Guar gum has mannose and galactose and form gels by cooling in the presence of salts via aggregation of helices [39]. At high temperature gellan gum turned into disordered coil while after cooling turns into double helix [40].

10.5 Characterization of Gel

Formation of gel is primarily conversion from a solid to a gel form in which elasticity altered rapidly while solid characteristics develop simultaneously. In many other words, during gel formation, continuous and discontinuous stages interact. As a consequence, assessing the viscoelastic (rheological) properties of materials is required, which is frequently achieved through a variety of experiments. Table 10.2 shows the non-rheological ways to evaluating characteristics of gel [9].

The geometry of the sample or the instrument used have no effect on the essential tests. Parameters such as stress/strain are commonly assessed at large deformations and are generally examined by uni-axial density and tension. To identify the basic factors and mechanisms involved in gelling and to define gel texture, features and

Table 10.2 Methods for the measuring the characteristics of gel

Measurement Type	Used Instrument	Measurement parameters	Applications	Ref.
Structural characterization	DSC	Rate of heat flow	Polyvinyl alcohol and gellan blend film	[41]
	XRD	Analysis of particle size	Food nano delivery system	[42]
Microscopic characterization	SEM	Components structural arrangement.	The size, number and distribution of particles in gel	[43]
	TEM	Structural distribution of particles.	Studies of mixed gel characteristic	[44]
Molecular characterization	NMR	Conformation changes.	Determination of particles structural features.	[42]
	FTIR	Molecular structure/ functional group identification	Determine components	[43]
			Infrared spectra	

rheological aspects of starch gels have been researched generally on functional studies [45]. Rayleigh scattering is a technique for determining the size of molecules in suspension. It can enlighten about the various agglomeration mechanisms that happen in gel formation. Dynamic light diffraction may be used to analyze the rheology of gelatin dispersion by measuring the diffusion co-efficient of dispersed molecules and the efficient viscosity of the gel [46].

The presence of a molecular network affects rheological properties. Measurements may be made on a gel under compression to demonstrate the correlation among strain and stress. Gel rheological measurements are commonly divided into small and big deformation testing small deformation tests commonly indicate structure, whereas huge deformation trials examine the condition and properties of a gel [47].

10.6 Nanotechnological Application of Gelling Agents

The most important thing for the food industry is to deliver quality, soft food and healthy foods. Foodstuffs with texture modification tends to be more demanding. In the production of soft foods, nanotechnology offer an appreciable input [48]. In order to soften the food, some conventional techniques can be used such as freeze-thawing, enzyme impregnation [49], high-pressure processing, pulsed electric fields and sonication [50] while some unique techniques are used to preserving the color and flavor such as microfluidics, 3D printing electrospinning and electro-spraying [51]. Moreover, lipids, proteins and carbohydrates are simple constituents in texture modified food. While heating, globular proteins unfold and denature and increase the viscosity of fluids while further heating, globular proteins assemble and form aggregates, fibrils and network chains of gels [52]. Polysaccharides are used to condense and stabilize the fluids as gelling agents [40] while gums and starches are used as thickening agents in enzymes and colorants [53].

10.7 Microgels and Nanogels

Microgels and nanogels are among the three types of hydrogels, the third being macrogels according to their size.

Microgels are polymer chains intra-molecularly crosslinked in small dimensions (from hundreds of nanometers to some micrometers) dispersed in colloidal solutions. Their structure is very close to solid particles, once their surface is well established [53, 54]. These kinds of gels have a high capacity for water content, large surface area, and an interior network useful for drug delivery systems. Biopolymer-based microgels are of great interest for food, drug delivery and tissue engineering systems because of its roles such as biodegradability, nontoxicity, and relatively low cost, beyond being abundant in nature.

Nanogels are innovative systems on the nanometer-scale of great potential in nanomedicine, nutraceutics, pharmaceutics, and bionanotechnology. Their internal structure is similar to that of microgels; however, there is variation in size (up to 100 nm) and responsiveness leading to several advantages. The nanoscale size improves the solubility of hydrophobic drugs, increases drug accumulation in tumors, makes the therapeutic agents very stable against enzymatic and chemical degradation, and decreases cytotoxic side effects. The nanogels helps in drug encapsulation, large surface area, and stable interior network structure [55].

10.7.1 Methods of Synthesis of Microgels and Nanogels

10.7.1.1 Methods of Synthesis of Microgels

Essentially, microgels may be prepared by physical or chemical crosslinking of hydrophilic polymers. Physical crosslinking is reversible upon external stimuli once it involves non-covalent attractive forces such as ionic and hydrophobic interactions. Biodegradable physically crosslinked microgels can encapsulate drugs, cells, and proteins and release them by their degradation process.

The production of microgels can be divided into homogeneous nucleation and polymerization, emulsification, or complexation methods. Basically, microgel is obtained from homogeneous solutions; in the second, aqueous droplets are dispersed in an oil phase followed by crosslinking; and, in the last, two water-soluble polymers are put together to form complexes with each other [56]. Homogeneous nucleation and polymerization methods are important to describe because of their relevance. These methods consist of mixing water-soluble monomers with crosslinking agents and an initiator. This type of process can be performed by emulsion polymerization using a water-soluble monomer, a radical initiator, and surfactants in an aqueous medium. To produce core-shell microgels, this emulsion step is followed by a second polymerization to form the shell. Microgels can also be made by water-in-oil heterogeneous emulsifications from the combination of a continuous oil phase, with oil-soluble surfactants and droplet emulsions of water-soluble polymers.

10.7.1.2 Methods of Synthesis of Nanogels

Different synthetic routes have been used for the development of nanogels that can be classified into two main categories: Chemically crosslinked nanogels and physically crosslinked nanogels, according to their crosslinked structure. The chemically-cross-linked nanogels present covalent bonds linking the polymer network and making them stable, rigid, and permanent, while the physically crosslinked nanogels present non-covalent bonds, which are weaker linkages, thus allowing sol-gel phase transitions as a result of the environmental stimuli [57]. The main methods of synthesis of nanogels are divided into two groups, one of which is known as crosslinking polymerization and involves techniques based on simultaneous polymerization and crosslinking, using monomers or their mixtures as substrates. The other group covers methods based on the crosslinking of macromolecules from polymer precursors which are polymers such as amphiphilic copolymers capable of forming nanogels by self-assembly or polymers with many reactive sites which can be directly used for chemical crosslinking.

Apart from these two groups, nanogels can also be prepared by controlled aggregation by physical self-assembly of hydrophilic polymers and template-assisted fabrication of nanogel particles.

The first method is a simple, and low-cost process conducted in dilute aqueous media that implies controlled association of hydrophilic or amphiphilic polymers linked by hydrogen bonds, van der Waals forces, hydrophobic forces, and/or electrostatic interactions.

The second method involving photolithography or micro-molding techniques, photolithography, uses exposure to ultraviolet (UV) radiation of UV cross-linkable polymers with direct collect of fabricated particles by the dissolution of the substrate in water.

The crosslinking of polymer precursors provides excellent properties relevant to many applications, especially when using ionizing radiation for crosslinking. It is known that the reaction of intra-molecular crosslinking can be obtained by using water-soluble polymers in dilute solutions and a cross-linker capable of reacting with the chain's functional groups. The ionizing radiation is an alternative method of intra-molecular crosslinking initiation which avoids the addition of any additives, allowing the reaction to be carried out in a pure polymer-solvent system, and in this way, one can produce nanogels for biomedical applications free from monomers, crosslinking agents, or surfactants, eliminating the purification step [58].

10.7.2 Applications of Microgels and Nanogels in the Food Industry

10.7.2.1 Application of Microgels

There are many important application of microgels in food system depending on intended goal. Some of their usefulness include for texture control, encapsulation, as delivery system and protection agent [59, 60]. Microgels have the ability to increase their viscosity properties by swelling in an appropriate solvent [61]. With the need to replace excess fats from food, microgels have seen application in the food industry as it helps improve texture and mouth feel of foods [61]. As an emerging encapsulation agent, microgels have in recent years been experimented for their ability to protect and release bioactive compounds such as macronutrients, phytochemicals, nutraceuticals, vitamins, minerals, antimicrobials, antioxidants, enzymes, probiotics and flavors, in a controlled manner [61]. Microgels importance as an encapsulating agent is arguably because of some negative sensory properties linked to many bioactives such as unstable chemical nature with potentials to undergo physical or chemical changes in the gastrointestinal tract that may negatively impact bioavailability and bioactivity of the bioactive compound. In other words, microgels can be designed to serve as good delivery system or also designed to prevent food spoilage.

In biomedical or pharmaceutical applications, microgels can be explored when stability is of concern or degradation, or even their ability to dissolve according to the purpose of the application that may be for wound dressings, tissue engineering, contact lenses, drug delivery systems, and others [62]. Among the several applications of microgels, delivery of chemotherapy for cancer treatment is one of great research interest. Although chemo treatments are effective in treating tumors, they have many limitations such as low specificity which is the main reason for their toxicity [63].

10.7.2.2 Application of Nanogels

Just like microgels, nanogels have found application in the food industry for food/nutrient encapsulation, delivery and protection of bioactive compounds. One of the widely used nanogel is protein nanogel obtained from milk proteins such as casein and whey proteins. The use of nanogels is on the rise because of the need to develop products that will resist degradation/spoilage or products with enhanced property of controlled release of bioactives for target delivery. Nanogels are used for the delivery of poorly water-soluble substance due to their single construction formed by a hydrophobic core, micelles, and hydrophilic exterior. A study by Hu et al. [64] explored the encapsulation of curcumin using acylated ovalbumin nanogels (AOVA) produced through acylation modification and heat-induced self-assembly as novel delivery system. Their study found that at gastrointestinal conditions, curcumin encapsulated in AOVA nanogels displayed 93.64% higher encapsulation efficiency with slower sustained release compared to native ovalbumin (NOVA) nanogels [64]. This exploration is necessary because curcumin is hydrophobic in nature and poorly absorbed [65, 66].

In biomedical and pharmaceutical applications, nanogels present a great potential of use in chemotherapy, diagnosis of diseases, the release of bioactive substances and vaccines, antimicrobials [64, 65], cell culture systems, contrast agents, biocatalysis, in the generation of bioactive scaffolds in regenerative medicine [65], besides being able to act as sensors, nanoreactors, nanodevices, superabsorbents, and biomimetic mechanical devices, such as artificial muscles. The interpenetrating network structure of the nanogels allows better encapsulation of drugs that can be delivered by various routes of administration such as oral, nasal, intraocular, and pulmonary pathways. It is also possible that nanogels entrap two drugs simultaneously, an important feature for the co-administration of two or more anticancer drugs.

10.8 Conclusion

Gels are elastic material, and the stability of food gelling agents is important for commercial application. As a result, recognizing the solid to the gel conversion is important for creating gelled materials. A number of elements, including as material type, concentration, time, pH, temperature, and so on, can greatly impact the gel formation process and, as a consequence, its consistency, which is the most essential feature for customer approval. The presence of cations, as well as the presence of hydrocolloids and proteins, influences the structure of gel. Gelling agents are used in the manufacture of restructured foods and innovative forms of foods products that have acceptable mechanical integrity, a prolonged storage value, good nutrient content, and customer acceptance.

References

1. Aguilera JM, Baffico P. Structure–mechanical properties of heat induced whey protein/cassava starch gels. J Food Sci. 1997;62:1048–66.
2. de Vries J. Hydrocolloid gelling agents and their applications. In: Gums and stabilisers for the food industry, vol. 12. London: RSC; 2004. p. 23–31.
3. Sutherland IW. Biotechnology of microbial polysaccharides in food. In: Food biotechnology. 2nd ed. Boca Raton, FL: CRC Press; 2007. p. 193–220.
4. Oakenfull D. Gelling agents. Crit Rev Food Sci Nutr. 1987;26:1–25.
5. Burey P, Bhandari BR, Howes T, Gidley MJ. Hydrocolloid gel particles: formation, characterization, and application. Crit Rev Food Sci Nutr. 2008;48(5):361–77.
6. Phillips G, Williams P, editors. Handbook of hydrocolloids. Amsterdam: Elsevier; 2009. p. 168–9.

7. Draget KI, Smidsrød O, Skjak-Braek G. Alginates from algae. In: Polysaccharides and polyamides in the food industry: properties, production, and patents. Weinheim: Wiley-VCH Verlag GMBH & Co. KGaA; 2005. p. 1–30.
8. Nishinari K, Zhang H. Recent advances in the understanding of heat set gelling polysaccharides. Trends Food Sci Technol. 2004;15:305–12.
9. Banerjee S, Bhattacharya S. Food gels: gelling process and new applications. Crit Rev Food Sci Nutr. 2012;52(4):334–46.
10. Dutta J, Tripathi S, Dutta PK. Progress in antimicrobial activities of chitin, chitosan and its oligosaccharides: a systematic study needs for food applications. Food Sci Technol Int. 2012;18(1):3–34.
11. Das N, Triparthi N, Basu S, Bose C, Maitra S, Khurana S. Progress in the development of gelling agents for improved culturability of microorganisms. Front Microbiol. 2015;6:698.
12. Becker A, Katzen F, Pühler A, Ielpi L. Xanthan gum biosynthesis and application: a biochemical/genetic perspective. Appl Microbiol Biotechnol. 1998;50(2):145–52.
13. Nazir A, Asghar A, Maan A. Food gels: gelling process and new applications. Amsterdam: Elsevier; 2017. p. 335–53.
14. Clark AH, Schwartzberg HG, Hartel RW. Gels and gelling. In: Physical chemistry of food. New York, NY: Marcel Dekker; 1992. p. 263–83.
15. Totosaus A, Montejano JG, Salazar JA, Guerrero I. A review of physical and chemical protein-gel induction. Int J Food Sci Technol. 2002;37(6):589–601.
16. Lauber S, Krause I, Klostermeyer H, Henle T. Microbial transglutaminase crosslinks β-casein and β-lactoglobulin to heterologous oligomers under high pressure. Eur Food Res Technol. 2003;216(1):15–7.
17. Ames JM. Applications of the Maillard reaction in the food industry. Food Chem. 1998;62(4):431–9.
18. Lucey JA, Singh H. Formation and physical properties of acid milk gels: a review. Food Res Int. 1997;30:529–42.
19. Park S, Okada T, Takeuchi D, Osakada K. Cyclopolymerization and copolymerization of functionalized 1,6-heptadienes catalyzed by Pd complexes: mechanism and application to physical-gel formation. Chem Eur J. 2010;16(29):8662–78.
20. Thakur VK, Thakur MK, Gupta RK. Graft copolymers of natural fibers for green composites. Carbohydr Polym. 2014;104:87–93.
21. Thakur VK, Thakur MK. Recent advances in graft copolymerization and applications of chitosan: a review. ACS Sustain Chem Eng. 2014;2(12):2637–52.
22. Ito K. Novel cross-linking concept of polymer network: synthesis, structure, and properties of slide-ring gels with freely movable junctions. Polym J. 2007;39(6):489–99.
23. Hurtado PI, Berthier L, Kob W. Heterogeneous diffusion in a reversible gel. Phys Rev Lett. 2007;98(13):135503.
24. Yadav S, Mehrotra GK, Bhartiya P, Singh A, Dutta PK. Preparation, physicochemical and biological evaluation of quercetin based chitosan-gelatin film for food packaging. Carbohydr Polym. 2020;227:115348.
25. Roy S, Rhim JW. Preparation of antimicrobial and antioxidant gelatin/curcumin composite films for active food packaging application. Colloids Surf B: Biointerfaces. 2020;188:110761.
26. Roy S, Rhim JW. Preparation of gelatin/carrageenan-based color-indicator film integrated with shikonin and propolis for smart food packaging applications. Am Constit Soc Appl Bio Mater. 2020;4(1):770–9.
27. Tang CH. Nanostructured soy proteins: fabrication and applications as delivery systems for bioactives (a review). Food Hydrocoll. 2019;91:92–116.
28. Wróblewska B, Juśkiewicz J, Kroplewski B, Jurgoński A, Wasilewska E, Złotkowska D, et al. The effects of whey and soy proteins on growth performance, gastrointestinal digestion, and selected physiological responses in rats. Food Funct. 2018;9(3):1500–9.
29. Montowska M, Fornal E. Detection of peptide markers of soy, milk and egg white allergenic proteins in poultry products by Qualitative tandem liquid chromatography quadrupole time of

flight mass spectrometry Qualitative tandem liquid chromatography quadrupole time of flight mass spectrometry (LC-Q-TOF-MS/MS). Lebensm-Wiss Technol. 2018;87:310–7.
30. Henchion M, Moloney AP, Hyland J, Zimmermann J, McCarthy S. Trends for meat, milk and egg consumption for the next decades and the role played by livestock systems in the global production of proteins. Animal. 2021;15:100287.
31. Said MI. Role and function of gelatin in the development of the food and non-food industry: a review. Inst Phys Conf Ser Earth Environ Sci. 2020;492(1):12086.
32. Kyselová J, Ječmínková K, Matějíčková J, Hanuš O, Kott T, Štípková M, et al. Physiochemical characteristics and fermentation ability of milk from Czech Fleckvieh cows are related to genetic polymorphisms of β-casein, κ-casein, and β-lactoglobulin. Asian Aust J Anim Sci. 2019;32(1):14.
33. Ippoushi K, Tanaka Y, Wakagi M, Hashimoto N. Evaluation of protein extraction methods for β-conglycinin quantification in soybeans and soybean products. LWT - Food Sci Technol. 2020;132:109871.
34. Sun C, Liu J, Yang N, Xu G. Egg quality and egg albumen property of domestic chicken, duck, goose, turkey, quail, and pigeon. Poult Sci. 2019;98(10):4516–21.
35. Bean SR, Akin PA, Aramouni FM. Zein functionality in viscoelastic dough for baked food products. J Cereal Sci. 2021;84:103270.
36. Cao L, Lu W, Mata A, Nishinari K, Fang Y. Egg-box model-based gelation of alginate and pectin: a review. Carbohydr Polym. 2020;242:116389.
37. Madni A, Khalid A, Wahid F, Ayub H, Khan R, Kousar R. Preparation and applications of guar gum composites in biomedical, pharmaceutical, food, and cosmetics industries. Curr Nanosci. 2021;17(3):365–79.
38. De Avelar MHM, Efraim P. Alginate/pectin cold-set gelation as a potential sustainable method for jelly candy production. LWT - Food Sci Technol. 2020;123:109119.
39. Mahuwala AA, Hemant V, Meharwade SD, Deb A, Chakravorty A, Grace AN, et al. Synthesis and characterisation of starch/agar nanocomposite films for food packaging application. IET Nanobiotechnol. 2020;14(9):809–14.
40. Imeson A. Food stabilisers, thickeners and gelling agents. New York, NY: John Wiley & Sons; 2011.
41. Sudhamani SR, Prasad MS, Sankar KU. DSC and FTIR studies on gellan and polyvinyl alcohol (PVA) blend films. Food Hydrocoll. 2003;17(3):245–50.
42. Luykx DM, Peters RJ, van Ruth SM, Bouwmeester H. A review of analytical methods for the identification and characterization of nano delivery systems in food. J Agric Food Chem. 2008;56(18):8231–47.
43. Moritaka H, Kimura S, Fukuba H. Rheological properties of matrix-particle gellan gum gel: effects of calcium chloride on the matrix. Food Hydrocoll. 2003;17(5):653–60.
44. Aguilera JM, Stanley DW. Microstructural principles of food processing and engineering. New York, NY: Springer Science & Business Media; 1999.
45. Jena R, Bhattacharya S. Viscoelastic characterization of rice gel. J Texture Stud. 2003;34(4):349–60.
46. Vittadini E, Carini E, Barbanti D. The effect of high pressure and temperature on the macroscopic, microscopic, structural and molecular properties of tapioca starch gels. In: Water properties of food, pharamaceutical and biological materials. London: Routledge; 2006. p. 471–83.
47. Van Vliet T. Mechanical properties of concentrated food gels. In: Dickinson E, Lorient D, editors. Proc. Int. Symp. Food macromolecules and colloids, Dijon; 1994. p. 447–55.
48. Kiss É. Nanotechnology in food systems: a review. Acta Aliment. 2020;49(4):460–74.
49. Eom S, Chun Y, Park C, Kim B, Lee S, Park D. Application of freeze–thaw enzyme impregnation to produce softened root vegetable foods for elderly consumers. J Texture Stud. 2018;49(4):404–14.
50. Nowacka M, Wiktor A, Dadan M, Rybak K, Anuszewska A, Materek L, et al. The application of combined pre-treatment with utilization of sonication and reduced pressure to accelerate

the osmotic dehydration process and modify the selected properties of cranberries. Foods. 2019;8(8):283.

51. Nielsen AV, Beauchamp MJ, Nordin GP, Woolley AT. 3D printed microfluidics. Annu Rev Anal Chem. 2020;13:45–65.
52. Jiang Y, Liu L, Wang B, Yang X, Chen Z, Zhong Y, et al. Polysaccharide-based edible emulsion gel stabilized by regenerated cellulose. Food Hydrocoll. 2019;91:232–7.
53. Barroso L, Viegas C, Vieira J, Pego C, Costa J, Fonte P. Lipid-based carriers for food ingredients delivery. J Food Eng. 2020;295:110451.
54. Funke W, Okay O, Joos-Müller B. Microgels-intramolecularly crosslinked macromolecules with a globular structure. Adv Polym Sci. 1998;136:139–234.
55. IUPAC. Compendium of chemical terminology (the "Gold Book"). Oxford: Scientific Publications; 1997.
56. Yallapu MM, Jaggi M, Chauhan SC. Design and engineering of nanogels for cancer treatment. Drug Discov Today. 2011;16:457–63.
57. Li D, Nostrum C, Mastrobattista E, Vermonden T, Hernnink W. Nanogels for intracelular delivery of biotherapeutics. J Control Release. 2017;259:16–28.
58. Sutekin S, Guven O. Application of radiation for the synthesis of poly(n-vinyl pyrrolidone) nanogels with controlled sizes from aqueous solutions. Appl Radiat Isot. 2019;145:161–9.
59. McClements DJ. Recent progress in hydrogel delivery systems for improving nutraceutical bioavailability. Food Hydrocoll. 2017;68:238.
60. McClements DJ. Designing biopolymer microgels to encapsulate, protect and deliver bioactive components: physicochemical aspects. Adv Colloid Interf Sci. 2017;240:31–59. https://doi.org/10.1016/j.cis.2016.12.005.
61. Stokes JR. Food biopolymer gels, microgel and nanogel structures, formation and rheology. In: Food materials science and engineering. New York, NY: Wiley; 2012. p. 151–76. https://doi.org/10.1002/9781118373903.ch6.
62. Zhang H, Zhai Y, Wang J, Zhai G. New progress and prospects: the application of nanogel in drug delivery. Mater Sci Eng. 2016;60:560–8.
63. Neamtu I, Rusu A, Diaconu A, Nita L, Chiriac A. Basic concepts and recent advances in nanogels as carriers for medical applications. Drug Deliv. 2017;24:539–57.
64. Hu G, Batool Z, Cai Z, Liu Y, Ma M, Sheng L, Jin Y. Production of self-assembling acylated ovalbumin nanogels as stable delivery vehicles for curcumin. Food Chem. 2021;355:129635. https://doi.org/10.1016/j.foodchem.2021.129635. PMID: 33780798
65. Khan J, Rudrapal M, Bhat EA, Ali A, Alaidarous M, Alshehri B, Banwas S, Ismail R, Egbuna C. Perspective insights to bio-nanomaterials for the treatment of neurological disorders. Front Bioeng Biotechnol. 2021;9:724158. https://doi.org/10.3389/fbioe.2021.724158.
66. Lee WH, Loo CY, Bebawy M, Luk F, Mason RS, Rohanizadeh R. Curcumin and its derivatives: their application in neuropharmacology and neuroscience in the 21st century. Curr Neuropharmacol. 2013;11(4):338–78. https://doi.org/10.2174/1570159X11311040002.

Chapter 11
Smart Use of Nanomaterials as Sensors for Detection and Monitoring of Food Spoilage

Aksa Fathima, Tafadzwa Justin Chiome, Archer Ann Catherine, Chukwuebuka Egbuna, Raghu Ram Achar, and Asha Srinivasan

11.1 Introduction

According to the data by World Health Organization (WHO) in 2020, an estimated 600 million—almost 1 out of 10 people are affected by foodborne illnesses, generally infectious or toxic and caused by microorganisms(bacteria, parasite, and virus) or chemical substances. These foodborne pathogens can cause diseases such as severe diarrhea, the most common illness causing 550 million people to fall sick and 2.3 lakhs (0.23 million) deaths every year. The more profound foodborne diseases

A. Fathima · T. J. Chiome · A. Srinivasan (✉)
Division of Nanoscience and Technology, School of Life Sciences, JSS Academy of Higher Education and Research, Mysuru, Karnataka, India
e-mail: justinchiome@jssuni.edu.in; asha.srinivasan@jssuni.edu.in

A. A. Catherine
Department of Microbiology, School of Life Sciences, JSS Academy of Higher Education and Research, Mysuru, Karnataka, India
e-mail: archerann@jssuni.edu.in

C. Egbuna
Department of Biochemistry, Faculty of Natural Sciences, Chukwuemeka Odumegwu Ojukwu University, Anambra State, Nigeria

Nutritional Biochemistry and Toxicology Unit, Africa Centre of Excellence, Centre for Public Health and Toxicological Research (ACE-PUTOR), University of Port-Harcourt, Rivers State, Nigeria

R. R. Achar (✉)
Department of Biochemistry, Faculty of Natural Sciences, Chukwuemeka Odumegwu Ojukwu University, Anambra State, Nigeria

Division of Biochemistry, School of Life Sciences, JSS Academy of Higher Education and Research, Mysuru, Karnataka, India
e-mail: rracharya@jssuni.edu.in

C. Egbuna et al. (eds.), *Application of Nanotechnology in Food Science, Processing and Packaging*, https://doi.org/10.1007/978-3-030-98820-3_11

are caused by bacteria and viruses [1, 2]. Food safety is one of the significant objectives of food safety laws [3]. The microbiological safety of food has become of substantial importance for the consumers and the food industry. Several lab methods exist to detect microorganisms contamination in food, such as traditional culturing methods, molecular methods, immunological protocols, and recently introducing spectroscopic techniques such as Raman and MALDI-TOF, during food processing [4]. Besides this, conventional packaging techniques have been handed-down to facilitate the proper product handling and preservation of nutrition value, increase shelf life, and decrease spoilage [5].

Food safety monitoring is vital throughout the food production and supply network [6]. An increase in public health issues worldwide and demand for international trading and complex food chains increases the chance of food spoilage and the supply of contaminated food products across national borders. Along with these developing cities, changes in climate, emigration, and international travel compounds also contribute to food contamination [1, 2]. With this globalization and dynamism of importation and exportation of food products and to meet the demand of public health, food safety, cost-effectiveness, traceability, tamper indication, and sustainability are the main objections to facilitating the development of new, improved packaging concepts, active packaging and smart packaging [7–9].

Smart packaging consists of nanomaterial-based sensors which have unique electrical and optical properties which can be devised to detect the presence of pathogens, gases, pesticides, and others in response to environmental changes. Smart packaging helps in the quality check of the food product to consumers, potentially reducing foodborne diseases worldwide. Though few nanotechnology-based indicators and sensors are already in the market, most of the task on nanosensors for food analysis is still in the initial stages of progression [10].

11.2 Factors Influencing Food Spoilage

Food spoilage is the procedure in which the edibleness of food decreases. A decrease in food edibility is related to food safety and can be identified by the appearance, texture, stench of the food. Food spoilage depends on numerous factors such as physical, microbial, or chemical actions. These mechanisms are interrelated to each other, which can cause food decay [11]. Physical factors which affect food decay are physical variations such as moisture content (water activity and water content), temperature- which eventually leads to microbial growth, glass transition temperature, crystal growth, and crystallization. *Microbial* spoilage is the typical cause of food degradation/ spoilage, which occurs due to bacteria, yeasts, and molds. The out-most growth of the microorganisms can be obstructed by storing at optimum temperature, lowering water activity, reducing pH, adding appropriate preservatives, and using appropriate packaging [12].

Chemical and biochemical responses occur naturally in foods which may lead to unpleasant tastes and smells in food products. Fresh foods may undergo undesirable quality changes caused by

1. pH changes due to microbial growth and metabolism
2. Virulent compounds, and
3. Rancid and discoloration of food because of lipid and pigment oxidation in fat.

Chemical deterioration is related to microbial activity. Nevertheless, the oxidation phenomenon is completely chemical and relay upon temperature fluctuations [13, 14].

11.3 Role of Nanotechnology in Food Packaging

Universally, the packaging of food products is practiced to protect the quality and improve the shelf life period by ensuring protection from various factors affecting food degradation such as physical (temperature, moisture content, pH), chemical, microbial, and any other environmental contaminants. A well-packaged food product undergoes a series of laboratory tests such as hot air drying and sterilization before the food reaches the supply chain [15]. In the past few decades, the food industry has witnessed a prominent growth in packaging technology, including food-packaging materials, food-packaging methods, food-packaging equipment, and auxiliary devices to ensure the safety of food [16].

Nanomaterials and their nonpareil properties, such as mechanical strength, solubility, diffusivity, optical features, have been intensively studied. Researchers and industry associates have recognized the future applications of nanotechnology in the field of food packaging [17]. Nanotechnology-based food packaging can enhance the storage period, freshness, and help in the development of devices/tools for rapid contaminant detection [6]. Nanotechnology-based food packaging is broadly classified as improved food packaging, active food packaging, and smart/intelligent food packaging.

11.3.1 Improved Food Packaging

Incorporation of nanoparticles to act as resistant to humidity and temperature to increase the physical and mechanical properties of packaging. Nanoparticles used are metal oxide NPs, nanoclays, carbon nanotubes, and metallic NPs. The most commonly used nanoparticle is nanoclay. Nanoclay acts as an oxygen scavenger or oxygen absorber; this can inhibit the invasion of oxygen upto 50% and moisture content up to 90% [18, 19].

11.3.2 Active Food Packaging

Active packaging incorporates nanoparticles that provide antimicrobial, antioxidant effects and absorb oxygen or moisture content into or from the food package. Active packaging consists of various nanoparticles such as AgNPs, AuNPs, metal oxides NPs, antimicrobial or antioxidant NPs, functionalized NPs. The most commonly used nanoparticles in active food packaging are gold nanoparticles, silver nanoparticles, antimicrobial/antioxidant nanoparticles, metal oxide nanoparticles, and functionalized nanoparticles. For instance, the combination of active compounds that are antimicrobial agents, preservatives, oxygen or moisture scavenger, ethylene removers with polymer provides better shelf life and food quality [19, 20].

11.3.3 Smart/Intelligent Food Packaging

Food packaging that provides information on the food quality of the food product using an internal or external sensor is known as Intelligent/ smart food packaging. The food industry regularly carries out microbial and chemical tests to ensure food products' quality assurance during production and delivery, but this is not possible once the product has reached the supply chain. Besides this, most of the food analyses carried out in the centralized laboratory with conventional techniques are limited to some samples that can be tested [21, 22]. These innovative food contact materials and items are endowed with the ability to sense any inner or atmospheric changes of the food package and communicate the unveiled information to consumers. The nanotechnology-based innovative communication methods allow inserting nanosized intelligent function known as nanosensor in Smart Food Packaging. These nanosensors are based on a chemical transduction mechanism, primarily designed to measure volatile compounds or on a biomolecular recognition for better selectivity (biosensor). Biosensors contain biological receptors such as enzymes, aptamers, cells, antibodies, etc. which can precisely recognize a selected target, and the building process is then transduced by electrochemical, optical, mass, or thermal process. Most biosensors are based on an optical or electrochemical process, e.g. color and electrical changes [10].

Nanomaterials commonly used in smart food packaging are (1) Gold NPs, Silver NPs with surface plasmon resonance (SPR) features, and high conductivity allowing them to be used in electrochemical and optical sensors. (2) Magnetic nanoparticles (Fe_2O_3 NPs) provide a magnetic effect for the efficient detachment and enhancement of the targeted analyte. (3) Carbon nanotubes and graphene-based nanomaterials amplify electrochemical signals due to their high electrical conductivity [6, 23].

11.4 Sensors

During food processing, one among the 300–400 packages is tested randomly for gas detection. However, this is laborious, expensive, and invalidating that untested packages meet safety quality, giving rise to the ideology for the development of noninvasive gas sensing methods based on nanotechnology, which can continuously monitor the gas content in all food packages throughout the supply chain [6]. It is possible that the existence of microbial contamination can be detected indirectly by measuring the changes in gas composition within the package as a result of micro-organism growth utilizing a gas sensor [9]. Figure 11.1 depicts the various types of sensors in nanotechnology.

11.4.1 Oxygen Sensor

Oxygen is essential for combustion or certain biological activities, such as oxidation reaction, fat oxidation, browning reaction, and pigment oxidation, which induces bacterial growth, leading to food deterioration [6]. Recently, several studies have been carried out to exploit metal-based semiconductors such as TiO_2, ZnO, and

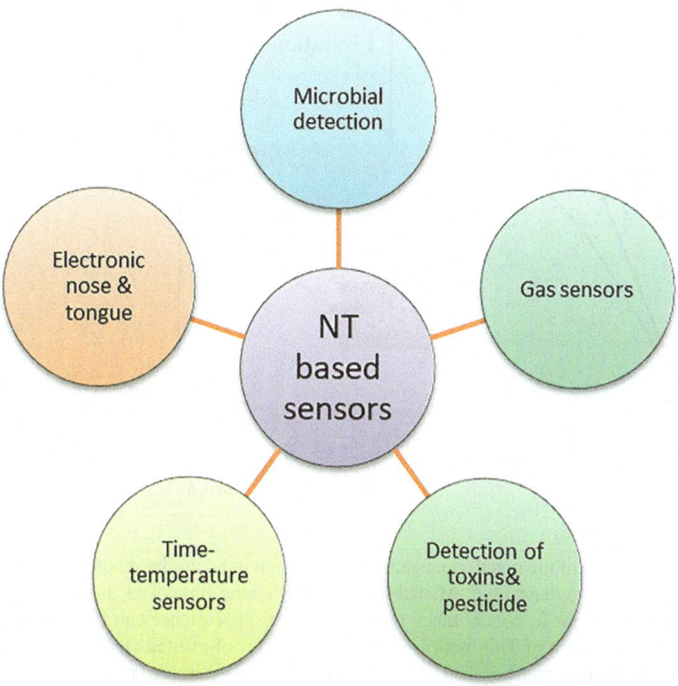

Fig. 11.1 Nanotechnology-based sensors

SnO_2 nanoparticles for oxygen sensor development. While these nanoparticle-based semiconductors are exposed to electromagnetic radiation with tremendous energy than their bandgap (in the UV range), electrons in their valence band bounce to the conduction band; their resulting excited electrons can be utilized for redox reactions, such as to activate redox dyes for oxygen detection.

Scientist Mills and coworkers developed a noninvasive sensor based on TiO_2 nanoparticles to detect oxygen in food packages. The TiO_2 nanoparticles are combined with a mild sacrificial electron donor [MSED] and a redox-active dye as an indicator. When TiO_2 nanoparticles are exposed to UV radiation, electron-hole pairs are formed on the surface of a semiconductor (Fig. 11.2). The MSED prevents the recombination of the electron-hole pairs of the semiconductor by donating electrons to the photogenerated holes, ensuring that the excited electrons are available for reducing the redox dye. As a result, the activation of dye occurs, the change from blue color oxidized state to colorless reduced state. On exposure to oxygen, the reduced dye is oxidized, and the blue color reappears. This UV-activated

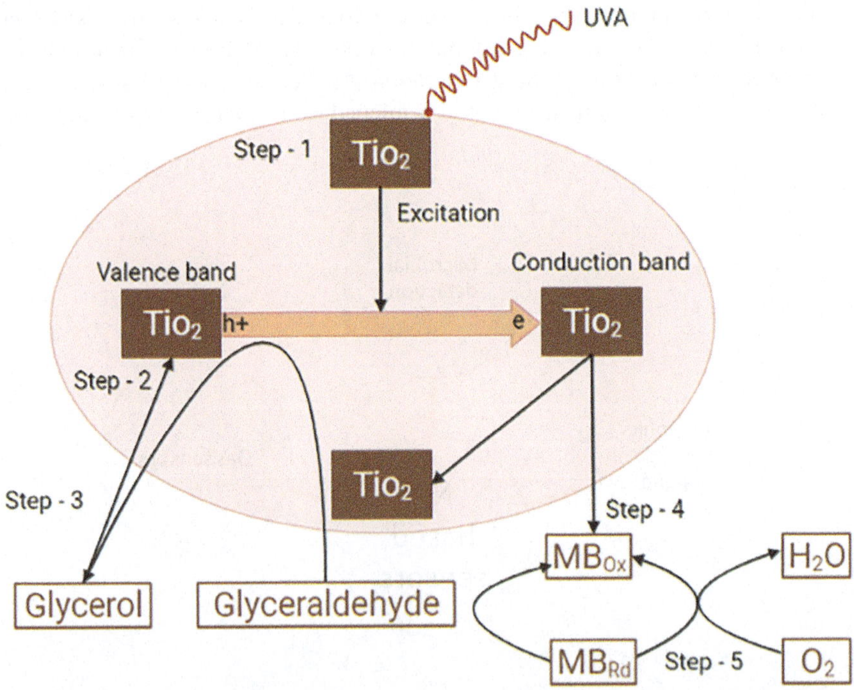

Fig. 11.2 Schematic of the reaction mechanism of oxygen indicator based on TiO_2 semiconductor. MBOX-oxidized methylene blue; MBRd-reduced methylene blue; TiO_2 holes in valence band: TiO_2—excited electrons in conductance band. The reaction sequence can be summarized as follows: Step 1—excitation of TiO_2 with UV; Step 2—transfer of excited electrons from the valence band to the conduction band of TiO_2; Step 3—photogenerated holes in valence band filled with electrons supplied by glycerol oxidation; Step 4—reduction of methylene blue by electrons from the conduction band; Step 5—oxidation of methylene blue by oxygen during end-use detection

mechanism is attractive for smart food packaging applications since the sensor can be triggered after it is placed within the package upon exposure to oxygen, which may occur due to leaky or damaged food packaging (Fig. 11.3) [17, 24].

Oxygen Sensor: SnO_2-ZnO nanocomposite sensor for trimethylamine: Trimethylamine (TMA) is an organic amine produced in the process of metabolism by animal organs and proteins TMA is one of the toxic gases in the foodstuff industry. SnO_2 and ZnO semiconductor oxides are the most frequently used materials, which have been widely studied due to their range of conductance variability and their response towards oxidative and reductive gases. ZnOmicrorods, along with SnO_2 nanoparticles, remarkably improve the gas sensing properties of TMA. This ZnO and SnO_2nanocomposite based sensor is used to monitor the freshness of a dead fish [25].

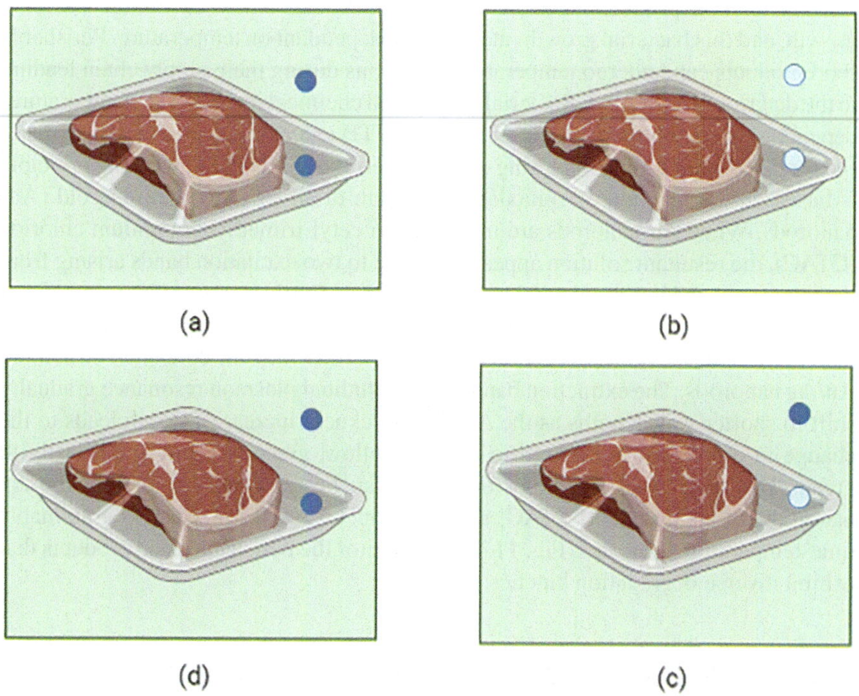

Fig. 11.3 Pictures of O_2 sensors which utilize UV-activated TiO_2 nanoparticles and methylene blue indicator dye, one placed inside of a food package flushed with CO_2 and one placed outside. In (**a**) the package is freshly sealed and both indicators are blue. The photograph in (**b**) shows the indicators immediately after activation with UVA light. After a few minutes, the indicator outside of the package returns to a blue color, whereas the indicator in an oxygen-free atmosphere remains white (**c**) until the package is opened, in which case the influx of oxygen causes it to change back to blue (**d**). This system could be used to easily and noninvasively detect the presence of leaks in every package immediately after production and at retail sites

11.4.2 Time-Temperature

Another factor influencing food spoilage is temperature. Time-temperature indicators based on nanotechnology can monitor and translate consumers about the quality of food products [26]. To ensure food product safety that requires specific temperature for storage such as extreme heat or freezing. Temperature fluctuations can occur at any point during the supply chain, such as loading, unloading, temperature cycling in walk-in coolers, store displays, and home transports of products. Monitoring of storage temperature of food products that need specific temperature is crucial since temperature largely determines the rate of microbial activity [2].

11.4.2.1 Time-Temperature Indicator Based on Kinetically Programmable Ag/Au Nanorods

Perishable food products such as dairy products can easily encourage bacterial growth, and this bacterial growth rate is highly dependant on temperature. Perishable food products can undergo temperature variations during their supply chain leading to the deterioration of food. Bacterial growth and chemical reactions are temperature-dependent. This time-temperature indicator (TTI) can act as a smart sensing material for the detection and monitoring of food deterioration. This TTI working principle is based on the reaction of epitaxial overgrowth of silver (Ag) shell on gold (Au) nanorods. When Au nanorods are immersed in cetyl-trimethylammonium chloride (CTAC), the resultant solution appears red; due to two excitation bands arising from the transverse and longitudinal plasmon resonance. Furthermore, $AgNO_3$ (precursor for Ag) and ascorbic acid (reducing agent) are instigated to produce Ag atoms, which are epitaxially deposited on the Au nanorods to form core/shell-structured Au/Ag nanorods. The extinction bands of longitudinal plasmon resonance gradually shift to shorter wavelengths as the Ag shell thickness increases, which leads to the change in color from initial red to orange, yellow, greenish-yellow, and lastly to green. The kinetics of this color deviation can be regulated by altering the Au nanorods, reducing agent, CTAC, and pH value. Such a low-cost programmable time-temperature indicator (Fig. 11.4) can monitor the perishable food products that exhibit diverse degradation kinetics [27].

11.4.2.2 Polydiacetylene (PDA) and Silica Nanoparticle-Based Time-Temperature Indicator

The PDA and silica nanoparticles can be used to develop a novel TTI. PDA molecules are well-polymerized vesicles that change color from deep blue to red under external stimuli like temperature, pH, and biological molecules. It can monitor, record, and translate the overall effect of temperature history on food quality in the chilled products and helps to assess the color change of all treatments upon temperature and time [28].

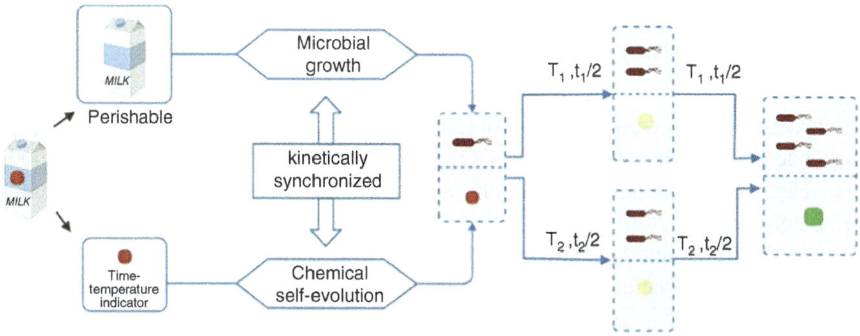

Fig. 11.4 Microbial growth in the perishable product and the chemical chronochromic self-evolution of the TTI is always kinetically synchronized, regardless of the temperature history. *T* temperature, *t* time

11.4.3 Detection of Microorganisms

Microbial growth can contribute to food deterioration in two possible ways- microbial growth of the organisms typically present in food which determines the food spoilage, and pathogenic microorganisms growth that can occur at any stage during logistics network [2].

Conventional biological detection methods are based upon immunological assays, which take advantage of Ab-Ag interaction. Nanomaterial-based sensors generally follow the same strategy. However, due to their unique optical and electrical properties combined with capacious, effortless functionalized surfaces, nanomaterials offer significant refinement in selectivity, speed, and sensitivity in contrast to chemical or biological methods based on macroscale materials. Along with this, one of the most outstanding advantages of the highly sensitive nanotechnology-based technique is the reduced incubation and evaluation time required to precisely detect bacterial pathogens [10, 17].

For instance, nanosized magnetic iron oxide particles attached with sugar molecule using a fluorescent dye (staining) were able to isolate ~88% *E.coli* in the sample with only 45mins of incubation time [17, 29]. Scientist Irudayaraj and his team improved this technique by using species and strain-specific antibodies to replace sugar molecules to isolate and detect target organisms in spinach extract and 2% milk. In another series of studies, they used magnetic nanoparticles in conjugation with gold nanorods (AuNRs) to separate and detect foodborne bacteria; in this case, magnetic nanoparticles facilitate separation, and AuNRs is used for optical detection. AuNRs have significant light-dependent absorption properties and efficient light-to-heat energy, hence offering simultaneous detection of multiple organisms [30, 31]. These optical calorimetric-based sensors are useful in the rapid detection of microorganisms.

11.4.4 Electrochemical Sensor for Detection of Microorganisms

Electrochemical detection of microorganisms using nanomaterials is impedance-based detection of bacteria- in this, TiO_2 nanowires bundles are coated with antibodies selective for targeted bacterium and placed between two gold electrodes protected by *n*-butythiol ligands. When the sensor is exposed to a complex matrix containing the targeted organism, a change in impedance (electrical) properties of the bundle due to bacterium-antibody binding can be observed [17] (Fig. 11.5).

11.5 Detection of Food-Contaminating Toxins, Pesticides and Chemicals

Toxins such as food-borne toxins, pesticides, chemicals and heavy metals in food are harmful to human health [10]. Nanotechnology-Based biosensors for toxin detection have been exploited (Table 11.1); the most common nanomaterials used for the detection are (1) Gold nanoparticles (GNPs), silver nanoparticles (AgNPs) as they possess surface platform resonance (SPR) characteristics and have high conductivity, which is essential in optical and electrochemical sensors, (2) magnetic NPs, e.g., Fe_2O_3 NPs which can provide magnetic functions that can be employed for the separation of the analytes reliably (3) carbon nanotubes (CNTs) and all the graphene-based nanomaterials which can amplify the electrochemical signals due to its high electrical conductivity [10, 32].

11.5.1 Gold Nanoparticles-Based Immunochromatographic Strip

Milk and dairy products are among the best sources of protein, fat, nutrients, and minerals for humans. The toxin involved in milk products is aflatoxin, *Bacillus cereus*, bisphenol A, and melamine. Aflatoxin B1 (AFB1) is a carcinogenic compound, the most common product of the *Aspergillus* species, and it is harmful to human health [33, 34]. The immunochromatographic strip that consists of AuNP's, along with anti-aflatoxin antibodies, has been used to detect aflatoxin B1, based on the classical indirect and direct competitive Enzyme-Linked Immunosorbent Assay (ELISA). The milk containing aflatoxin M1 appears to be a colorless zone on the strip, whereas the absence of aflatoxin M1 shows red bands on the strips. Likewise, using superparamagnetic beads consisting of aflatoxin M1 and gold nanoprobes reduced the detection time in milk samples containing aflatoxin M1 [32, 35]. Toxins such as food-borne toxins, pesticides, chemicals, and heavy metals in food are harmful to human health. Pesticides are intensely used in agriculture to protect

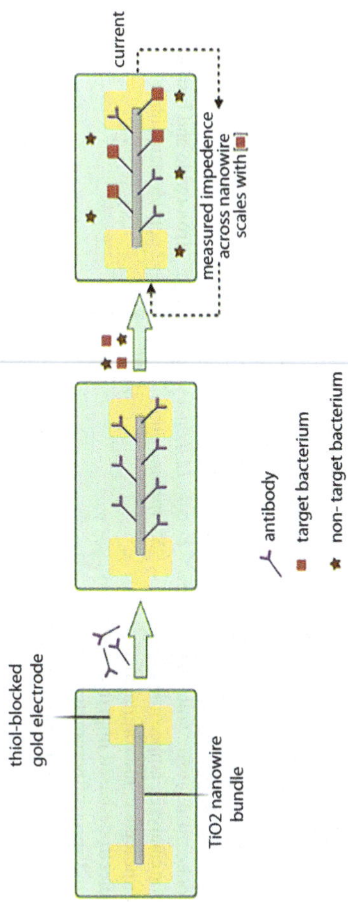

Fig. 11.5 Schematic representation of electrochemical sensor for detection of microorganism

Table 11.1 Nanomaterials based sensor used for detection and monitoring food spoilage

Types of sensors	Technique employed	Analyte	Nanomaterial employed	References
Gas sensor	Colorimetry and luminescence	Oxygen	Methylene bue-titanium doxide and poly-(styrene-block-vinylpyrrolidone), SnO_2	[35]
	Electrochemical	Hydrogen peroxide	Gold nanoparticles	
	Colorimetric: ultraviolet-visible spectroscopy and electrochemical: cyclic voltammetry	Xanthine and hypoxanthine	Gold	
Microbial detection	Electrochemical: linear sweep voltammetry	*Bacillus globigii*	Polypyrrole nanowires	[47]
	Field-effect transistor	*Salmonella infantis*	Single walled carbon nanotube	[48]
	Surface Plasmon resonance, electrochemical: cyclic voltammetry and different pulse voltammetry	*E. coli, Staphylococcus aureus, Vibrio parahaemolyticus* and *Salmonella* spp	Gold	[35]
	Electrochemical: cyclic voltammetry, flow injection analysis: amperometry, bioluminescence, interdigitated array microelectrode-based impedance analysis, PCR, and spectrofluorometry	*E. coli, L. monocytogenes,* and *S. typhimurium*	Iron oxide and bismuth nanofilm and peptide nanotubes	[35]
	Spectrofluorometry and flow cytometry	*E. coli, S. typhimurium* and *B. cereus*	Tris-(2,2-bipyridyl) dichlororuthenium(II) hexahydrate doped silica	[49]
	Fluorescence microscopy	*Escherichia coli* and *Salmonella typhimurium*	Zinc sulfite–cadmium selenide	[35]
	Electrochemical	*Salmonella*	AuNPs label	[6]
	Electrochemical	*Pathogens (E. coli O157:H7)*	Cd QDs encapsulated in ZIF-8 metal organic framework	
	Optical	*Staphylococcus aureus*	Gold nanorods $MnFe_2O_4$@ Au DNA-AuNPs	

Detection of food contaminants, toxins, pesticides, and chemicals	Electrochemical: cyclic voltammetry, immunochromatographic and enzyme-linked immunosorbent assay	Botulinum neurotoxin type B and brevetoxins	Gold	[35]
	Immunoassay and enzyme-linked immunosorbent assay; ion nanogating	Mycotoxin: Zearalenone and HT-2	Iron oxide, quartz nanopipettes	
	Immunoassay and enzyme-linked immunosorbent assay	Aflatoxins B1 and aflatoxin M1	Gold, iron oxide and superparamagnetic	
	Immunoassay and electrochemiluminescence	Microcystin-LR and palytoxin	SWCNT and MWCNT	
	Electrochemical	Toxins (BPA)	Cu NPs based ink	[6]
	Electrochemical	Adulterants (melamine) toxins (BPA)	Multiwalled carbon nanotube	
	Electrochemical	Adulterants, Sudan I	Graphene oxide/Graphenenanoribbons Pd/Au core-shell Nanocrystalline (Pd/Au CSNs) CdSe@CdS QDs	
	Electrochemical	Adulterants (dopamine), residual pesticides (chlorpyrifo, fenthion, and methyl parathion)	TiO_2/CeO_2 NP's CuO nanostructures	[6]
	Optical	Mycotoxin (Ochratoxin)	CeO_2 NP's	
	Optical	Adulterants (Melamine)	AuNP's	
	Optical	Adulterants (dopamine)	CeO_2 NP's	
	Electrochemical: square wave voltammetry; colorimetry, fluorescence, photoluminescence and ultraviolet-visible spectroscopy	Organophosphorus: parathion, paraoxon and carbamate	Liposome, gold, zirconium dioxide–gold and zinc sulfide–cadmium selenide and thioglycolicacid–cadmiumselenide	[35]
	Colorimetric, fluorescence and ultraviolet-visible spectroscopy	Melamine	Silver and gold	

(continued)

Table 11.1 (continud)

Types of sensors	Technique employed	Analyte	Nanomaterial employed	References
Time-temperature	Colorimetric: ultraviolet-visible	–	Silver shell, gold nanorods, and polydiacetylene-silica	
Electronic nose and electronic tongue	Electrical: source meter and field-effect transistor	Helional, hexanal, L-monosodium glutamate, phenylthiocarbamide, and propylthiouracil	Carbon nanotubes and carboxylated polypyrrole nanotube	
	Field-effect transistors	Trimethylamine, amyl butyrate, sucrose, phenylthiocarbamide and propylthiouracil	Single-walled carbon nanotube	
	Electrical: resistance measurement and dynamic headspace extraction analysis, linear discriminant analysis and electrodeless monolithic multichannel quartz crystal microbalance	Ethanol gas	Silver–tin dioxide, cobalt tetraoxide, molybdenum trioxide, magnesium–zinc oxide nanowires, gold–tungsten oxide, platinum–tin dioxide, tin dioxide, and zinc oxide	
	Electrical: resistance measurement	Ethylene gas	Tungsten oxide–tin oxide and silver–tin dioxide	[35]
	Surface XXXlasmon resonance and indirect competitive immunoassay	Benzaldehyde and olive oil	Gold	
	Electrical: resistance measurement	Chinese Vinegar	Zinc oxide	

crops and increase food production, but their toxicity causes human health risks. Pesticides are of various chemicals, such as fungicides, insecticides, herbicides, and plant growth regulators. After application, residual pesticides endure in soil, plants, and water, which may cause long-term health effects, including cancer, acute poisoning, and effect on reproduction health [1, 36]. Among the various pesticides, organophosphate pesticides such as chlorpyrifos, fenthion, and methyl parathion are the most common [10].

11.5.2 CuO Nanosensor for Detection of Organophosphate Pesticides

Organophosphate pesticides were detected by a non-enzymatic approach using CuO nanostructures grown in situ over indium tin oxide (ITO). The basic detection principle is the inhibition of pralidoxime chloride (PAM) coated over the pimelic acid-functionalized CuO nanostructures on the ITO surface. The in situ assemblies of the nanostructure provided enhanced surface area and loading, hence enhanced signal. Chlorpyrifos was detected using this method in cabbage and spinach extract [37, 38]. In recent advancement, disposable graphene-based biosensor has been developed for the detection of organophosphates [39].

11.6 Electronic Tongue and Electronic Nose for Artificial Taste and Smell Sensing

Electronic tongues and electronic noses have human-like efficiency used for artificial detection of taste and smell of food products, which helps monitor desired quality and taste food. Nanoparticles have been extensively studied for designing electronic noses and electronic tongues [35].

11.6.1 Electronic Tongue

The electronic tongue is defined as an analytical instrument comprising an array of nonspecific, low-selective, chemical sensors with high stability and cross-sensitivity to different species in solution and an appropriate method of PARC and/or multivariate calibration for data processing" [40]. Electronic tongues discriminate, identify or quantify the sample through chemometric methods and artificial intelligence [41, 42]. Usually, electronic tongue is composed of sensors such as electrochemical (amperometric, potentiometric, voltammetric, impedimetric and conductometric), gravimetric and optical (absorbance luminescence and reflectance). A wide range of sensor arrays has been developed to analyze and classify flavors of multicomponent mixtures making a versatile tool used in food analysis. The food quality and freshness can be continuously monitored using E-tongue, which can be carried out with the help of a sensory panel through evaluation of the aroma and taste properties [40].

11.6.2 Nanobioelectronics Tongue Sensor

Nanobiolectrnic tongue sensor has been developed by functionalization of carboxylated polypyrrole nanotube field-effect transistor along with human taste receptor, hTAS2R38 for better taste detection in a food sample. This device could detect target bitter tastants with high selectivity. Taster type (PAV) and nontaster type (AVI) hTAS2R38s are immobilized on a CPNT-FET sensor platform. PAV-CPNT-FET exclusively responded to the target bitterness compound. However, no notable changes were observed in the AVI-CPNT-FET with response to the target bitterness. Nanobioelectronics-tongue developed by the combination of human taste receptors with CPNT and allowed for high selectivity and sensitivity in the detection of target bitter tastants in mixtures and real food samples [43].

11.6.3 Electronic Nose

An electronic nose is an instrument that comprises an array of electrochemical sensors with partial specificity and an appropriate pattern recognition system capable of recognizing complex or straightforward odors. The more appropriate definition can be "an intelligent chemical-array sensor system that mimics the mammalian olfactory system." Electronic nose functions are less than human nose applications as they have been mainly developed for desired applications and consist of gas sensors that sense any odor. E-nose consists of components such as chemical array sensors, electronic circuitry, and software for data analysis. Besides this, it may also consist of a sampling, filtering, and conditioning system (for reproducible collection of mixture) [44, 45] (Fig. 11.6).

The incorporation of nanomaterials in sensors enhances the sensing ability as nanomaterial offer various advantages such as, (1) high surface area to volume ratio; (2) enhanced and tunable surface reactivity; (3) faster response kinetics; (4) size and effective Debye length comparable to the size of analytes; (5) crystalline dislocation-defect free structure with precise chemical composition, surface, and terminations; (6) high crystallinity implies superior stability; (7) easy to incorporate into microelectronic devices; (8) smaller size infers low power consumption; (9) and low cost. This nano-based EN sensor can be efficiently used to detect and monitor food spoilage during the packaging and storage of food products [44].

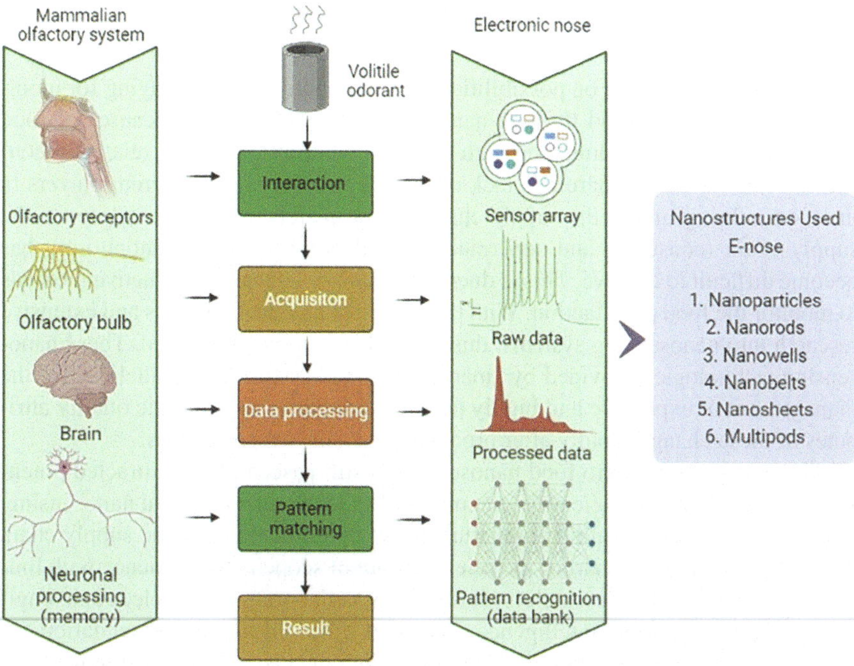

Fig. 11.6 Use of nanostructures in E-nose

11.6.4 Gold Nanoparticles Conjugated Oligopeptides Based E-Nose

The electronic nose comprises seven array quartz crystal micro balanced sensors modified with Gold Nanoparticles (GNP's) conjugated with oligopeptides used for quality assessment of extra virgin, virgin and non-edible lampante olive oil. Commercially olive oils are categorized as extra virgin, virgin, and lampante olive oils, which is strictly dependant on the sensory analysis that evaluates the presence and the level of sensory defects such as muddy, fusty, vinegary, moldy, and rancid. As the sensory perception depends upon the chemical composition of the olive oil sample and mainly on the headspace composition, the use of dynamic headspace (HS) high-resolution gas chromatography coupled with mass spectrometry (GC-MS) along with chemical and sensory evaluation has been used in the sensor. This gold nanoparticle-based electronic nose sensor can be used as a low-cost, easy-to-use, and rapid system for the quality check of extra virgin, virgin, and lampante (non-edible) olive oil [46].

11.7 Conclusion and Future Perspectives

This chapter has focused on possibilities to develop intelligent packaging for monitoring the quality of food through nanosensing technologies. The causes of food wastes and losses in medium and high-income countries are mainly related to consumer behavior. Often there is a lack of coordination between different players in the food industry and in the supply chain. Unfortunately, it is in these steps in the supply chain (consumer and supermarket retailer) where quality monitoring has become difficult to achieve. This is dues to a paucity in the non-destructive methods to monitor the food putrefaction. With the advent of nanotechnologies and extensive research into nanosensing systems, this loss is likely to be minimized. These nanosensing technologies provided by smart packaging systems give guidelines for the management of expensive and highly perishable foods as the intrinsic quality attributes of these change rapidly after processing causing economic loss.

Although, research into food nanosensors is still nascent, it has attracted potential interests among the scientific community for developing potential nanosensing. Further, nanosensor should be miniaturized, so that it could meet the supply chain demands. Although this part of the science is out of scope in food science with limited recommendations, research is warranted in collaboration with electronic engineering and artificial intelligence scientists. An ideal recommendation of miniaturized food sensor should be integrated into the package for food units, cost-effective relative to a food product, accurate and reliable, easy to scale up, and sustainable.

Acknowledgments All the authors thank JSS Academy of Higher Education & Research, Mysuru for the support towards research activities.

References

1. Food safety [Internet]. Who.int. 2021. https://www.who.int/NEWS-ROOM/FACT-SHEETS/DETAIL/FOOD-SAFETY
2. Yousefi H, Su H, Imani S, Alkhaldi KM, Filipe C, Didar T. Intelligent food packaging: a review of smart sensing technologies for monitoring food quality. ACS Sensors. 2019;4(4):808–21.
3. Resende-Filho M, Buhr B, Economics of traceability for mitigation of food recall costs, MPRA Paper 3650, University Library of Munich, Germany. 2007. Available: https://ideas.repec.org/p/pra/mprapa/3650.html. Retrieved: October 30, 2021.
4. Ferone M, Gowen A, Fanning S, Scannell A. Microbial detection and identification methods: bench top assays to omics approaches. Compr Rev Food Sci Food Saf. 2020;19(6):3106–29.
5. Rodriguez-Aguilera R, Oliveira J. Review of design engineering methods and applications of active and modified atmosphere packaging systems. Food Eng Rev. 2009;1(1):66–83.
6. Mustafa F, Andreescu S. Nanotechnology-based approaches for food sensing and packaging applications. RSC Adv. 2020;10(33):19309–36.
7. Anvar A, Ahari H, Ataee M. Antimicrobial properties of food nanopackaging: a new focus on foodborne pathogens. Front Microbiol. 2021.
8. Cooksey K. Food packaging, principles and practices, 2nd ed., by Gordon Robertson. CRC, Taylor and Francis: Boca Raton, FL, 2006. ISBN 0-8493-3775-5: 550 pages. Packag Technol Sci. 2006;21(1):57–9.

9. Kuswandi B, Wicaksono Y, Abdullah JA, Heng L, Ahmad M. Smart packaging: sensors for monitoring of food quality and safety. Sens Instrumen Food Qual. 2011;5(3–4):137–46.
10. Duncan T. Applications of nanotechnology in food packaging and food safety: barrier materials, antimicrobials and sensors. J Colloid Interface Sci. 2011;363(1):1–24.
11. Amit S, Uddin M, Rahman R, Islam S, Khan M. A review on mechanisms and commercial aspects of food preservation and processing. Agric Food Secur. 2017;6(1):51.
12. Tianli Y, Jiangbo Z, Yahong Y. Spoilage by Alicyclo bacillus bacteria in juice and beverage products: chemical, physical, and combined control methods. Compr Rev Food Sci Food Saf. 2014;13(5):771–97.
13. Van Boekel M. Kinetic modeling of food quality: a critical review. Compr Rev Food Sci Food Saf. 2008;7(1):144–58.
14. Huisin't VJ. Microbial and biochemical spoilage of foods: an overview. Int J Food Microbiol. 1996;33(1):1–18.
15. Pal M, Devrani M, Hadush A. Recent developments in food packaging technologies. Beverage Food World. 2019;48:21–5.
16. Meng X, Kim S, Puligundla P, Ko S. Carbon dioxide and oxygen gas sensors-possible application for monitoring quality, freshness, and safety of agricultural and food products with emphasis on importance of analytical signals and their transformation. J Korean Soc Appl Biol Chem. 2014;57(6):723–33.
17. Wahab A, Rahim AA, Hassan S, Egbuna C, Manzoor MF, Okere KJ, Walag AMP. Application of nanotechnology in the packaging of edible materials. In: Egbuna C, Mishra AP, Goyal MR, editors. Preparation of phytopharmaceuticals for the management of disorders. San Diego, CA: Academic; 2021. p. 215–25.
18. Kim S, Cha S. Thermal, mechanical, and gas barrier properties of ethylene-vinyl alcohol copolymer-based nanocomposites for food packaging films: effects of nanoclay loading. J Appl Polym Sci. 2013;131(11).
19. Alfei S, Marengo B, Zuccari G. Nanotechnology application in food packaging: a plethora of opportunities versus pending risks assessment and public concerns. Food Res Int. 2020;137:109664.
20. Sharma C, Dhiman R, Rokana N, Panwar H. Nanotechnology: an untapped resource for food packaging. Front Microbiol. 2017;8:1735.
21. Ghaani M, Cozzolino C, Castelli G, Farris S. An overview of the intelligent packaging technologies in the food sector. Trends Food Sci Technol. 2016;51:1–11.
22. Bouwmeester H, Dekkers S, Noordam M, Hagens W, Bulder A, de Heer C, et al. Review of health safety aspects of nanotechnologies in food production. Regul Toxicol Pharmacol. 2009;53(1):52–62.
23. Wang J, Wu X, Wang C, Rong Z, Ding H, Li H, et al. Facile synthesis of Au-coated magnetic nanoparticles and their application in bacteria detection via a SERS method. ACS Appl Mater Interfaces. 2016;8(31):19958–67.
24. Mihindukulasuriya S, Lim L. Oxygen detection using UV-activated electrospun poly(ethylene oxide) fibers encapsulated with TiO2 nanoparticles. J Mater Sci. 2013;48(16):5489–98.
25. Zhang W, Zhang W. Fabrication of SnO2–ZnO nanocomposite sensor for selective sensing of trimethylamine and the freshness of fishes. Sens Actuators B. 2008;134(2):403–8.
26. Fuertes G, Soto I, Carrasco R, Vargas M, Sabattin J, Lagos C. Intelligent packaging systems: sensors and nanosensors to monitor food quality and safety. J Sensors. 2016;2016:1–8.
27. Zhang C, Yin A, Jiang R, Rong J, Dong L, Zhao T, et al. Time–temperature indicator for perishable products based on kinetically programmable Ag overgrowth on Au nanorods. ACS Nano. 2013;7(5):4561–8.
28. Nopwinyuwong A, Kaisone T, Hanthanon P, Nandhivajrin C, Boonsupthip W, Pechyen C, et al. Effects of nanoparticle concentration and plasticizer type on colorimetric behavior of polydiacetylene/silica nanocomposite as time-temperature indicator. Energy Procedia. 2014;56:423–30.
29. El-Boubbou K, Gruden C, Huang X. Magnetic glyco-nanoparticles: a unique tool for rapid pathogen detection, decontamination, and strain differentiation. J Am Chem Soc. 2007;129(44):13392–3.

30. Wang C, Irudayaraj J. Gold nanorod probes for the detection of multiple pathogens. Small. 2008;4(12):2204–8.
31. Wang C, Irudayaraj J. Multifunctional magnetic-optical nanoparticle probes for simultaneous detection, separation, and thermal ablation of multiple pathogens. Small. 2010;6(2):283–9.
32. Girigoswami A, Ghosh MM, Pallavi P, Ramesh S, Girigoswami K. Nanotechnology in detection of food toxins—focus on the dairy products. Biointerface Res Appl Chem. 2021;11(6):14155–72.
33. Azad T, Ahmed S. Common milk adulteration and their detection techniques. Int J Food Contam. 2016;3(1):22.
34. Var I, Kabak B. Detection of aflatoxin M1 in milk and dairy products consumed in Adana, Turkey. Int J Dairy Technol. 2009;62(1):15–8.
35. Kumar V, Guleria P, Mehta S. Nanosensors for food quality and safety assessment. Environ Chem Lett. 2017;15(2):165–77.
36. Olsson A, Rutkowska A, Rachon D. Health risk of exposure to Bisphenol A (BPA). Roczniki Panstwowego Zaktada Higieny. 2015;66:5–11.
37. Aktar W, Sengupta D, Chowdhury A. Impact of pesticides use in agriculture: their benefits and hazards. Interdiscip Toxicol. 2009;2(1):1–12.
38. Tunesi M, Kalwar N, Abbas M, Karakus S, Soomro R, Kilislioglu A, et al. Functionalised CuO nanostructures for the detection of organophosphorus pesticides: a non-enzymatic inhibition approach coupled with nano-scale electrode engineering to improve electrode sensitivity. Sens Actuators B. 2018;260:480–9.
39. Hondred J, Breger J, Alves N, Trammell S, Walper S, Medintz I, et al. Printed graphene electrochemical biosensors fabricated by inkjet maskless lithography for rapid and sensitive detection of organophosphates. ACS Appl Mater Interfaces. 2018;10(13):11125–34.
40. Podrażka M, Bączyńska E, Kundys M, Jeleń P, Witkowska NE. Electronic tongue—a tool for all tastes? Biosensors. 2017;8(1):3.
41. Vlasov Y, Legin A. Non-selective chemical sensors in analytical chemistry: from "electronic nose" to "electronic tongue". Fresenius J Anal Chem. 1998;361(3):255–60.
42. Ciosek P, Wróblewski W. Sensor arrays for liquid sensing—electronic tongue systems. Analyst. 2007;132(10):963.
43. Song H, Kwon O, Lee S, Park S, Kim U, Jang J, et al. Human taste receptor-functionalized field effect transistor as a human-like nanobioelectronic tongue. Nano Lett. 2012;13(1):172–8.
44. Ramgir N. Electronic nose based on nanomaterials: issues, challenges, and prospects. ISRN Nanomater. 2013;2013:1–21.
45. El Kazzy M, Weerakkody J, Hurot C, Mathey R, Buhot A, Scaramozzino N, et al. An overview of artificial olfaction systems with a focus on surface Plasmon resonance for the analysis of volatile organic compounds. Biosensors. 2021;11(8):244.
46. Del Carlo M, Fusella G, Pepe A, Sergi M, Di Martino M, Mascini M, et al. Novel oligopeptides based e-nose for food quality control: application to extra-virgin olive samples. Qual Assur Saf Crops Foods. 2014;6(3):309–17.
47. Villamizar R, Maroto A, Rius F, Inza I, Figueras M. Fast detection of Salmonella Infantis with carbon nanotube field effect transistors. Biosens Bioelectron. 2008;24(2):279–83.
48. García-Aljaro C, Bangar M, Baldrich E, Muñoz F, Mulchandani A. Conducting polymer nanowire-based chemiresistive biosensor for the detection of bacterial spores. Biosens Bioelectron. 2010;25(10):2309–12.
49. Zhao X, Hilliard L, Mechery S, Wang Y, Bagwe R, Jin S, et al. A rapid bioassay for single bacterial cell quantitation using bioconjugated nanoparticles. Proc Natl Acad Sci. 2004;101(42):15027–32.

Chapter 12
Active, Smart, Intelligent, and Improved Packaging

Chinaza Godswill Awuchi and Terwase Abraham Dendegh

12.1 Introduction

Globally, food industries are under intense demand to attain customer needs for fresh, well, and safe foodstuff, besides the challenges of meeting updated and strict food safety regulations by the regulatory body(s). The binary broadcast organizations had stimulated and enlightened the end-users to ask for fresh, nutritious, safe, marginally prepared, and convenient food produce with properly-attached labels for easy access. To ensure the safety and credibility of foods all through the food supply chain, food producers, consumers, traders, manufacturers, and regulatory bodies must develop new methods that are fast, cost-effective, and consistent in observing the quality food packaged; more efficiently than its conventional submissive barrier abstraction, since packaging is a vital section of each part of the food processing industries [1]. The art, science, and technology of preparing food for sale, storage, distribution, and transportation from the manufacturing house to the end-users is known as food packaging. This conventional method tends to give preservation by changing the persistent and exceptional chemical, biological, and even physical needs. Conventional containers or packages like plastic, paperboard, metal, glassy materials, and a combination of substance of various physical and synthetic natures and shapes, was applied to make certain the idea (reason) and importance of food packaging depending on the type. The development of different types of packaging has been on the increase to help meliorate its efficacy in food standards with good

C. G. Awuchi (✉)
Department of Biochemistry, Kampala International University, Bushenyi, Uganda

School of Natural and Applied Sciences, Kampala International University, Kampala, Uganda

T. A. Dendegh
Department of Food Science and Technology, Federal University of Agriculture, Makurdi, Nigeria

C. Egbuna et al. (eds.), *Application of Nanotechnology in Food Science, Processing and Packaging*, https://doi.org/10.1007/978-3-030-98820-3_12

convenience of preparation and end-use [2]. These efforts are aimed towards nanotechnology to include active, smart/intelligent, and improved packaging material. Bio-nanotechnology has now become a vital practical application of science in food processing. The science of food packaging is a branch that deals with many aspects involving technology which also doubles as a way of manipulating, fabricating, and/or examining facts regarding nanoscale attributes [3].

Nanotechnology is the science that involves the fabrication, manipulation, and characterization of structures, devices, or materials of very small size (approximately 1–100 nm) that have at least one dimension of materials [4–7]. Nanotechnology is been adopted much more rapidly in the domain of food packaging [8]. However, challenges concerning the rate that these nanomaterials could possibly permeate into foods, and the possible health problems that these may have on the consumers are there. The efforts are quite motivating, and the assistance is extremely perceived—various nano-amplification for packages are readily available in markets, aid in prolonging the storage quality of the food, therefore eligibly ease of manufacture, preparation, and management [2]. There are many varieties of nanomaterials such as nano-clay, titanium nitride nanoparticles, and nano-titanium dioxide, nano-zinc oxide, and silver nanoparticles are introduced as functional additives to food packaging [9]. Nanotechnology-enabled food packaging can be divided into three major categories; improved packaging, active, and intelligent/smart packaging [10]. This chapter presents the latest information on the active, improved, intelligent, and smart packaging commonly used in food and nutraceutical industries, with an emphasis on nanotechnological applications.

12.2 Improved Packaging Through Nano-Composites

Nano-composites are a combination of both the conventional food packaging material and nanoparticles which are drawing high attention in the packaging industries. Furthermore, to the striking antimicrobial spectrum it has, it also tends to display great mechanical efficiency and tough resistant properties [1]. It also tends to provide better gas inhibition properties, and also to, moisture protection and temperature of the packaging. The USDA has authorized the application of nano-composites to be in close interaction with the food [7]. These materials (nano-composites) are composed of polymer matrices that are in a nonstop or discontinuous phase. Based on the nanomaterial, the nano-dimensional phase is generally made up of nanoparticles or nano-spheres, and/or nano-whisker(s), nano-sheets and nano-rods, nano-tubes, or nano-platelets. The mechanical attributes of polymers are made up of nano-phases in which its flexibility strain is selected to nano-reinforced material. Due to these attributes, nano-composite had been identified as a yardstick for imprecision of both the barrier and mechanical properties of nanocomposites (polymers). Asides from enhancing the mechanical and barrier properties, nano-particles do add smart/intelligent or active attributes to the packaging Scheme [1]. Bio-polymers are degradable compounds and are capable of improving the potential and protecting the foods while also ensuring their safety and quality. However, some great challenges of these types of packaging materials are their weak resistance barrier and weaker mechanical properties they have as it measured against petroleum-based equivalent. These shortcomings tend to hinder the application of these materials (biopolymer) in a variety of food packaging uses. Nano-composite technology has proven to help improve these materials (biopolymer) properties, as well as artificial thermoplastics [11]. The mechanical attributes such as toughness, fatigue, thermal stability, stiffness, delamination resistance, barrier features, shear strength, and tensile strength (TS) properties of polymeric materials could be improved by adding nanomaterials in their matrices [11].

12.3 Clay and Silicate Nano-Platelets (Nanoparticles and Nano-Crystals)

Sequel to the accessibility, low cost, and relatively easy processing ability, it has attracted the focus of scientists and technologists as potential nano-particles. The silicates layer is barely 1 nm thick to a couple of μ in length and is often engaged in nano-composites of a two-dimensional structure. However, the collation of these materials (polymers/silicates) tends to bestow good barrier properties to foods. The combination tends to enhance the diffusive pathway for infiltrating molecules [1]. Furthermore, to the tactoid structure of micro-composites, the relationship amid films of polymers and silicates could be an outcome in inserting or spreading nanocomposites. These added nano-composites occupy the space of multiple layered structures with cyclic inanimate films laid further apart by nano-meters. The forms occur when the foray of polymer chains into the interlayer region of the clay lead [11]. The exfoliated nanocomposites are composed of large polymer penetrations with haphazardly driving off of clay layers. The use of silicate clay as a nanocomponent in large polymers such as, polyvinyl chloride, starch, polyethylene, and nylon dates back to 1990s [12]. This group of silicate aluminum (montmorillonite) has been the major fillers. This is an octahedron layer of aluminum (OH) oxide within silica nine-sided (tetrahedral) bi-layers [11]. They are connected by delicate electrostatic pressure [11]. The disequilibrium within the outer anions charges is provided for by the existence of interchangeable cations, Ca^{2+} and Na^{+}. A Clay layer offers more barrier resistance to water vapor permeability and gases [11]. Furthermore, the addition of clays as nano-composite tends to improve the mechanical strength of biopolymers. Five (%w/w) applications of montmorillonite in thermosetting starch (TPS) improve the impulsive traits and decreases H_2O vapor transmission of the biopolymer starch. These montmorillonites also tend to influence the thermo-decomposition temperatures and glass transition [11]. The prospect of nano-clays materials with regards to their applications in intelligent/smart food packaging is great. Studies have suggested a nanotechnological packaging system that is formulated by the introduction of extracts of blueberries within silicate inter-layers of clays. These fruits are mostly characterized by anthocyanins which change color as a result of pH, are tied to a shift within flavylium and quinoidal forms. The addition of these fruits (blueberry) juice could transform these materials into smart/intelligent and active nano-composites [11]. Two types of ideal nano-scale composites are produced when silicates and polymer layered chains react together. The polymer chain raid occurs as a result of the insertion of nano-composites into the inter-layer region of the clay, and this, in turn, results in an organized complex form having a shifting inorganic/polymer sheet at recurring space of small nano-meters [13].

Barrier materials that are majorly used are the aluminum thin layer covering on thermoplastic films which are void-fixed, in packaging snacks, candy foods amongst others. Aluminum thin layer covering is made 50 nm thick and is proven capable as a nanomaterial, (a one-dimensional nanoscale). Also, the outer boundaries of glasses used as beverages containers are treated with organo-silanes, with plasma or high temperatures, which are very popular. The rate of accumulated molecules of gas in the material from surroundings/polymer border and the saturation rate of adsorbed gas molecules crossing the network makes the pervasion of polymeric compounds [14]. There are recent developments regarding the incorporating of nano clays (platelet) in polymers. They are used to decrease pervasion magnitude by preventing the movement of aroma and tainting substances, water, and oxygen into the package. Usually, incorporation of a small mass fraction, (small percent) into nano-composites, tends to improve barrier properties when compared to polymer materials alone [15]. Furthermore, these nanofillers enhance the barrier properties of the polymers by activating the polymer matrix in interfacial regions. If the reciprocal actions of polymer–nanoparticle are encouraging, the polymer patterns will be in part immobilized which are close to each nano-particle. These results could be that the molecules of gas going through the interfacial layer reduce their rate of flow between altered density, size of holes, and or free volume opening, a situation that is observed directly with the help of positron annihilation lifetime spectroscopy (PALS) [2]. Nanoplatelets can be illustrated through to the tortuous pathway [2]; with evolving gas molecules making a path that is straight to the film orientation; and, nanocomposite, evolving molecules tends to meander about tight platelets and passes the surface intentioned zones with a variety of absorbance attributes to the polymer. The path adds to the keeping quality of perishable foods through an increase in the time of the gas movement.

12.4 Nanotechnology in Active Packaging

The inculcation of nanomaterials in active packaging development is useful to relating directly with food and its environment to bring about excellent protection of the food products. Varieties of nanomaterials like carbon nanotubes, nano-magnesium oxide, nano-titanium dioxide, nano-silver and nano-copper oxide gives good inhibition traits against microorganisms. The application of silicate nano-particles as an anti-bacteria substance in the packaging of food is currently an increase [7]. Unlike the traditional packaging systems, active packages are purposefully made package system to inculcate components (bioactive NPs or bio-activated NPs) that can discharge substances the destroys microorganisms, substances that inhibit oxidation, imbued water vapor or oxygen substance in or from the food packaged and its environment. A combination of active forms such as ethylene removers, absorbers, preservatives, water vapor, and O_2, antimicrobial agents, etc. with polymer causes the package more capable of boosting the keeping quality and quality of the food product [1, 16]. The existence of active packaging concepts occurs where packages are visualized to alter the content, the kind of food, the surrounding condition the heritage of the food in packages [2]. The presences of nanoparticles are very high in surface areas and hold a lofty likelihood for containing or removing chemicals. For example, when enhancing agents like preservatives, colors are added, nanoencapsulation is applied to remove these enhancing agents from the food surface. The amalgamation of food packaging materials and active substances are new approaches to prevent surface microbial food contaminations. Without a doubt, some nanomaterials, like gold, zinc, and silver show antimicrobial effects. Silver nanoparticles have many applications commercially. Owning to its high-temperature stability and low volatility, silver at the nano-scale is established as a good anti-fungal and antimicrobial agent and claimed to be effective against many different bacteria [2].

12.4.1 Inorganic Nanoparticles

Metals such as zinc (Zn), silver (Ag), and gold (Au), and metal oxides, such as magnesium oxide (MgO), silicon oxide (SiO_2), zinc oxide (ZnO), and titanium dioxide (TiO_2), are nanomaterials which have seen application in various active package usage [17, 18]. These nanomaterials may perform on direct contact with the food sample or move consciously and respond to choice food components. Antimicrobiological property of nanomaterials might be because of one of the following mechanisms: oxidizing cell components; contact with microbial cells (bursting/breakthrough the cell-wall or membrane, temporarily stopping trans-membrane electron transfer) and foliation by-products (e.g., dissolved heavy metal ions or reactive oxygen species—ROS), leading to cell damage [1].

12.4.2 Silver Nanoparticles (AgNPs)

They are one of the most used nanomaterials, due to its strong antimicrobial properties against various pathogenic and commensal strains [1]. In addition to inhibiting bacteria strains, nanoparticles, e.g., AgNPs, are known to inhibit a variety of fungi and probably several viruses. Silver nanoparticles bind to proteins, enzymes, and DNA during the microbial metabolic process; resulting in bacterio-static effects

[19]. It tends to become unstable and break both the cytoplasmic membranes and cell walls. The respiratory chain enzymes are inhibited leading to the production of reactive oxygen species [1, 20]. Films containing silver microparticles tend to kill up to 80% of bacteria cells. Busolo et al. [21] reported an earlier study, that recorded the effective antibacterial activity of silver-OMMT/PLA nano-composite as it affects *Salmonella spp*. Antibacterial properties of PVA combined with cellulose nanocrystals and silver (Ag) nano-particles had been shown to act against *Staphylococcus aureus* and *E. coli* [1, 22–24]. Silver glassy materials are employed also to assemble anti-bacterial polymer composites. They are resistant and provide a gradual discharge of silver-ion into stored foods, resulting in antimicrobial activity. Other study shows the antimicrobial properties of AgNP-having cellulose sheets were shown in freshly foods, with gradual respiration rates and longer shelf-life [25]. Furthermore, these cellulose sheets having silver-NPs has been used with success for beef thin-coating, resulting to important decline in the micro-organisms load as reported [26].

12.4.3 Enzyme Immobilization

Enzymes generally play multiple roles in the food manufacturing industry but have a few challenges which include enzyme inhibitors and/or intolerance to production conditions such as temperature and pH amongst others may sometimes reduce their applicability in food systems. Fixation (Immobilization) and inclusion of these organic catalysts in packages give options to straight application of these organic catalysts to food systems [27]. This at a nano-scale tends to increase the surface and improve the effectiveness by enhancing it to stabilize to temperature, resistance to proteases, and pH; besides aiding to control the discharge of catalysts into the food system [28, 29]. Lopez-Rubio et al. [30] conducted research that shows that fixation is a very effective technique to enhance enzyme order to temperature and pH, opposed to protease-enzymes, and also gives ample condition for re-use and other denaturing components or restrained execution in foods. Fernandez et al. [31] reported that introducing catalyst such as lactase or cholesterol reductase into containing materials might improve the product (food) positively to also meet the end-user standard that has enzyme deficiencies. Rhim and Ng [32] made an inquiry toward enzyme adsorption into nano-clays addition to polymers were further aided by Gopinath and Sugunan [33] who also reported that nano-clay has excellent traits for protein uptake which may be applied as the basis of enzyme immobilization. However, Sharma et al. [34] established this through the fixation of glucose oxidase onto layers of anilineco- fluoroaniline. Modification of SiO_2 nano-particles as reported by Qhobosheane et al. [35] to fix lactate dehydrogenase and glutamate dehydrogenase shows greater enzymatic characteristics aft fixation.

12.4.4 Titanium (TiO_2) Dioxide

Titanium (TiO_2) dioxide exists in nature in the major forms of brookite, anatase, and rutile respectively; it possesses varied size crystals. It contains photo-catalytic abilities and at nano-scale, it shows surface reactivity which tends to fasten it with biological molecules such as phosphorylated proteins and peptides, and DNA. The surface energy of TiO_2 nano-particles also tends to amplify with mass and it is an important aspect in filler interaction/polymer. The outer strength consisting of titanium particles is larger than those of anatase particles of the same size. Titanium dioxide is largely inquired in the arrangement of many nanomaterials such as mesoporous, nanotubes, nanowires, nanorods, nanoparticles, and nano-porous TiO_2 accommodating materials [1]. Titanium dioxide antibacterial properties are well known as reported by [36, 37]. However, it is restricted to the risk of ultraviolent emission [1]. Though, Titanium (TiO_2) dioxide biocidal activity is not fully understood yet. It could be linked to its inceptive oxidative strike on the inner/outer cell membrane of bacterial, destruction to the DNA by means of hydroxyl radicals, and altering of Co-enzyme A-dependent enzyme activity. Examination of the photoactive biocidal attributes of Titanium (TiO_2) dioxide NPs based EVOH layers against some organisms such as *E. coli, L. plantarum, P. fluorescens, S. aureus, Bacillus* sp., *E. caratovora, P. jadinii, B. stearothermophilus, Z. rouxii* was done by Cerrada et al. [38]. The nano-particles of Titanium (TiO_2) dioxide were uniformly dispersed rapidly. Five (5) logarithm reductions for *P. jadinii, L. plantarum, Bacillus* sp, and *B. stearothermophilus* were reported in duration of 30 min of emission exposure within the aura of EVOH/Titanium (TiO_2) dioxide materials.

In other studies, Titanium (TiO_2) dioxide nanoparticles which are thin-layered with plastic films were measured counter to spoilage in apples, lemons, and tomatoes by *Penicillium expansum*. The result shows growth performance of *P. expansum* was suppressed because of the photocatalytic activity of Titanium dioxide particles when subjected to light [39]. An amalgamation of Titanium dioxide nanoparticles and silver is proven to improve anti-microbiological activities at a significant level [1]. New materials used for food packaging with enhanced functional properties tend to render longer keeping quality of food commodity. Novel active packaging materials used for foods involve a variety of absorbers, coatings, emitters, and scavengers. The substances could likely be enveloped in the traditional non-degradable (plastics, etc.) packaging. However, they are employed in connection with bio-degradable parts. Table 12.1 shows the functional Characteristics of active food packaging to prolong the keeping quality and enhance the safety of foods.

Table 12.1 Functional characteristics of active food packaging to prolong the keeping quality and enhance the safety of foods

Functional agents	Types	Functions	References
Organic (ascorbic acid, catechol, tocopherol), metallic (iron powder, Zn, activated iron), inorganic (ZnO, thiosulfate, sulfite,), enzyme-based (glucose oxidase, laccase), polymer-based (polymer metallic complex), $KMnO_4$, SiO_2, Ag, TiO_2	Oxygen scavengers	Preventing lipid oxidation	[3]
$PdCl_2$, Pd-impregnated zeolite, polyvinyl chloride films containing ZnO nanoparticles	Ethylene scavengers	Fruit and vegetables ripening reduction	[8]
Inorganic (non-synthetic clay (zeolite, montmorillonite), silica gel, chlorides (Mg, Na, Al, Ca, K,), organic (fructose, sorbitol, cellulose, xylitol, and their derivatives), oxides (Ca, Ba), bentonite, polymer-based (polyvinyl alcohol, absorbent resin, starch copolymers)	Moisture absorbers	Microbial growth reduction	[40, 41]
Ascorbate and citric acid, sodium bicarbonate	Carbon dioxide (CO_2) emitters	Inhibition of microbial spoilage	[3]

12.4.5 Various Antimicrobial Nanoparticles

Nano-particles of Copper were also shown to have an inhibitory effect on expansion of *E. coli, Saccharomyces cerevisiae, L. monocytogenes,* and *S. aureus* on polymeric materials after subjecting it to a 4 h culture test [42]. Reports as observed by Sheikh et al. [43] show that copper nanoparticles possess good bacterio-static properties when tested on B. *subtilis* and *E.coli* in polyurethane nano-fibers with copper nanoparticles. Copper nanoparticles result in various poisonous characteristics which include lipid peroxidation, DNA degradation, protein oxidation, and generation of reactive oxygen species (ROS); this could possibly be the reason for the antimicrobial properties [44]. Zinc nano-crystals are used also as an antifungal and antimicrobial agents. This happens if it is combined along with plastic materials [45]. Various nano-particles oxide(s) including zinc oxide (ZnO), magnesium oxide (MgO), titanium dioxide (TiO_2), and silicon oxide (SiO_2) are useful in the packaging of food, partly because of their capability to behave as a UV barrier and disinfecting agents (photo-catalytic) [46]. Titanium (TiO_2) dioxide particles are very much promising [47, 48]. Titanium dioxide nanoparticles' antimicrobial activities are photo-catalyzed and are active only in the existence of UV rays. It also shows good inhibitory properties when applied to *L. monocytogenes, Vibrio parahaemolyticus*, and *S. choleraesuis* in the existence of UV illumination [49].

12.5 Smart and Intelligent Packaging System

Intelligent/smart packages are designed basically in sensing microbial or biochemical variation in food products. An intelligent/smart package has the ability to detect specific pathogens that are growing in the food and/or some gases produced during food spoilage. Few smart packaging had been made to use RFID gadgets for food security. Supermarkets such as MonoPrix Supermarket and other organizations like British Airways and Nestle are applying sensors (e.g., chemical sensors) that could easily discover color changes. Nanomaterial integrated packages could be "smart", that is, they can respond to their surrounding situations, mend themselves or give impairs alert to consumers and pathogens that may present [2, 41, 50]. According to reports, smart or intelligent packaging tends to protect food quality along the food distribution chain [50].

However, because of the variation in the optical nature and greater outer irritability of nanomaterials which includes photonic nanocrystals or metal nanoparticles, its high act as compared to conventional colorimetric indicators could be observed. The use of optical marks, among nano-sensors, is mostly seen in the commercial market because of their suitability and ease of use [3]. With nano-sensors, the reaction inside or outside framework tends to change within the food package or in its surrounding is carried out with report back data to the customers, to guarantee food safety and quality. Various nanomaterials are applied to ameliorate and reform the performance of the package. These nano-sensors have great potentials for fast quantification, identification, and detection of allergy-causing proteins, decaying substances, and pathogenic microorganisms. E Custom-made nano-sensors used in intelligent/smart packages are applied for food analysis (detecting food pathogens, toxins, unwanted chemicals) and in the identification of colors, flavors, etc. [3, 50]. The Food packaging material could be equipped with these nano-sensors that are perceptive to gas formation, temperature, or humidity variations and for instance, gasses which are formed as a result of festering of food, the package will then tends to change the color of the indicator and therefore alerting consumers to the unsuitability of the product(s). Packages that are equipped with nano-sensors, could be used successfully to cut down the standard for deciding the keeping quality of the food and also, real-time monitoring of food freshness since the nano-sensors could react to some chemical indicators, toxins, and pathogens in food [1, 3, 51]. Bio-nano-sensors are a combination of biosensors and nanotechnology. There are various uses of bio-nano-sensors in food packaging. Nano-sensors also are designed to trace either the internal or external conditions of food commodities, pellets, and containers, all through the supply chain. These packages have the ability to monitor temperature or humidity over time and provide feedback on these food commodities such as color change. Nano-sensors can sense gases evolving from food when it is spoilt. The packaging itself changes color which alerts consumers. Recent studies on active, intelligent, smart, and improved packaging are shown in Table 12.2.

Table 12.2 Recent studies on active, smart, and improved packaging

Reference	Title of study	Type of study	Relevance
Wu et al. [52]	"Development and Characterization of an Enzymatic Time-Temperature Indicator (TTI) Based on *Aspergillus niger* Lipase"	Research	Presents interesting area of smart packaging
Shivraj et al. [53]	"Nanotechnologies in Food Science: Applications, Recent Trends, and Future Perspectives"	Review	Relevant to areas of active, improved, and smart packaging
Egbuna et al. [18]	"Toxicity of Nanoparticles in Biomedical Application: Nanotoxicology"	Review	This review summarizes the toxicity of nanomaterials
Awuchi et al. [8]	"Nanotechnology Application in Food Science and Nutrition and Its Safety Issues; A Review"	Review	The review presents interesting areas of nanotechnology application in food and nutrition, including active and smart packaging
Spielman [54]	"Smart Packaging: Connecting the Physical with the Digital"	Online resource	Presents interesting area of smart packaging using digitalization
Cammarelle et al. [55]	"Intention to Purchase Active and Intelligent Packaging to Reduce Household Food Waste: Evidence from Italian Consumers"	Research	Presents interesting area of active and intelligent packaging
Salgado et al. [56]	"Recent Developments in Smart Food Packaging Focused on Biobased and Biodegradable Polymers"	Review	This review summarizes the advancement in smart packaging based on polymeric materials
Eghbal et al. [57]	"Antimicrobial Films Based on Pectin and Sodium Caseinate for the Release of Antifungal Natamycin"	Research	Presents interesting area of smart packaging using antimicrobial films
Ehsani et al. [58]	"Comparative Evaluation of Edible Films Impregnated with Sage Essential Oil or Lactoperoxidase System: Impact on the Chemical and Sensory Quality of Carp Burgers"	Research	Presents a comparison between edible films infused with sage essential oils or lactoperoxidases
Bhargava et al. [59]	"Active and Intelligent Biodegradable Packaging Films Using Food and Food Waste-Derived Bioactive Compounds: A Review"	Review	Presents interesting area of active and smart packaging
Assis et al. [60]	"Active Biodegradable Cassava Starch Films Incorporated Lycopene Nanocapsules"	Research	Presents the application of lycopene nanocapsules and cassava in active and smart packaging

12.6 Conclusion

Packaging in recent times has not just been the enclosure of foods in a package to make certain that the product is delivered in its highest form to consumers. Incorporating nano-sensors amongst other technology has provided useful feedback on the condition of the food inside the packages. These packages are made in such a way that it releases antioxidants, nutraceuticals, flavors, antimicrobials, and enzymes to prolong the keeping quality of the food products. Active and smart/intelligent packaging has become more important in adding value to foods. The importance of preserving the nutritional components of food, to counteract the deterioration of food, and to prevent bacterial contamination has brought about the development of these novel preservative methods and also conceptualization of natural or artificial adjuvant, and also to the manufacturing of ingenious packages. The introduction of nanomaterials in the formulation of food packaging has presented a new preformat and the likelihood in keeping food quality and prolonging the keeping quality food and substantial useful effect its organoleptic acumen of end-users, thereby leading to a decrease in post-harvest losses and hence, improvement in the economy.

References

1. Sharma C, Dhiman R, Rokana N, Panwar H. Nanotechnology: an untapped resource for food packaging. Front Microbiol. 2017;8:1735. https://doi.org/10.3389/fmicb.2017.01735.
2. Gokularaman S, Stalin Cruz A, Pragalyaashree MM, Nishadh A. Nanotechnology approach in food packaging—review. J Pharm Sci Res. 2017;9(10):1743–9.
3. Primožič M, Knez Ž, Leitgeb M. (Bio) Nanotechnology in food science—food packaging. Nanomaterials. 2021;11:292. https://doi.org/10.3390/nano11020292.
4. Ravichandran R. Nanoparticles in drug delivery: potential green nanobiomedicine applications. Int J Nanotechnol Biomed. 2010;1:108–30.
5. Wahab A, Rahim AA, Hassan S, Egbuna C, Manzoor MF, Okere KJ, Walag AMP. Application of nanotechnology in the packaging of edible materials. In: Egbuna C, Mishra AP, Goyal MR, editors. Preparation of phytopharmaceuticals for the management of disorders. San Diego, CA: Academic; 2021. p. 215–25.
6. Kuswandi B. Nanotechnology in food packaging. Chemo and Biosensors Group, Faculty of Pharmacy, University of Jember; 2016.
7. Pal M. Nanotechnology: a new approach in food packaging. J Food Microbiol Saf Hyg. 2017;2:2. https://doi.org/10.4172/2476-2059.1000121.
8. Awuchi CG, Twinomhwezi H, Choudghal S, Khan MG, Yezdani U, Akram MV. Nanotechnology application in food science and nutrition and its safety issues; a review. Adv Biores. 2020;11(6):23–35. https://doi.org/10.15515/abr.0976-4585.11.6.2335.
9. Tager J. Nanomaterials in food packaging: FSANZ fails consumers gain. Chain Reaction. 2014;122:16–7.
10. Duncan TV. Applications of nanotechnology in food packaging and food safety: barrier materials, antimicrobials and sensors. J Colloid Interface Sci. 2011;363:1–24.
11. Mihindukulasuriya SDF, Lim LT. Nanotechnology development in food packaging: a review. Trends Food Sci Technol. 2014;40(2014):149–67.
12. Montazer M, Harifi T. New approaches and future aspects of antibacterial food packaging: from nanoparticles coating to nanofibers and nanocomposites, with foresight to address the regulatory uncertainty. In: Grumezescu AM, editor. Food package. San Diego, CA: Academic; 2017. p. 533–59.

13. Weiss J, Takhistov P, McClements DJ. Functional materials in food nanotechnology. J Food Sci. 2006;71(9):107–16.
14. Mercea P. Models for diffusion in polymers. In: Piringer OG, Baner AL, editors. Plastic packaging. 2nd ed. Weinheim: Wiley-VCH GmbH and Co. KGaA; 2008.
15. Bradley EL, Castle L, Chaudhry Q. Applications of nanomaterials in food packaging with a consideration of opportunities for developing countries. Trends Food Sci Technol. 2011;22:604–10.
16. Alfei S, Marengo B, Zuccari G. Nanotechnology application in food packaging: a plethora of opportunities *versus* pending risks assessment and public concerns. Food Res Int. 2020;137:109664.
17. Bikiaris DN, Triantafyllidis KS. HDPE/cu-nanofiber nanocomposites with enhanced antibacterial and oxygen barrier properties appropriate for food packaging applications. Mater Lett. 2013;93:1–4. https://doi.org/10.1016/j.matlet.2012.10.128.
18. Egbuna C, Parmar VK, Jeevanandam J, Ezzat SM, Patrick-Iwuanyanwu KC, Adetunji CO, Khan J, et al. Toxicity of nanoparticles in biomedical application: nanotoxicology. J Toxicol. 2021;2021:9954443. https://doi.org/10.1155/2021/9954443.
19. Cavaliere E, De Cesari S, Landini G, Riccobono E, Pallecchi L, et al. Highly bactericidal ag nanoparticle films obtained by cluster beam deposition. Nanomed Nanotechnol Biol Med. 2015;11:1417–23. https://doi.org/10.1016/j.nano.2015.02.023.
20. Emamifar A, Kadivar M, Shahedi M, Soleimanian-Zad S. Evaluation of nanocomposites packaging containing Ag and ZnO on shelflife of fresh orange juice. Innov Food Sci Emerg Technol. 2011;11:742–8. https://doi.org/10.1016/j.ifset.2010.06.003.
21. Busolo MA, Fernandez P, Ocio MJ, Lagaron JM. Novel silver based nanoclay as an antimicrobial in polylactic acid food packaging coatings. Food Addit Contam. 2010;27:1617–26. https://doi.org/10.1080/19440049.2010.506601.
22. Fortunati E, Peltzer M, Armentano I, Jimenez A, Kenny JM. Combined effects of cellulose nanocrystals and silver nanoparticles on the barrier and migration properties of PLA nanobiocomposites. J Food Eng. 2013a;118:117–24. https://doi.org/10.1016/j.jfoodeng.2013.03.025.
23. Fortunati E, Puglia D, Luzi F, Santulli C, Kenny JM, Torre L. Binary PVA bio-nanocomposites containing cellulose nanocrystals extracted from different natural sources: part I. Carbohydr Polym. 2013b;97:825–36. https://doi.org/10.1016/j.carbpol.2013.03.075.
24. Sadeghnejad A, Aroujalian A, Raisi A, Fazel S. Antibacterial nano silver coating on the surface of polyethylene films using corona discharge. Surf Coat Technol. 2014;245:1–8. https://doi.org/10.1016/j.surfcoat.2014.02.023.
25. Fernandez A, Picouet P, Lloret E. Cellulose-silver nanoparticle hybrid materials to control spoilage-related microflora in absorbent pads located in trays of fresh-cut melon. Int J Food Microbiol. 2010;142:222–8. https://doi.org/10.1016/j.ijfoodmicro.2010.07.001.
26. Smolkova B, El Yamani N, Collins AR, Gutleb AC, Dusinska M. Nanoparticles in food. Epigenetic changes induced by nanomaterials and possible impact on health. Food Chem Toxicol. 2015;77:64–73. https://doi.org/10.1016/j.fct.2014.12.015.
27. Ranjan S, Dasgupta N, Chakraborty AR, Samuel SM, Ramalingam C, Shanker R, et al. Nanoscience and nanotechnologies in food industries: opportunities and research trends. J Nanopart Res. 2014;16:2464. https://doi.org/10.1007/s11051-014-2464-5.
28. Rhim JW, Park HM, Ha CS. Bio-nanocomposites for food packaging applications. Prog Polym Sci. 2013;38:1629–52. https://doi.org/10.1016/j.progpolymsci.2013.05.008.
29. Brandelli A, Brum LFW, dos Santos JHZ. Nanostructured bioactive compounds for ecological food packaging. Environ Chem Lett. 2017;15:193–204. https://doi.org/10.1007/s10311-017-0621-7.
30. Lopez-Rubio A, Gavara R, Lagaron JM. Bioactive packaging: turning foods into healthier foods through biomaterials. Trends Food Sci Technol. 2006;17:567–75. https://doi.org/10.1016/j.tifs.2006.04.012.
31. Fernandez A, Cava D, Ocio MJ, Lagaron JM. Perspectives for biocatalysts in food packaging. Trends Food Sci Technol. 2008;19:198–206. https://doi.org/10.1016/j.tifs.2007.12.004.

32. Rhim JW, Ng PKW. Natural biopolymer-based nanocomposite films for packaging applications. Crit Rev Food Sci Nutr. 2007;47:411–33. https://doi.org/10.1080/10408390600846366.
33. Gopinath S, Sugunan S. Enzymes immobilized on montmorillonite K 10: effect of adsorption and grafting on the surface properties and the enzyme activity. Appl Clay Sci. 2007;35:67–75. https://doi.org/10.1016/j.clay.2006.04.007.
34. Sharma AL, Singhal R, Kumar A, Rajesh Pande KK, Malhotra BD. Immobilization of glucose oxidase onto electrochemically prepared poly (aniline-co-fluoroaniline) films. J Appl Polym Sci. 2004;91:3999–4006. https://doi.org/10.1002/app.13553.
35. Qhobosheane M, Santra S, Zhang P, Tan WH. Biochemically functionalized silica nanoparticles. Analyst. 2001;126:1274–8. https://doi.org/10.1039/b101489g.
36. Macwan D, Dave PN, Chaturvedi S. A review on nano- TiO2 sol–gel type syntheses and its applications. J Mater Sci. 2011;46:3669–86. https://doi.org/10.1007/s10853-011-5378-y.
37. Montazer M, Seifollahzadeh S. Enhanced self-cleaning, antibacterial and UV protection properties of nano TiO2 treated textile through enzymatic pretreatment. Photochem Photobiol. 2011;87:877–83. https://doi.org/10.1111/j.1751-1097.2011.00917.x.
38. Cerrada ML, Serrano C, Sánchez-Chaves M, Fernández-García M, Fernández-Martín F, de Andrés A. Self-sterilized EVOHTiO2 nanocomposites: effect of TiO2 content on biocidal properties. Adv Funct Mater. 2008;18:1949–60. https://doi.org/10.1002/adfm.200701068.
39. Maneerat C, Hayata Y. Antifungal activity of TiO2 photocatalysis against Penicillium expansum in vitro and in fruit tests. Int J Food Microbiol. 2006;107:99–103. https://doi.org/10.1016/j.ijfoodmicro.2005.08.018.
40. Awuchi CG, Ondari EN, Ofoedu CE, Chacha JS, Rasaq WA, Morya S, Okpala COR. Grain processing methods' effectiveness to eliminate mycotoxins: an overview. Asian J Chem. 2021a;33(10):2267–75. https://doi.org/10.14233/ajchem.2021.23374.
41. Awuchi CG, Ondari EN, Ogbonna CU, Upadhyay AK, Baran K, Okpala COR, Korzeniowska M, Guiné RPF. Mycotoxins affecting animals, foods, humans and plants: types, occurrence, toxicities, action mechanisms, prevention and detoxification strategies—a revisit. Foods. 2021b;10:1279. https://doi.org/10.3390/foods10061279.
42. Cioffi N, Torsi L, Ditaranto N, Tantillo G, Ghibelli L, Sabbatini L, et al. Copper nanoparticle/polymer composites with antifungal and bacteriostatic properties. ChemMater. 2005;17:5255–62. https://doi.org/10.1021/cm0505244.
43. Sheikh FA, Kanjawal MA, Saran S, Chung WJ, Kim H. Polyurethane nanofibers containing copper nanoparticles as future materials. Appl Surf Sci. 2011;257:3020–6. https://doi.org/10.1016/j.apsusc.2010.10.110.
44. Chatterjee AK, Chakraborty R, Basu T. Mechanism of antibacterial activity of copper nanoparticles. Nanotechnology. 2014;25:135101. https://doi.org/10.1088/0957-4484/25/13/135101.
45. Vermeiren L, Devlieghere F, Debevere J. Effectiveness of some recent antimicrobial packaging concepts. Food Addit Contam. 2002;19:163–71.
46. Fujishima A, Rao TN, Tryk DA. Titanium dioxide photocatalysis. J Photochem Photobiol Phytochem Rev. 2000;1:1–21. https://doi.org/10.1016/S1389-5567(00)00002-2.
47. Kong H, Song J, Jang J. Photocatalytic antibacterial capabilities of TiO(2)-biocidal polymer nanocomposites synthesized by a surfaceinitiated photopolymerization. Environ Sci Technol. 2010;44:5672–6. https://doi.org/10.1021/es1010779.
48. Farhoodi M. Nanocomposite materials for food packaging applications: characterization and safety evaluation. Food Eng Rev. 2016;8:35–51. https://doi.org/10.1007/s12393-015-9114-2.
49. Robertson JMC, Robertson PKJ, Lawton LA. A comparison of the effectiveness of TiO2 photocatalysis and UVA photolysis for the destruction of three pathogenic micro-organisms. J Photochem Photobiol. 2005;175:51–6. https://doi.org/10.1016/j.jphotochem.2005.04.033.
50. Li Z, Sheng C. Nanosensors for food safety. J Nanosci Nanotechnol. 2014;14:905–12. https://doi.org/10.1166/jnn.2014.8743.
51. Pramanik PKD, Solanki A, Debnath A, Nayyar A, El-Sappagh S, Kwak K. Advancing modern healthcare with nanotechnology, nanobiosensors, and internet of nanothings: taxonomies, applications, architecture, and challenges. IEEE Access. 2020;8:65230–66.

52. Wu D, Hou S, Chen J, Sun Y, Ye X, Liu D, Wang Y. Development and characterization of an enzymatic time-temperature indicator (TTI) based on *Aspergillus niger* lipase. LWT- Food Sci Technol. 2015;60(2):1100–4. https://doi.org/10.1016/j.lwt.2014.10.011.
53. Nile SH, Baskar V, Selvaraj D, Nile A, Xiao J, Kai G. Nanotechnologies in food science: applications, recent trends, and future perspectives. Nano Micro Lett. 2020;12:45. ISSN 2311-6706.
54. Spielman S. Smart packaging: connecting the physical with the digital. 2021. https://www.foodengineeringmag.com/articles/99310-smart-packaging-connecting-the-physical-with-the-digital. Accessed 4 Oct 2021.
55. Cammarelle A, Viscecchia R, Bimbo F. Intention to purchase active and intelligent packaging to reduce household food waste: evidence from Italian consumers. Sustainability. 2021;13:4486. https://doi.org/10.3390/su13084486.
56. Salgado PR, Di Giorgio L, Musso YS, Mauri AN. Recent developments in smart food packaging focused on biobased and biodegradable polymers. Front Sustain Food Syst. 2021;5:630393. https://doi.org/10.3389/fsufs.2021.630393.
57. Eghbal N, Dumas E, Yarmand MS, Mousavi ME, Oulahal N, Gharsallaoui A. Antimicrobial films based on pectin and sodium caseinate for the release of antifungal natamycin. J Food Process Preserv. 2019;43:e13953. https://doi.org/10.1111/jfpp.13953.
58. Ehsani A, Hashemi M, Aminzare M, Raeisi M, Afshari A, Mirza Alizadeh A, et al. Comparative evaluation of edible films impregnated with sage essential oil or lactoperoxidase system: impact on chemical and sensory quality of carp burgers. J Food Process Preserv. 2019;43:e14070. https://doi.org/10.1111/jfpp.14070.
59. Bhargava N, Sharanagat VS, Mor RS, Kumar K. Active and intelligent biodegradable packaging films using food and food waste-derived bioactive compounds: a review. Trends Food Sci Technol. 2020;105:385–401. https://doi.org/10.1016/j.tifs.2020.09.015.
60. Assis RQ, Lopes SM, Costa TMH, Flôres SH, de Rios AO. Active biodegradable cassava starch films incorporated lycopene nanocapsules. Ind Crop Prod. 2017;109:818–27. https://doi.org/10.1016/j.indcrop.2017.09.043.

Chapter 13
Application of Nanoformulations in Improving the Properties of Curcuma (*Curcuma longa* L.).

Sirley González Laime, Claudia Chávez Hernández, Ariel Martínez García, and Juan Abreu Payrol

13.1 Introduction

Turmeric (*Curcuma longa* L.) belongs to *the Zingiberaceae* family, native to India, and currently cultivated in various other parts of the world, including Southeast Asia, China, and Latin America [1, 2]. Turmeric is a common spice used in curry preparation in India and other Asian countries due to its taste and color [3], it has been recognized for its medicinal properties. Aside from its use as a curry spice and dietary supplement, it has historically been used as a natural colorant (food, cosmetics, and textiles), an insect repellent, and an antimicrobial agent. According to Ayurvedic medicine, turmeric has been used for various medical purposes, such as wound healing, respiratory problems, liver and dermatological disorders. It has been used as a component of many traditional medicines in the Eastern world for many centuries [3–6].

Curcumin (CUR), demethoxycurcumin (DMC) and bisdemethoxycurcumin (BMC) are the main bioactive polyphenolic compounds identified in turmeric (Fig. 13.1) [7, 8], they are collectively called curcuminoids (CCM). Commercial

S. G. Laime · C. C. Hernández
ECTI Sierra Maestra, Complejo Barlovento Ave 5ta. y 246, La Habana, Cuba
e-mail: sirley@bionaturasm.cu; cchavez@bionaturasm.cu

A. M. García
ECTI Sierra Maestra, Complejo Barlovento Ave 5ta. y 246, La Habana, Cuba

Instituto de Ciencia y Tecnología de Materiales, Universidad de La Habana, La Habana, Cuba
e-mail: arielmg@imre.uh.cu

J. A. Payrol (✉)
ECTI Sierra Maestra, Complejo Barlovento Ave 5ta. y 246, La Habana, Cuba

Escuela Latinoamericana de Medicina, Carretera Panamericana, La Habana, Cuba
e-mail: japayrol@gmail.com

C. Egbuna et al. (eds.), *Application of Nanotechnology in Food Science, Processing and Packaging*, https://doi.org/10.1007/978-3-030-98820-3_13

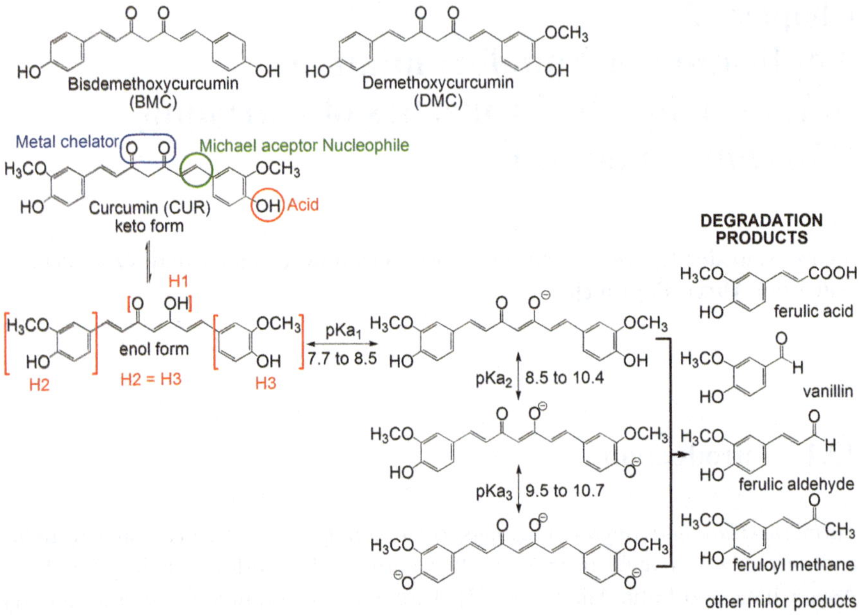

Fig. 13.1 Chemistry of curcumin

curcuminoid powder is a mixture of curcumin, demethoxycurcumin, and bisdemethoxycurcumin in a ratio of approximately 77:17:3 [5]. Curcumin is a yellow polyphenol extracted from the rhizome of the turmeric root (*Curcuma longa*). According to the US Food and Drug Administration (FDA), curcuminoids are generally recognized as safe (GRAS, Generally Recognized as Safe) [3, 8].

Curcumin has attracted much attention in recent decades due to its therapeutic potential, supported by several *in vitro* studies, clinical trials, and *in vivo* [1, 7, 9–11]. Curcumin has shown promise in treating wound healing, arthritis, Alzheimer's and has been shown to exhibit effects as antibacterial, anti-fungal, etc. [12–15].

However, the therapeutic potential of curcumin is limited by its extremely low solubility in aqueous media, its low bioavailability, and its pharmacokinetic profiles, characterized by its instability in body fluids and its rapid metabolism [16–18]. To address these problems, several different formulations (materials/mixtures that combine curcumin with other elements, including polymers, lipids, and nanoparticles in appropriate proportions) have been produced and used in multiple studies [1, 7, 9–11]. In particular, the researchers encapsulate curcumin in nanocarriers, such as liposomes, polymeric micelles, polymeric nanoparticles, mesoporous silica nanoparticles, protein-based nanocarriers, solid lipid nanoparticles, cyclodextrins, nanogels, nanocrystals, etc. [19]

These facts have motivated a significant increase in research on curcumin. The Pubmed database shows the progress of published research in the last 30 years, which shows that research on this topic is an active field at the moment (Fig. 13.2).

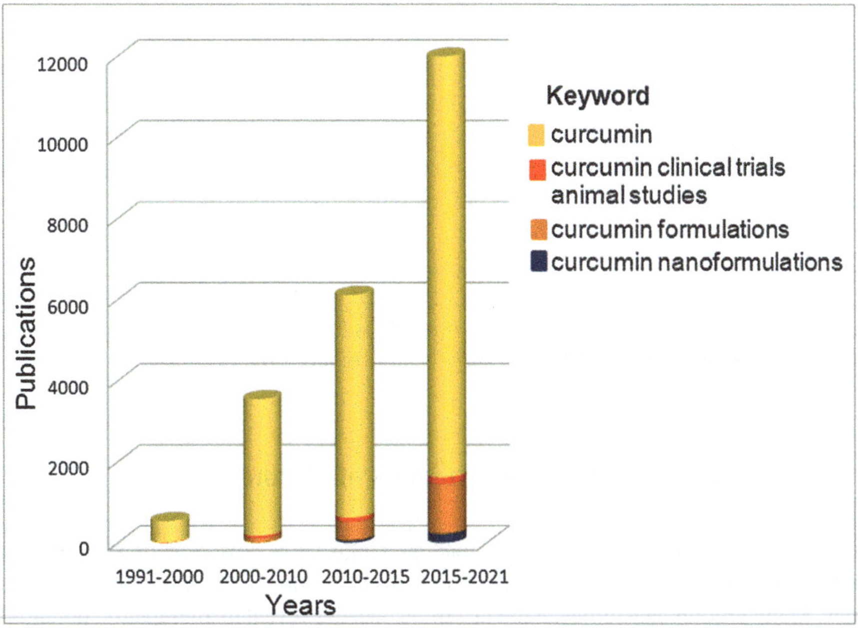

Fig. 13.2 Publications in Pubmed in the period 1990–2021 on curcumin

13.2 Curcuminoids

The chemical composition of turmeric consists of approximately 70% carbohydrates, 13% moisture, 2 to 7% fiber, 6% protein, 6% essential oils (phelandrene, sabinene, cineole, borneol, zingiberene, and sesquiterpenes), 5% fat, 3% minerals (potassium, calcium, phosphorus, iron, and sodium), 3 to 5% curcuminoids and traces of vitamins (B1, B2, C and niacin) [18].

Curcumin is a symmetric molecule, also known as diferuloylmethane. Its name according to IUPAC rules is (1E, 6E)-1,7-bis (4-hydroxy-3-methoxyphenyl)-1,6-heptadiene-3,5-diona. Three chemical units stand out in its structure: two aromatic ring systems containing o-methoxyphenolic groups, connected by a chain of seven carbon atoms consisting of an α, β-unsaturated β-diketone residue (Fig. 13.1).

The dikete group exhibits keto-enolic tautomerism (Fig. 13.1), which can exist in different types of conformers depending on the environment. In the crystalline state, it takes a cis-enol configuration, stabilized by resonance-assisted hydrogen bonds, in a structure of three substituted planar groups conjugated through two double bonds. Curcumin is practically insoluble at room temperature in aqueous solutions at neutral and acidic pH. It has a ground state dipole moment of 10.77 D, it is a hydrophobic molecule with a logP value of ~3.0; It is easily soluble in polar organic solvents such as methanol, ethanol, acetone, dimethylsulfoxide, acetonitrile, chloroform, and ethyl acetate. It is poorly soluble in hydrocarbon solvents such as cyclohexane and hexane. Curcumin is a weak acid, it has a pKa of 8.54 and three acidic protons at neutral pH, one enolic and two phenolics, which provide three pKas, corresponding to three ionic equilibria (Fig. 13.1). At neutral and acidic pH the keto form predominates, under alkaline conditions exclusively the enolic tautomer is present. The solubility of curcumin in an aqueous solution increases at alkaline pH, but it degrades rapidly (greater reactivity) both at that pH and under neutral conditions. At pH > 10, curcumin is completely deprotonated, is red in color, shows a maximum absorbance at 467 nm and a molar extinction coefficient is 53,000 $dm^3\ mol^{-1}\ cm^{-1}$ [4, 20].

With the first pKa, pH range 7.5 to 8.5, curcumin changes from yellow to red. It is not clear whether enolic OH or phenolic OH is the more acidic, although calculations indicate that enolic OH is the more acidic group [20].

13.3 Pharmacological Potentialities of Turmeric

Curcumin is a significantly pleiotropic molecule, it is a polyphenol with anti-inflammatory, hypoglycemic, antioxidant, wound healing, and antimicrobial activities. Many preclinical studies in the last three decades reveal its therapeutic potential against a wide range of human diseases. How a single agent can have so many diverse effects has been an enigma over the years for both basic scientists and physicians [4] Clinical trials conducted so far have indicated the therapeutic potential of curcumin against a wide range of human diseases, including cancer, cardiovascular disease, neurological and autoimmune diseases [9] (Fig. 13.3).

The combination of hydrophobic interactions, including $\pi - \pi$ interactions, extensive hydrogen bonding, metal chelation, and covalent bonds, covering such a large area of the molecule, provides curcumin with many possible mechanisms for interacting with target proteins. Curcumin has been shown to modulate a wide range of signaling molecules at the molecular level. It can cause positive or negative regulation, depending on the target and the cellular context. These goals fall into two categories [21, 22]:

1. those to which it binds directly (inflammatory molecules, cell survival proteins, protein kinases, protein reductases, histone acetyltransferase, histone deacetylase, glyoxalase I, xanthine oxidase, proteasomes, HIV1 integrase, HIV1 protease, sarco/reticulum endoplasmic Ca^{2+}-ATPase, DNA methyltransferase 1, FtsZ, carrier proteins and metal ions);
2. those whose activity is indirectly modulated (transcription factors, enzymes, inflammatory mediators, protein kinases, drug resistance proteins, adhesion molecules, growth factors, receptors, cell cycle regulatory proteins, cell survival proteins, chemokines, and chemokine receptors.

A full review of the topic can be found in numerous published articles. The safety, tolerability, and non-toxicity of high-dose curcumin are well established by human clinical trials. Although it has shown efficacy against numerous diseases, low

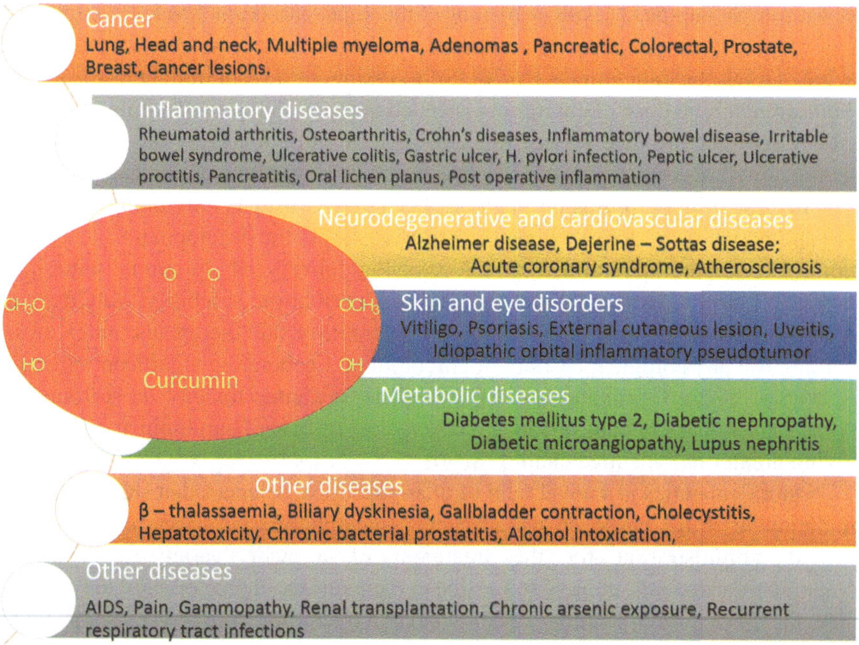

Fig. 13.3 Therapeutic potentialities of curcumin

bioavailability resulting from malabsorption, rapid metabolism, and rapid systemic elimination limit its therapeutic efficacy. That is why many efforts have been made to improve bioavailability by altering these characteristics [8, 21, 22].

It should also be noted that there is still uncertainty about the efficacy of curcumin, due to the poor quality of primary clinical trials (in relation to the number of trials for each review, design, participants and systematic duration of treatment) and the methodology of analysis used [4, 23].

13.4 Curcumin and Nanoformulations

Curcumin has the following limitations for its use in clinical settings: (1) low hydrophilicity and dissolution rates, (2) physicochemical instability, (3) rapid metabolism, (4) low bioactive absorption, (5) pharmacokinetics and bioavailability deficient, and (6) low penetration and targeting efficiency. All of these factors significantly affect the efficient use of curcumin as a therapeutic molecule [24, 25].

Curcumin's poor bioavailability (demonstrated by many preclinical and clinical studies in mice and humans), its extremely low solubility in aqueous buffers (only 0.0004 mg/mL at pH 7.3), its instability in body fluids, and its rapid metabolism have hampered its therapeutic applications. After orally administering 12 g of curcumin to humans, a still safe range, the final curcumin level in serum was approximately only 50 ng/mL, a minimal availability in the systemic circulation to achieve its therapeutic effects [5, 26]. So to maintain the required therapeutic concentration, a high dose of curcumin is needed along with repeated administration; a person will need to take between 12 and 20 g of curcumin every day to reach the therapeutic threshold, this is not feasible from the perspective of clinical trials [27].

The best solution to this problem has been to encapsulate curcumin in nanoformulations (nanocurcumin). Numerous studies address the development and evaluation of nanocurcumin aimed at the specific delivery of curcumin with high permeability, a longer circulation, greater biodistribution, solubility, stability, cellular absorption, internalization efficiency, tolerability, and therapeutic index, overcoming the physiological barriers of curcumin and generating significant effective responses. These formulations can also reduce unwanted toxicity on surrounding normal cells or tissues and enhance the sustained release of curcumin at cell targets giving greater therapeutic benefit [28–30].

Nanotherapeutics is a novel field of pharmaceutical sciences that studies the formulation of drugs based on nanoparticles (nanopharmaceuticals) and their controlled administration for the treatment of a wide variety of diseases. Nanopharmaceuticals are very stable due to their surface characteristics and optimal size, they have a high loading capacity and can overcome limitations such as those mentioned for curcumin [27].

The most used techniques for the synthesis of nanocurcumin include nanoprecipitation, single emulsion, microemulsion, spray drying, emulsion polymerization, solvent evaporation, antisolvent precipitation, ultrasound, coacervation technique, ionic gelation, wet milling, solid dispersion, hydration of thin film and Fessi method, each with its advantages and peculiarities [31, 32]. The synthesis of curcumin nanoparticles by ionic gelation and antisolvent precipitation have been the most efficient, since they show better solubility and stability than with other techniques.

The ionic gelling technique is based on the ability of polymers to crosslink in the presence of counter ions, it is used in the preparation of non-toxic, biocompatible, and biodegradable natural polymers (chitosan/alginate) [33–35]. Antisolvent precipitation, a promising and profitable technique, simple and easy to apply, is widely used to prepare curcumin nanoparticles, its effectiveness depends on the time interval, temperature, and stirring speed, providing better solubility and stability of the nanoparticles [36].

Nanocarriers are generally pharmaceutical materials and products between 1 and 100 nm, although the US Food and Drug Administration (FDA) often considers 1000 nm as an upper limit [37]. They offer a high surface/volume ratio and increase the solubility and dissolution rate of drugs. Furthermore, the small particle size can prolong the maintenance of the drug in the systemic circulation, modify the distribution of the drug and allow the transport and selection of drugs across barriers. Therefore, nanoformulations can improve the solubility and bioavailability of curcumin [38]. Curcumin is encapsulated in nanocarriers such as liposomes, polymeric micelles, polymeric nanoparticles, mesoporous silica nanoparticles, protein-based nanocarriers, solid lipid nanoparticles, cyclodextrins, nanogels, nanocrystals, dendrimers, etc. [5, 19, 24, 36].

Different curcumin nanoformulations have been designed, produced as emulsion, cream, solution, pill, gel, band-aid, etc. [8] for conventional or exploratory administration in *vitro* and *in vivo* studies, using both active and non-invasive, ineffective local, circulatory, affinity, or active targeting applications, some have improved the results of clinical applications against various diseases. Many of these efforts were initially aimed at improving bioavailability, but later attention is being paid to the efficient targeting of curcumin in the diseased area with the help of the mediation of antibodies, aptamers, and peptides [24, 36, 39].

13.5 Nanocarriers Platforms in Curcumin Nanoformulations

There are numerous strategies for creating carrier nanoparticles that improve drug delivery regimens [5, 36]. Impressive pharmacokinetic data are accumulating for drug-loaded nanoparticles ranging from 10 to 200 nm in diameter, noting multiple enhancements in the prolonged systemic circulation, biodistribution, sustained drug release profile, and specific orientation [27].

13.5.1 Physical and Chemical Properties of Nanocurcumin

Not only does the chemical composition of nanocurcumin determine its characteristics, its physical and chemical properties also influence it. Particle size, surface area, surface charge, and hydrophobicity are key physicochemical properties as they influence solubility and bioavailability, pharmacokinetic profile, and active targeting [40].

The smaller the particle size, the greater the efficiency of nanocurcumin, due to its greater intracellular absorption capacity, high systemic bioavailability, and *in vivo* distribution in plasma and tissues. It is generally used between 10–100 nm for medicinal applications and clinical trials, managing to enter organs that free curcumin hardly achieves [27, 41, 42]. Its circulation time, retention time, and mean residence time within the body are also increased.

Surface area is another primary property of nanoparticles. A greater relative surface area increases its degradation rate and aqueous solubility, the drug is exposed to the surface of the particle, which promotes its rapid release, inducing an increase in bioavailability and therefore a better response of the drug to a specific molecular target and increased pharmacological activity [43].

The electrical potential of the nanoparticles is defined by the surface charge, it depends on their chemical composition. Negative and positive zeta potentials prevent nanoparticle aggregation, therefore nanoparticles are very stable in suspension. Nanocurcumin dissolves completely in aqueous media without forming aggregates thanks to its zeta potential, while curcumin, poorly soluble in water, forms aggregates and is susceptible to opsonization. Small positive zeta potential is very good because the nanoparticle can penetrate deep into cell membranes and have a high absorption rate compared to negatively charged particles. A higher positive charge leads to toxicity to cells [24]. The negatively charged nanoparticle does not enter the cell wall, but prevents it from being destroyed under certain conditions and increases the stability of the particle in circulation.

The hydrophobicity of nanomaterials intervenes in the stability and biodistribution of nanocarriers. Curcumin, hydrophobic in nature, tries to reach the cell membrane and bind via hydrogen bonds and hydrophobic interactions to the fatty acid chains of membrane lipids. Many biological processes depend on this property: protein adsorption and denaturation, immune cell activation, cell uptake, and toxicity depend on the hydrophobicity of the surface. The method of preparation and the type of carrier system used to produce nanocurcumin also greatly influence the loading and entrapment efficiencies of the nanopharmaceutical.

13.5.2 Liposomes

Liposomes were the first nanocarriers to be approved by the FDA for clinical application. They consist of spherical vesicles composed of single or multiple phospholipid bilayers that surround aqueous units that closely resemble the structure of the cell membrane. They are classified into single unilamellar vesicles (SUV), large unilamellar vesicles (LUV), and multilamellar vesicles (MLV) according to their structure. Among the methods to prepare them are the thin film dispersion method, ultrasonic dispersion method, injection method, reverse phase evaporation method, etc. [44].

They are ideal delivery systems as targeted drug carriers, both *in vitro* and *in vivo*, since their composition is identical to that of human cell membranes. They have many advantages: high biocompatibility and biodegradability, non-immunogenicity, high stability, low toxicity, better solubility, target to specific cells, controlled distribution, varied loading capacity, flexibility, and easy preparation [45, 46]. Therefore, they are the solid drug carrier system of choice for researchers, although low encapsulation efficiency, easy melting, and leakage have been obstacles to their application.

The size of the vesicle varies between 25 and 1000 nm [47], it is an important factor in deciding the circulation time of the liposomes and impacts, together with the number of bilayers, on the amount of the drug to be encapsulated. The liposome solubilizes curcumin in the phospholipid bilayer, facilitating its distribution in an aqueous medium and increasing its effect [48], the therapeutic index, and reducing side effects.

Many studies have revealed that liposomes could be the best carrier of curcumin, with effective effects on cancer [49–51], in addition, they are effective nanocarriers for combined drug administration due to the ability to encapsulate both hydrophilic and lipophilic drugs [52, 53]. Liposomes accumulate mainly in the liver, spleen, lungs, bone marrow, or other tissues and organs, Chen et al. (2020) cites numerous examples of the effectiveness of these nanocarriers [5].

13.5.3 Polymeric Nanoparticles

Polymeric nanoparticles are solid colloidal particles of approximately 1 to 100 nm in diameter, formed from polymeric materials designed at the atomic or molecular level. They are 1000 times smaller than an average human cell. There is a growing interest in them due to their controllable structure and easy industrial production, their good biodegradability, and excellent biocompatibility, thus they can last longer in the blood circulation [40, 54–57]. They improve pharmacokinetics and drug solubility which makes them suitable for both controlled and targeted drug delivery systems [58]. Various natural and synthetic polymers have been used to prepare curcumin nanoparticles.

13.5.3.1 Poly (Lactic-Co-Glycolic) Acid, PLGA

PLGA, polymerized from lactic acid and glycolic acid, has good biocompatibility and biodegradability [59]. There are multiple reports of PLGA nanoparticles loaded with curcumin. They have been obtained by various methods: solvent evaporation, nanoprecipitation using stabilizers such as PVA, a particle size of up to 76.2 nm has been achieved; zeta potential value close to 0 mV (better stability), curcumin loading capacity 5.75%, and encapsulation efficiency of 91.96%; the solubility of CUR-PLGA-NP was 4.35 mg/mL (that of free curcumin was 6.79 μg/mL, 500 times lower). The half-life was longer and oral bioavailability improved between 5 and 9 times.

PLGA nanoparticles loaded with curcumin have been widely evaluated in cancer therapy due to their good biocompatibility and stability, showing good efficacy, with better bioactivity under *in vitro* conditions and higher bioavailability under *in vivo* conditions than native curcumin [60–62].

13.5.3.2 *N*-Isopropylacrylamide, NIPAAM

It is a temperature-sensitive polymer, with a low critical solution temperature (LCST) at 32 °C [63], hydrophobic at a temperature above the LCST, and hydrophilic at a temperature below the LCST; this reversible sensitivity to ambient temperature stimuli makes it very useful in drug administration, chromatographic analysis, and other fields [64, 65]. Good results have been reported on its utility, it significantly inhibits the growth of brain tumor cells, reduces the clonogenicity of brain tumor cells [66], it could modulate the signaling pathways that affect both survival and the phenotype of neoplastic stem cells, increased cytotoxicity against breast cancer cells [67], etc.

13.5.3.3 Other Formulations

PBCA nanoparticles [(poly (butyl cyanoacrylate)] have been widely used due to their good biocompatibility, biodegradability and non-immunogenicity, prepared mainly by the emulsion polymerization method and the interfacial polycondensation method, loading the drugs into the structure porous nanoparticles. They can enhance curcumin transport, inhibit the growth of hepatocellular carcinoma in mice, and suppress tumor angiogenesis.

Chitosan is a semi-synthetic linear polysaccharide derived from chitin. Due to its properties: good biodegradability, biocompatibility, non-immunogenicity, non-irritating, non-polygenic and non-toxic, it is valuable as an ideal carrier material for drug delivery [68], there are multiple reports of chitosan nanoparticles, including curcumin, for pharmaceutical and biomedical applications [69, 70].

13.5.4 Polymeric Micelles

They are thermodynamically stable colloidal solutions, formed by self-assembly of amphiphilic compounds in water over the critical micellar concentration (CMC), the micelles are formed of a hydrophilic layer [polyethylene glycol (PEG), chitosan, polyvinylpyrrolidone (PVP)] and a hydrophobic nucleus [distearoylphosphoethanolamine (DSPE), poly-lactic acid (PLA), dioleoyphosphatidylethanolamine, poly-(ε-caprolactone) (PCL), poly-(lactic-co-glycolic) (PLGA)] that can load hydrophobic drugs and its size ranges between 20 and 100 nm.

Numerous curcumin-loaded polymeric micelles have been developed [71], with good results in terms of drug loading capacity, zeta potential, encapsulation efficiency, and sustained release behavior [72]. Furthermore, pharmacokinetic parameters, chemotherapeutic efficacy [73], inhibition of cell proliferation, induced apoptosis, and stimulation of antitumor immunity were significantly improved.

Systems were synthesized to increase intracellular uptake and improve therapeutic efficacy [74], the surface of the micelles was modified with several specific ligands to recognize the receptors on the tumor surface [75, 76], significantly improving the anticancer effect both *in vitro* and *in vivo* [77–81].

13.5.5 Mesoporous Silica Nanoparticles

Porous materials, such as silica mesoporous nanoparticles (SMNs), metal-organic systems [82], and iron oxide-gold nanoparticles, have been widely used in drug delivery. SMNs have been studied extensively due to their particular properties: good biocompatibility, large specific surface area, adjustable pore sizes, and volumes, and easily modifiable surface area. They are usually prepared by microwave synthesis, hydrothermal synthesis, sol-gel method, and template synthesis using surfactants, the most widely used due to convenience and milder conditions.

The characteristics of SMNs: high loading efficiency, biocompatibility, pH sensitivity, penetrability, better solubility, *in vitro* release, and cytotoxicity make them a valuable system for curcumin release [83–88]. Modifying the surface with polymers plays an effective role in improving biocompatibility, providing controlled drug release, improving transport and therapeutic effect, and decreasing systemic toxicity [89–91], while increasing efficiency. Curcumin anticancer.

13.5.6 Protein-Based Nanocarriers

Proteins are a natural biomaterial with unique advantages, excellent non-immunogenicity, biocompatibility, biodegradability and low cost. Protein-based nanocarriers have been widely used in the field of biomedicine [92] for the administration of curcumin, many forms have been developed from bovine serum albumin (BSA), human serum albumin (HSA), ovalbumin (OVA), zein, casein and silk fibroin (SF). There are reports of stable long circulation and better solubility, stable controlled release, reduced cytotoxicity in normal cells, and strong anticancer effect [5, 93, 94].

13.5.7 Various Nanoformulations

Other formulations have been used to enhance curcumin activity: nanogels, nanodisks, yeast cells, and metal complexes. A nanogel is a hydrogel synthesized by physical or chemical crosslinking of polymers under controlled conditions, with a size between 10–100 nm. The release of drugs to the cells is achieved with better stability, maintaining their activity and avoiding their immunogenicity [95]. Solubility and cytotoxicity have been improved, killing tumor cells [96], with antimalarial activity in mice, always with values higher than free curcumin [97, 98].

Nanodisks are disc-shaped bilayers, stabilized by apolipoproteins and self-assembled, achieving an increase in the solubility of curcumin. It is an effective strategy to treat mantle cell lymphoma or other cancers, some combinations showed potent activity against Alzheimer's disease *in vivo* [99] and a strong antitumor effect against tumor cells *in vitro* [100].

13.6 Conclusions

Systems for delivering curcumin as nanopharmaceuticals has yielded very encouraging results. Nanocurcumin disperses in water much more than free curcumin, giving greater systemic availability. Nanoencapsulated curcumin favors better circulation and retention of the drug, allowing lower doses to be applied and maintaining the threshold level of curcumin. Food grade formulations, including proprietary formulas, have been developed, changing the image of turmeric from a nutritional spice to clinical medicine. Larger and well-controlled human studies are required to demonstrate the safety and efficacy of this polyphenol and the nanoformulations based on it. That should be the focus of future research on this topic.

References

1. Amalraj A, Pius A, Gopi S, Gopi S. Biological activities of curcuminoids, other biomolecules from turmeric and their derivatives—a review. J Tradit Complement Med. 2017;7(2):205–33.
2. Carolina Alves R, Perosa Fernandes R, Fonseca-Santos B, Damiani Victorelli F, Chorilli M. A critical review of the properties and analytical methods for the determination of curcumin in biological and pharmaceutical matrices. Crit Rev Anal Chem. 2019;49(2):138–49.
3. Hewlings SJ, Kalman DS. Curcumin: a review of its effects on human health. Foods. 2017;6(10):92.
4. Kotha RR, Luthria DL. Curcumin: Biological, pharmaceutical, nutraceutical, and analytical aspects. Molecules. 2019;24(16):2930.
5. Chen Y, Lu Y, Lee RJ, Xiang G. Nano encapsulated curcumin: and its potential for biomedical applications. Int J Nanomedicine. 2020;15:3099–120.
6. Huang Y-S, Hsieh T-J, Lu C-Y. Simple analytical strategy for MALDI-TOF-MS and nanoUPLC–MS/MS: quantitating curcumin in food condiments and dietary supplements and screening of acrylamide-induced ROS protein indicators reduced by curcumin. Food Chem. 2015;174:571–6.
7. Gopi S, Jacob J, Varma K, Jude S, Amalraj A, Arundhathy CA, et al. Comparative oral absorption of curcumin in a natural turmeric matrix with two other curcumin formulations: an open-label parallel-arm study. Phytother Res. 2017;31(12):1883–91.
8. Gupta SC, Patchva S, Aggarwal BB. Therapeutic roles of curcumin: lessons learned from clinical trials. AAPS J. 2013;15(1):195–218.
9. Kunnumakkara AB, Bordoloi D, Padmavathi G, Monisha J, Roy NK, Prasad S, et al. Curcumin, the golden nutraceutical: multitargeting for multiple chronic diseases. Br J Pharmacol. 2017;174(11):1325–48.
10. Hussain Z, Thu HE, Ng S-F, Khan S, Katas H. Nanoencapsulation, an efficient and promising approach to maximize wound healing efficacy of curcumin: a review of new trends and state-of-the-art. Colloids Surf B Biointerfaces. 2017;150:223–41.
11. Mirzaei H, Shakeri A, Rashidi B, Jalili A, Banikazemi Z, Sahebkar A. Phytosomal curcumin: a review of pharmacokinetic, experimental and clinical studies. Biomed Pharmacother. 2017;85:102–12.
12. Wang J, Wang H, Zhu R, Liu Q, Fei J, Wang S. Anti-inflammatory activity of curcumin-loaded solid lipid nanoparticles in IL-1β transgenic mice subjected to the lipopolysaccharide-induced sepsis. Biomaterials. 2015;53:475–83.
13. Peng K-T, Chiang Y-C, Huang T-Y, Chen P-C, Chang P-J, Lee C-W. Curcumin nanoparticles are a promising anti-bacterial and anti-inflammatory agent for treating periprosthetic joint infections. Int J Nanomedicine. 2019;14:469–81.
14. Yavarpour-Bali H, Ghasemi-Kasman M, Pirzadeh M. Curcumin-loaded nanoparticles: a novel therapeutic strategy in treatment of central nervous system disorders. Int J Nanomedicine. 2019;14:4449–60.
15. Chen Y, Chen C, Zhang X, He C, Zhao P, Li M, et al. Platinum complexes of curcumin delivered by dual-responsive polymeric nanoparticles improve chemotherapeutic efficacy based on the enhanced anti-metastasis activity and reduce side effects. Acta Pharm Sin B. 2020;10(6):1106–21.

16. Schneider C, Gordon ON, Edwards RL, Luis PB. Degradation of curcumin: from mechanism to biological implications. J Agric Food Chem. 2015;63(35):7606–14.
17. Ghosh S, Banerjee S, Sil PC. The beneficial role of curcumin on inflammation, diabetes and neurodegenerative disease: a recent update. Food Chem Toxicol. 2015;83:111–24.
18. Nelson KM, Dahlin JL, Bisson J, Graham J, Pauli GF, Walters MA. The essential medicinal chemistry of curcumin. J Med Chem. 2017;60(5):1620–37.
19. Wong KE, Ngai SC, Chan K-G, Lee L-H, Goh B-H, Chuah L-H. Curcumin nanoformulations for colorectal cancer: a review. Front Pharmacol. 2019;10:152.
20. Priyadarsini KI. The chemistry of curcumin: from extraction to therapeutic agent. Molecules. 2014;19(12):20091–112.
21. Gupta SC, Prasad S, Kim JH, Patchva S, Webb LJ, Priyadarsini IK, et al. Multitargeting by curcumin as revealed by molecular interaction studies. Nat Prod Rep. 2011;28(12):1937–55.
22. Gupta SC, Patchva S, Koh W, Aggarwal BB. Discovery of curcumin, a component of golden spice, and its miraculous biological activities. Clin Exp Pharmacol Physiol. 2012;39(3):283–99.
23. Pagano E, Romano B, Izzo AA, Borrelli F. The clinical efficacy of curcumin-containing nutraceuticals: an overview of systematic reviews. Pharmacol Res. 2018;134:79–91.
24. Yallapu MM, Nagesh PK, Jaggi M, Chauhan SC. Therapeutic applications of curcumin nanoformulations. AAPS J. 2015;17(6):1341–56.
25. Burgos-Morón E, Calderón-Montaño JM, Salvador J, Robles A, López-Lázaro M. The dark side of curcumin. Int J Cancer. 2010;126(7):1771–5.
26. Siviero A, Gallo E, Maggini V, Gori L, Mugelli A, Firenzuoli F, et al. Curcumin, a golden spice with a low bioavailability. J Herb Med. 2015;5(2):57–70.
27. Flora G, Gupta D, Tiwari A. Nanocurcumin: a promising therapeutic advancement over native curcumin. Crit Rev Ther Drug Carrier Syst. 2013;30(4):331–68.
28. Bhatia A, Flamer D, Shah PS, Cohen SP. Transforaminal epidural steroid injections for treating lumbosacral radicular pain from herniated intervertebral discs: a systematic review and meta-analysis. Anesth Analg. 2016;122(3):857–70.
29. Fonseca-Santos B, Dos Santos AM, Rodero CF, Gremião MPD, Chorilli M. Design, characterization, and biological evaluation of curcumin-loaded surfactant-based systems for topical drug delivery. Int J Nanomedicine. 2016;11:4553–62.
30. Gera M, Sharma N, Ghosh M, Huynh DL, Lee SJ, Min T, et al. Nanoformulations of curcumin: an emerging paradigm for improved remedial application. Oncotarget. 2017;8(39):66680–98.
31. Rai M, Pandit R, Gaikwad S, Yadav A, Gade A. Potential applications of curcumin and curcumin nanoparticles: from traditional therapeutics to modern nanomedicine. Nanotechnol Rev. 2015;4(2):161–72.
32. Dhivya S, Rajalakshmi DA. A review on the preparation methods of curcumin nanoparticles. Pharma Tutor. 2018;6:6–10.
33. Giri TK. 20 - alginate containing Nanoarchitectonics for improved cancer therapy. In: Holban AM, Grumezescu AM, editors. Nanoarchitectonics for smart delivery and drug targeting. Burlington, MA: William Andrew Publishing; 2016. p. 565–88.
34. Bhunchu S, Muangnoi C, Rojsitthisak P, Rojsitthisak P. Curcumin diethyl disuccinate encapsulated in chitosan/alginate nanoparticles for improvement of its in vitro cytotoxicity against MDA-MB-231 human breast cancer cells. Pharmazie. 2016;71(12):691–700.
35. Bhunchu S, Rojsitthisak P, Rojsitthisak P. Effects of preparation parameters on the characteristics of chitosan–alginate nanoparticles containing curcumin diethyl disuccinate. J Drug Deliv Sci Technol. 2015;28:64–72.
36. Karthikeyan A, Senthil N, Min T. Nanocurcumin: a promising candidate for therapeutic applications. Front Pharmacol. 2020;11:487.
37. Parthasarathi S, Muthukumar SP, Anandharamakrishnan C. The influence of droplet size on the stability, in vivo digestion, and oral bioavailability of vitamin E emulsions. Food Funct. 2016;7(5):2294–302.

38. Peng S, Li Z, Zou L, Liu W, Liu C, McClements DJ. Improving curcumin solubility and bioavailability by encapsulation in saponin-coated curcumin nanoparticles prepared using a simple pH-driven loading method. Food Funct. 2018;9(3):1829–39.
39. Yallapu M, Jaggi MC, Chauhan S. Curcumin nanomedicine: a road to cancer therapeutics. Curr Pharm Des. 2013;19(11):1994–2010.
40. Biswas AK, Islam MR, Choudhury ZS, Mostafa A, Kadir MF. Nanotechnology based approaches in cancer therapeutics. Adv Nat Sci Nanosci Nanotechnol. 2014;5(4):043001.
41. Zou P, Zhang J, Xia Y, Kanchana K, Guo G, Chen W, et al. ROS generation mediates the anti-cancer effects of WZ35 via activating JNK and ER stress apoptotic pathways in gastric cancer. Oncotarget. 2015;6(8):5860–76.
42. Dende C, Meena J, Nagarajan P, Nagaraj VA, Panda AK, Padmanaban G. Nanocurcumin is superior to native curcumin in preventing degenerative changes in experimental cerebral malaria. Sci Rep. 2017;7(1):10062.
43. Mohanty C, Sahoo SK. The in vitro stability and in vivo pharmacokinetics of curcumin prepared as an aqueous nanoparticulate formulation. Biomaterials. 2010;31(25):6597–611.
44. Feng T, Wei Y, Lee RJ, Zhao L. Liposomal curcumin and its application in cancer. Int J Nanomedicine. 2017;12:6027–44.
45. Moballegh Nasery M, Abadi B, Poormoghadam D, Zarrabi A, Keyhanvar P, Khanbabaei H, et al. Curcumin delivery mediated by bio-based nanoparticles: a review. Molecules. 2020;25(3):689.
46. He C, Zhang X, Yan R, Zhao P, Chen Y, Li M, et al. Enhancement of cisplatin efficacy by lipid–CaO2 nanocarrier-mediated comprehensive modulation of the tumor microenvironment. Biomater Sci. 2019;7(10):4260–72.
47. Zhang W, Ma W, Zhang J, Song X, Sun W, Fan Y. The immunoregulatory activities of astragalus polysaccharide liposome on macrophages and dendritic cells. Int J Biol Macromol. 2017;105:852–61.
48. Chang M, Wu M, Li H. Antitumor activities of novel glycyrrhetinic acid-modified curcumin-loaded cationic liposomes in vitro and in H22 tumor-bearing mice. Drug Deliv. 2018;25(1):1984–95.
49. Tefas LR, Sylvester B, Tomuta I, Sesarman A, Licarete E, Banciu M, et al. Development of antiproliferative long-circulating liposomes co-encapsulating doxorubicin and curcumin, through the use of a quality-by-design approach. Drug Des Devel Ther. 2017;11:1605–21.
50. Huang M, Liang C, Tan C, Huang S, Ying R, Wang Y, et al. Liposome co-encapsulation as a strategy for the delivery of curcumin and resveratrol. Food Funct. 2019;10(10):6447–58.
51. Vetha BSS, Kim E-M, Oh P-S, Kim SH, Lim ST, Sohn M-H, et al. Curcumin encapsulated micellar Nanoplatform for blue light emitting diode induced apoptosis as a new class of cancer therapy. Macromol Res. 2019;27(12):1179–84.
52. Ruttala HB, Ko YT. Liposomal co-delivery of curcumin and albumin/paclitaxel nanoparticle for enhanced synergistic antitumor efficacy. Colloids Surf B Biointerfaces. 2015;128:419–26.
53. Cheng Y, Zhao P, Wu S, Yang T, Chen Y, Zhang X, et al. Cisplatin and curcumin co-loaded nano-liposomes for the treatment of hepatocellular carcinoma. Int J Pharm. 2018;545(1):261–73.
54. Ferrari R, Sponchioni M, Morbidelli M, Moscatelli D. Polymer nanoparticles for the intravenous delivery of anticancer drugs: the checkpoints on the road from the synthesis to clinical translation. Nanoscale. 2018;10(48):22701–19.
55. Wu W, Chen M, Luo T, Fan Y, Zhang J, Zhang Y, et al. ROS and GSH-responsive S-nitrosoglutathione functionalized polymeric nanoparticles to overcome multidrug resistance in cancer. Acta Biomater. 2020;103:259–71.
56. Kamaly N, Yameen B, Wu J, Farokhzad OC. Degradable controlled-release polymers and polymeric nanoparticles: mechanisms of controlling drug release. Chem Rev. 2016;116(4):2602–63.
57. El-Say KM, El-Sawy HS. Polymeric nanoparticles: promising platform for drug delivery. Int J Pharm. 2017;528(1):675–91.

58. Rudramurthy GR, Swamy MK, Sinniah UR, Ghasemzadeh A. Nanoparticles: alternatives against drug-resistant pathogenic microbes. Molecules. 2016;21(7):836.
59. Pan Q, Li W, Yuan X, Rakhmanov Y, Wang P, Lu R, et al. Chondrogenic effect of cell-based scaffold of self-assembling peptides/PLGA-PLL loading the hTGFβ3 plasmid DNA. J Mater Sci Mater Med. 2015;27(1):19.
60. Bowerman CJ, Byrne JD, Chu KS, Schorzman AN, Keeler AW, Sherwood CA, et al. Docetaxel-loaded PLGA nanoparticles improve efficacy in Taxane-resistant triple-negative breast cancer. Nano Lett. 2017;17(1):242–8.
61. Kennedy PJ, Sousa F, Ferreira D, Pereira C, Nestor M, Oliveira C, et al. Fab-conjugated PLGA nanoparticles effectively target cancer cells expressing human CD44v6. Acta Biomater. 2018;81:208–18.
62. Reardon PJ, Parhizkar M, Harker AH, Browning RJ, Vassileva V, Stride E, et al. Electrohydrodynamic fabrication of core-shell PLGA nanoparticles with controlled release of cisplatin for enhanced cancer treatment. Int J Nanomedicine. 2017;12:3913–26.
63. Wang Y, Yang N, Wang D, He Y, Chen L, Zhao Y. Poly (MAH-β-cyclodextrin-co-NIPAAm) hydrogels with drug hosting and thermo/pH-sensitive for controlled drug release. Polym Degrad Stab. 2018;147:123–31.
64. Wu D-Q, Zhu J, Han H, Zhang J-Z, Wu F-F, Qin X-H, et al. Synthesis and characterization of arginine-NIPAAm hybrid hydrogel as wound dressing: in vitro and in vivo study. Acta Biomater. 2018;65:305–16.
65. Aguilar LE, GhavamiNejad A, Park CH, Kim CS. On-demand drug release and hyperthermia therapy applications of thermoresponsive poly-(NIPAAm-co-HMAAm)/polyurethane core-shell nanofiber mat on non-vascular nitinol stents. Nanomedicine. 2017;13(2):527–38.
66. Lim KJ, Bisht S, Bar EE, Maitra A, Eberhart CG. A polymeric nanoparticle formulation of curcumin inhibits growth, clonogenicity and stem-like fraction in malignant brain tumors. Cancer Biol Ther. 2011;11(5):464–73.
67. Zeighamian V, Darabi M, Akbarzadeh A, Rahmati-Yamchi M, Zarghami N, Badrzadeh F, et al. PNIPAAm-MAA nanoparticles as delivery vehicles for curcumin against MCF-7 breast cancer cells. Artif Cells Nanomed Biotechnol. 2016;44(2):735–42.
68. Zhang X, He C, Yan R, Chen Y, Zhao P, Li M, et al. HIF-1 dependent reversal of cisplatin resistance via anti-oxidative nano selenium for effective cancer therapy. Chem Eng J. 2020;380:122540.
69. He Z, Santos JL, Tian H, Huang H, Hu Y, Liu L, et al. Scalable fabrication of size-controlled chitosan nanoparticles for oral delivery of insulin. Biomaterials. 2017;130:28–41.
70. Esfandiarpour-Boroujeni S, Bagheri-Khoulenjani S, Mirzadeh H, Amanpour S. Fabrication and study of curcumin loaded nanoparticles based on folate-chitosan for breast cancer therapy application. Carbohydr Polym. 2017;168:14–21.
71. Chen S, Yang K, Tuguntaev RG, Mozhi A, Zhang J, Wang PC, et al. Targeting tumor microenvironment with PEG-based amphiphilic nanoparticles to overcome chemoresistance. Nanomedicine. 2016;12(2):269–86.
72. Kheiri Manjili H, Ghasemi P, Malvandi H, Mousavi MS, Attari E, Danafar H. Pharmacokinetics and in vivo delivery of curcumin by copolymeric mPEG-PCL micelles. Eur J Pharm Biopharm. 2017;116:17–30.
73. Manjili HK, Sharafi A, Danafar H, Hosseini M, Ramazani A, Ghasemi MH. Poly(caprolactone)–poly(ethylene glycol)–poly(caprolactone) (PCL–PEG–PCL) nanoparticles: a valuable and efficient system for in vitro and in vivo delivery of curcumin. RSC Adv. 2016;6(17):14403–15.
74. Phan QT, Le MH, Le TTH, Tran THH, Xuan PN, Ha PT. Characteristics and cytotoxicity of folate-modified curcumin-loaded PLA-PEG micellar nano systems with various PLA:PEG ratios. Int J Pharm. 2016;507(1):32–40.
75. Guan J, Zhou Z-Q, Chen M-H, Li H-Y, Tong D-N, Yang J, et al. Folate-conjugated and pH-responsive polymeric micelles for target-cell-specific anticancer drug delivery. Acta Biomater. 2017;60:244–55.

76. Zhong Y, Goltsche K, Cheng L, Xie F, Meng F, Deng C, et al. Hyaluronic acid-shelled acid-activatable paclitaxel prodrug micelles effectively target and treat CD44-overexpressing human breast tumor xenografts in vivo. Biomaterials. 2016;84:250–61.
77. Jin H, Pi J, Zhao Y, Jiang J, Li T, Zeng X, et al. EGFR-targeting PLGA-PEG nanoparticles as a curcumin delivery system for breast cancer therapy. Nanoscale. 2017;9(42):16365–74.
78. Yang C, Chen H, Zhao J, Pang X, Xi Y, Zhai G. Development of a folate-modified curcumin loaded micelle delivery system for cancer targeting. Colloids Surf B Biointerfaces. 2014;121:206–13.
79. Zhang J, Li J, Shi Z, Yang Y, Xie X, Lee SM, et al. pH-sensitive polymeric nanoparticles for co-delivery of doxorubicin and curcumin to treat cancer via enhanced pro-apoptotic and anti-angiogenic activities. Acta Biomater. 2017;58:349–64.
80. Zhao S, Ma L, Cao C, Yu Q, Chen L, Liu J. Curcumin-loaded redox response of self-assembled micelles for enhanced antitumor and anti-inflammation efficacy. Int J Nanomedicine. 2017;12:2489–504.
81. Zhao D, Zhang H, Yang S, He W, Luan Y. Redox-sensitive mPEG-SS-PTX/TPGS mixed micelles: an efficient drug delivery system for overcoming multidrug resistance. Int J Pharm. 2016;515(1):281–92.
82. Molavi H, Zamani M, Aghajanzadeh M, Kheiri Manjili H, Danafar H, Shojaei A. Evaluation of UiO-66 metal organic framework as an effective sorbent for Curcumin's overdose. Appl Organomet Chem. 2018;32(4):e4221.
83. Bollu VS, Barui AK, Mondal SK, Prashar S, Fajardo M, Briones D, et al. Curcumin-loaded silica-based mesoporous materials: synthesis, characterization and cytotoxic properties against cancer cells. Mater Sci Eng C. 2016;63:393–410.
84. Kotcherlakota R, Barui AK, Prashar S, Fajardo M, Briones D, Rodríguez-Diéguez A, et al. Curcumin loaded mesoporous silica: an effective drug delivery system for cancer treatment. Biomater Sci. 2016;4(3):448–59.
85. Chen C, Sun W, Wang X, Wang Y, Wang P. Rational design of curcumin loaded multifunctional mesoporous silica nanoparticles to enhance the cytotoxicity for targeted and controlled drug release. Mater Sci Eng C. 2018;85:88–96.
86. Sun X, Wang N, Yang L-Y, Ouyang X-K, Huang F. Folic acid and PEI modified mesoporous silica for targeted delivery of curcumin. Pharmaceutics. 2019;11(9):430.
87. Li N, Wang Z, Zhang Y, Zhang K, Xie J, Liu Y, et al. Curcumin-loaded redox-responsive mesoporous silica nanoparticles for targeted breast cancer therapy. Artif Cells Nanomed Biotechnol. 2018;46(sup2):921–35.
88. Elbialy NS, Aboushoushah SF, Sofi BF, Noorwali A. Multifunctional curcumin-loaded mesoporous silica nanoparticles for cancer chemoprevention and therapy. Microporous Mesoporous Mater. 2020;291:109540.
89. Kong Z-L, Kuo H-P, Johnson A, Wu L-C, Chang KLB. Curcumin-loaded mesoporous silica nanoparticles markedly enhanced cytotoxicity in hepatocellular carcinoma cells. Int J Mol Sci. 2019;20(12):2918.
90. Datz S, Engelke H, Cv S, Nguyen L, Bein T. Lipid bilayer-coated curcumin-based mesoporous organosilica nanoparticles for cellular delivery. Microporous Mesoporous Mater. 2016;225:371–7.
91. Ahmadi Nasab N, Hassani Kumleh H, Beygzadeh M, Teimourian S, Kazemzad M. Delivery of curcumin by a pH-responsive chitosan mesoporous silica nanoparticles for cancer treatment. Artif Cells Nanomed Biotechnol. 2018;46(1):75–81.
92. Karimi M, Bahrami S, Ravari SB, Zangabad PS, Mirshekari H, Bozorgomid M, et al. Albumin nanostructures as advanced drug delivery systems. Expert Opin Drug Deliv. 2016;13(11):1609–23.
93. Xie M, Fan D, Li Y, He X, Chen X, Chen Y, et al. Supercritical carbon dioxide-developed silk fibroin nanoplatform for smart colon cancer therapy. Int J Nanomedicine. 2017;12:7751–61.

94. Thadakapally R, Aafreen A, Aukunuru J, Habibuddin M, Jogala S. Preparation and characterization of PEG-albumin-curcumin nanoparticles intended to treat breast cancer. Indian J Pharm Sci. 2016;78(1):65–72.
95. Wang S, Ha Y, Huang X, Chin B, Sim W, Chen R. A new strategy for intestinal drug delivery via pH-responsive and membrane-active Nanogels. ACS Appl Mater Interfaces. 2018;10(43):36622–7.
96. Reeves A, Vinogradov SV, Morrissey P, Chernin M, Ahmed MM. Curcumin-encapsulating Nanogels as an effective anticancer formulation for intracellular uptake. Mol Cell Pharmacol. 2015;7(3):25–40.
97. Amanlou N, Parsa M, Rostamizadeh K, Sadighian S, Moghaddam F. Enhanced cytotoxic activity of curcumin on cancer cell lines by incorporating into gold/chitosan nanogels. Mater Chem Phys. 2019;226:151–7.
98. Priya P, Mohan Raj R, Vasanthakumar V, Raj V. Curcumin-loaded layer-by-layer folic acid and casein coated carboxymethyl cellulose/casein nanogels for treatment of skin cancer. Arab J Chem. 2020;13(1):694–708.
99. Bicer N, Yildiz E, Yegani AA, Aksu F. Synthesis of curcumin complexes with iron(iii) and manganese(ii), and effects of curcumin–iron(iii) on Alzheimer's disease. New J Chem. 2018;42(10):8098–104.
100. Li Y, Gu Z, Zhang C, Li S, Zhang L, Zhou G, et al. Synthesis, characterization and ROS-mediated antitumor effects of palladium(II) complexes of curcuminoids. Eur J Med Chem. 2018;144:662–71.

[illegible]

Chapter 14
Bioavailability of Nano Nutrients, Potential Safety Issues, and Regulations

Jayashree V. Hanchinalmath, R. Surabhi, Nevaj Jain, Megha Banerjee, P. Lochana, Alekhya Batchu, Kirankumar Shivasharanappa, M. S. Sheeja, and Snehva Roy

Abbreviations

Cu	Copper
DNA	Deoxyribonucleic acid
NMs	Nanomaterials
pH	Potential of hydrogen
PUFA	Poly unsaturated fatty acid
USA	United States of America

14.1 Introduction

The majority of the nutrients that feed the world's population come from agricultural systems, and nutrient sufficiency is the foundation of excellent health, productive life, and longevity for everyone. In 2020, the global population reached 7.7 billion people, with a projected increase to 9.7 billion by 2050. As a result, total demand for agricultural products is predicted to rise by 50% to 80%. Meanwhile, traditional agricultural technology's efficiency has been steadily declining since 2000, with flattening yield curves, decreased crop resiliency, increased environmental harm, and rising global food demand necessitating the development of new agricultural technologies [1].

J. V. Hanchinalmath (✉) · R. Surabhi · N. Jain · M. Banerjee · P. Lochana · A. Batchu
Department of Life Sciences, School of Sciences, Jain (Deemed To Be University), Bangalore, Karnataka, India

K. Shivasharanappa · M. S. Sheeja · S. Roy
Department of Life Sciences, Garden City University, Bangalore, Karnataka, India

C. Egbuna et al. (eds.), *Application of Nanotechnology in Food Science, Processing and Packaging*, https://doi.org/10.1007/978-3-030-98820-3_14

More engineered nanomaterials (NMs), such as nano-fertilizers, nano-pesticides, and nano-sensors, have been successfully applied in agricultural production and management for the enhancement of soil nutrient bioavailability and food quality for the past 10 years due to their high ratio of surface-volume area, unique physico-chemical, mechanical, and electronic properties in comparison to their bulk counterparts. Nanotechnology-based agricultural inputs will reach $75.8 billion by 2020, contributing to $3.4 trillion in global economic growth. According to Kah et al. [2] the efficacy of nano-agrochemicals is 30% higher than that of conventional goods, and nanotechnology has the potential to change the agricultural output. Despite a large number of articles on the interaction of NMs with food plants, our understanding of how NMs alter the nutritional value of edible food plant tissue is still restricted [2].

The potential influence of NMs translocating into edible food tissues on food quality and safety has been a major societal concern. NMs interact with soil components (such as minerals and organic matter) before entering plants, and hence their bioavailability may be influenced by soil physicochemical parameters, particularly the pH and ionic strength of soil pore water. In the rhizosphere, large numbers of microbes (e.g., bacteria and fungi) and soil fauna (e.g., earthworms) congregate around plant roots, ultimately shaping a unique biotic and abiotic environment through rhizosphere processes, such as root secretion, microbial growth, and earthworm activity in soils. These rhizosphere mechanisms could affect NM geochemical processes as aggregation, anti-aggregation, redox reaction, and transformation, and hence influence NM translocation into plants and NM-mediated physiological and biochemical responses in edible crop tissue.

Up to the present time, there is a dearth of understanding of the rhizosphere processes that regulate geochemical activity and NM bioavailability, thus boosting food nutritional quality. By doing so, we hope to examine the main geochemical behaviors caused by rhizosphere processes and assess the enhancing impacts of NMs on nutritional quality in food plants by meta-analysis. Nanoencapsulation-based technologies are a new and innovative topic of research in the food and pharmaceutical industries, with advantages such as enhanced bioavailability, shelf stability, and controlled release of active substances as shown in Fig. 14.1.

Any novel technological approaches have the potential to cause ethical dilemmas and raise ethical concerns among stakeholders, albeit whether or not this happens depends on the technology's qualities. Food safety, dangers and advantages (to human health, the environment, and socioeconomic repercussions), and consumer choice are all important concerns to consider when discussing ethics, food, and new technology. It's important to consider how equitably the advantages (and hazards, if any) are spread (for example, among food producers in various countries or between farmers and consumers). These can be viewed from an ethical standpoint in terms of how they affect basic ethical concepts such as non-malfeasance, beneficence, autonomy, and fairness. The purpose of this review is to examine potential ethical and regulatory challenges linked with existing and future nanotechnology-related food and agriculture applications [3].

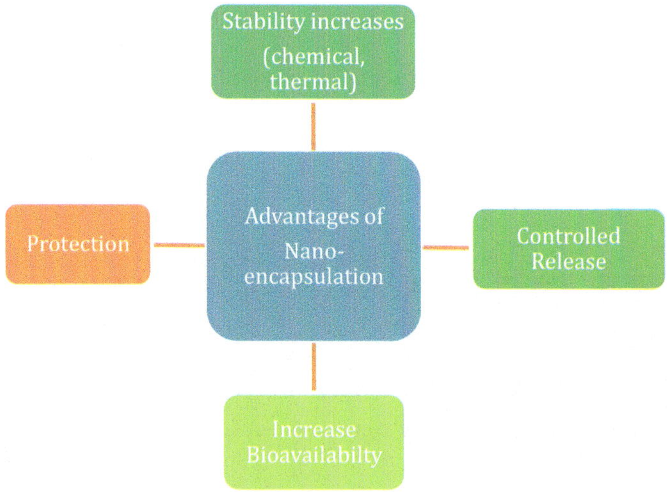

Fig. 14.1 Applications of nano-encapsulation

14.2 Classification of Nano Nutrients

Nutraceuticals are a part of Nano nutrients which are categorized based on foods available in the market [4, 5]. Figure 14.2 shows the classification of nutraceuticals.

- Traditional nutraceuticals
- Non-traditional nutraceuticals.

14.2.1 Traditional Nutraceuticals

Traditional nutraceuticals are merely natural with no changes to the food. Food contains several natural components that deliver benefits beyond basic nutrition, such as lycopene in tomatoes, omega-3 fatty acids in salmon, or saponins in soy.

14.2.1.1 Nutrients

Chemical constituents such as vitamins, minerals, amino acids, and fatty acids are examples of substances with well-established nutritional activities. Vitamins are found in most vegetables, wholegrain cereals, dairy products, fruits, and animal products such as meat and chicken, and are beneficial in the treatment of heart disease, stroke, cataracts, osteoporosis, diabetes, and cancer is illustrated in Fig. 14.3. Minerals found in the plant, animal, and dairy products aid in the treatment of

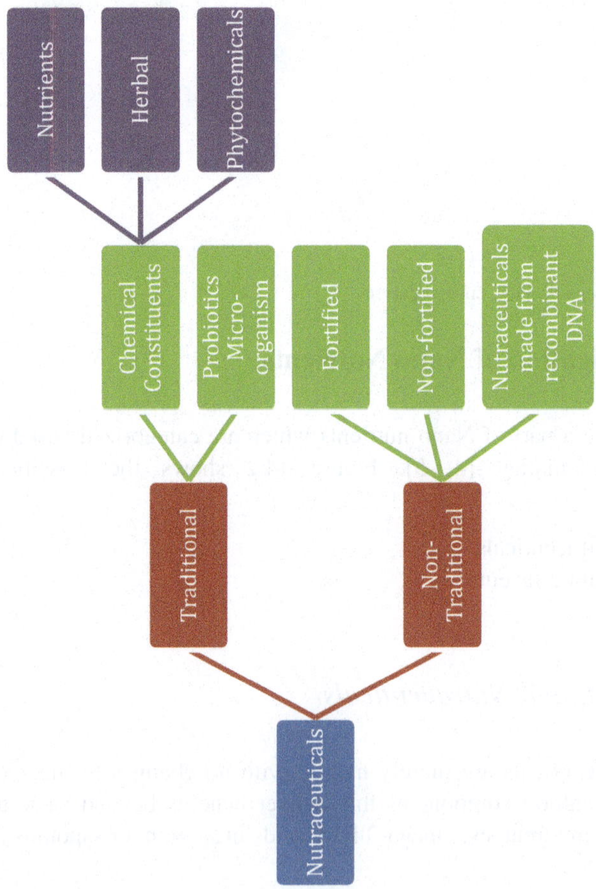

Fig. 14.2 Classification of nutraceuticals

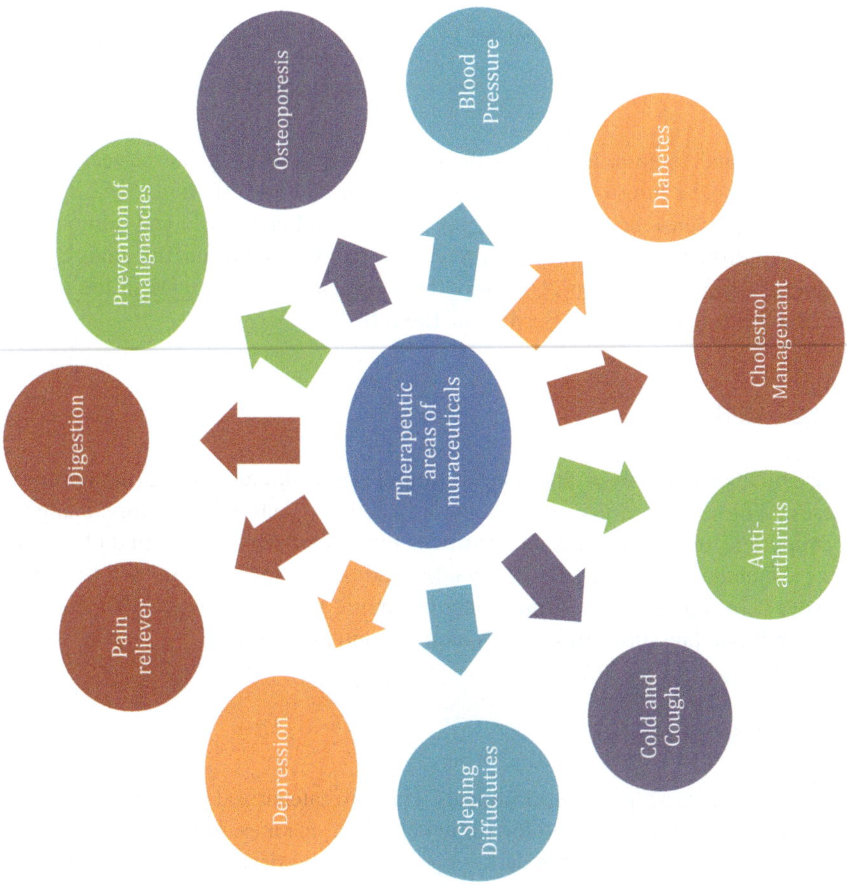

Fig. 14.3 Therapeutic areas of nutraceuticals

Table 14.1 Important nutrients, their sources, and their role

Nutrients	Source	Used in treatment of
Vitamins	Most vegetables, wholegrain cereals, dairy products, fruits, and animal products like meat and chicken	Heart disease, stroke, cataracts, osteoporosis, cancer, and diabetes [4]
Minerals	Plant, animal, and dairy products	Osteoporosis and anemia, development of strong bones, teeth, and muscles, improvement of nerve impulses, and heart rhythm [5]
Omega-3 polyunsaturated fatty acids (PUFAs)	Flaxseed and salmon	Effective controllers of inflammatory processes, maintain brain function and lower cholesterol deposits [5]

osteoporosis and anemia, as well as the development of strong bones, teeth, and muscles, also in the improvement of nerve impulses and heart rhythm [5].

Omega-3 polyunsaturated fatty acids (PUFAs) found in flaxseed and salmon are effective controllers of inflammatory processes, maintain brain function, and lower cholesterol deposits. The source of important nutrients and their role in the treatment of various diseases are described in Table 14.1.

14.2.1.2 Herbals

Nutraceuticals have a lot of promise for improving health and preventing chronic diseases. Willow bark (*Salix nigra*) contains salicin, an anti-inflammatory, analgesic, antipyretic, astringent, and antiarthritic active component. Peppermint (*Mentha piperita*) includes the key ingredient menthol and is used to treat colds and flu [18]. Lavender (*Lavandula angustifolia*) contains tannin, which can assist with depression, high blood pressure, stress, colds, coughs, and asthma [6].

14.2.1.3 Phytochemicals

Phytochemicals are a type of nutraceuticals. They are categorized based on the chemical names assigned to them and their phytochemical features. Non-flavonoid polyphenolics, which can be found in dark grapes, raisins, berries, peanuts, and turmeric roots, are powerful anti-inflammatory, antioxidant, and anti-clotting agents that help lower cholesterol. The antioxidant activity of phenolic acids, which can be found in blueberries, tomatoes, and bell peppers, reduces the mutagenicity of polycyclic aromatic hydrocarbons. Nutraceutical Enzymes, Part III Enzymes are vital components of life; without them, our bodies would stop working. Those suffering from medical illnesses such as hypoglycemia, blood sugar imbalances, digestive issues, and obesity might reduce their symptoms by adding enzyme supplements to their diet. These enzymes come from a variety of sources, including bacteria, plants, and animals [7].

14.2.2 Non-Traditional Nutraceuticals

Biotechnology-assisted artificial foods are known as non-traditional nutraceuticals. Bioactive components in food samples have been developed to create products for human wellness. They're divided into two categories:

(a) Fortified nutraceuticals and
(b) Non-fortified nutraceuticals.
(c) Nutraceuticals made from recombinant DNA.

Nutraceuticals with added nutrients consist of food that has been fortified with increased nutrients and/or additives as a result of agricultural breeding. For example, calcium-fortified orange juice, vitamin- and mineral-fortified cereals, and folic acid-fortified flour. Some examples include cholecalciferol-fortified milk, which is used to treat vitamin D insufficiency. In children 24, prebiotic and probiotic fortified milk with *Bifidobacterium lactis* HN019 was utilized to treat diarrhea, respiratory infections, and severe illnesses [8]. Kumar 25 identified banana enhanced with soybean ferritin gene in iron deficit. Recombinant nutraceuticals Biotechnology is used to generate energy-giving foods such as bread, alcohol, fermented starch, yogurt, cheese, vinegar, and others. Biotechnology allows for the manufacture of probiotics and the extraction of bioactive components using enzyme/fermentation methods, as well as genetic engineering.

14.3 Commercial Nutraceuticals

Finding a new chemical is difficult, expensive, and risky than ever before. Many pharmaceutical companies are now attempting to create nutraceuticals due to the enormous and rapidly growing market. Anti-arthritic, cold and cough, sleeping difficulties, digestion, and the prevention of some malignancies, osteoporosis, blood pressure, cholesterol management, pain relievers, depression, and diabetes are just a few of the therapeutic areas covered by nutraceuticals. One of the most promising advances in human nutrition and disease prevention research in the last three decades is the recognition of health benefits from eating omega-3 rich seafood.

14.4 Importance of Measuring Bioavailability

The importance of bioavailability in determining dietary requirements and using those requirements in food labeling is critical. The amount of a nutrient in a diet that the body can use varies by age and physiologic situation, such as pregnancy. In addition to infant foods, nutritional supplements, and enteral formulas (for patients who can't digest solid foods), nutrient availability is critical in the testing and marketing of these products [9].

Knowing bioavailability is particularly important since consumers' eating patterns change frequently for a variety of reasons, including health, economics, and personal preference, and knowledge of nutrient bioavailability may affect their decisions. Furthermore, as the number of food products available to customers grows (particularly with the introduction of new and unusual convenience meals), the food processing sector is increasingly interested in the impact of food processing and preparation on nutrient bioavailability [10].

Food options are also expanding as a result of demographic shifts henceforth determining the nutrient availability and adequacy of ethnic foods has become more important. With as many as 50% of Americans using vitamin and mineral supplements, there is a need for precise data on the nutrients available in these supplements. Finally, nutrient-drug interactions can influence nutrient absorption and consequently nutritional status in those taking specific medications.

14.5 Bioavailability Analysis and Measurements

The amount of a nutrient in a diet that the body can use to execute specified physiological processes is referred to as bioavailability.

The bioavailability of a nutrient is influenced by several factors.

1. Digestion
2. Absorption
3. Distribution of the nutrient by circulating blood and
4. Entry of the nutrient into the specific bodily tissues and fluids where it may be physiologically useful

Bioavailability can be measured to some extent by assessing the quantity of the nutrient in multiple body tissues and fluids, or the nutrient-dependent growth or enzyme activity. However, because a nutrient is rarely retained in a single human tissue, measuring nutrient levels in individual tissues may not correctly reflect real bioavailability. For example, nutrient levels in the blood, which may be measured, may not reflect nutrient levels in other tissues that serve as primary storage, such as the liver.

Modifications in response factors like growth, immunological competence, or enzyme activity must be confirmed by comparing them to other conditions, as they may not indicate actual bioavailability on their own. The activity of selenium-dependent glutathione peroxidase in the liver may not reflect selenium bioavailability for other proteins that require it. Furthermore, none of these functional responses talk much about how a nutrient is processed at different stages of digestion, absorption, and use [11].

Each step in the bioavailability process is influenced by a multitude of elements in the food as well as the individual's nutritional state. When nutrients are present in a variety of forms in foods and tissues, determining bioavailability is very difficult.

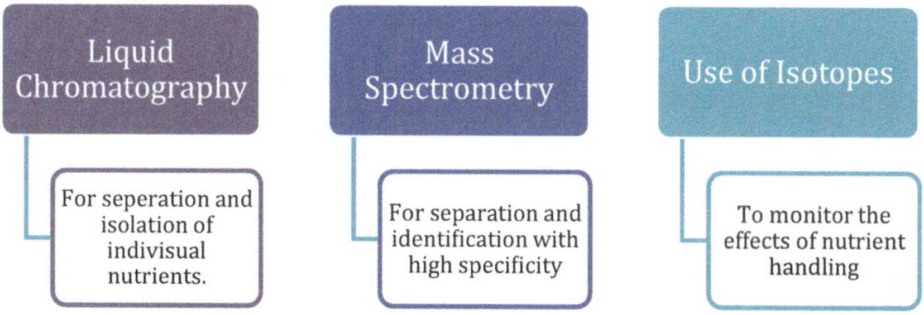

Fig. 14.4 Techniques used for analyzing nutrients

The evaluation of nutrient bioavailability, as difficult as it may look, is still vital to our insights into how people utilize vital nutrients from eaten foods and our appreciation of how foods meet our nutritional needs.

Researchers have discovered new analytic tools that allow for more precise identification and quantification of nutrients in foods and tissues, and they've used these techniques to increase our knowledge of observed variations in nutrient bioavailability from different types of food.

Methods to measure vitamins and minerals include affinity and high-performance liquid chromatography for separating and isolating individual nutrients; mass spectrometry for separation and identification with high specificity; and the use of "tagged" nutrients (or isotopes that can be chemically identified at various stages) as tracers to monitor the effects of nutrient handling at each step that may affect bioavailability. Different techniques used for the analysis of nutrients are outlined in Fig. 14.4.

In certain cases, food can be intrinsically labeled with tagged nutrients by cultivating plants and animals under the influence of tagged nutrients. This empirical strategy is more relevant or practical for investigating nutrient bioavailability than one that involves adding the tracer form of the nutrient to ingested foods [11].

14.6 Individual Nutrients and Food Factors That Affect Bioavailability

The bioavailability of nutrients can be reduced or increased by a variety of food components. Some components can create complexes with nutrients, preventing them from being digested or absorbed, or even degrading them, as is the case with foods containing an enzyme that breaks down the vitamin B (thiamine) as depicted in Fig. 14.5. Cooking generally destroys protein inhibitors, which restrict nutritional absorption. Other complexes can help with absorption by increasing solubility [12]. The following is a summary of recent changes in the availability of specific nutrients:

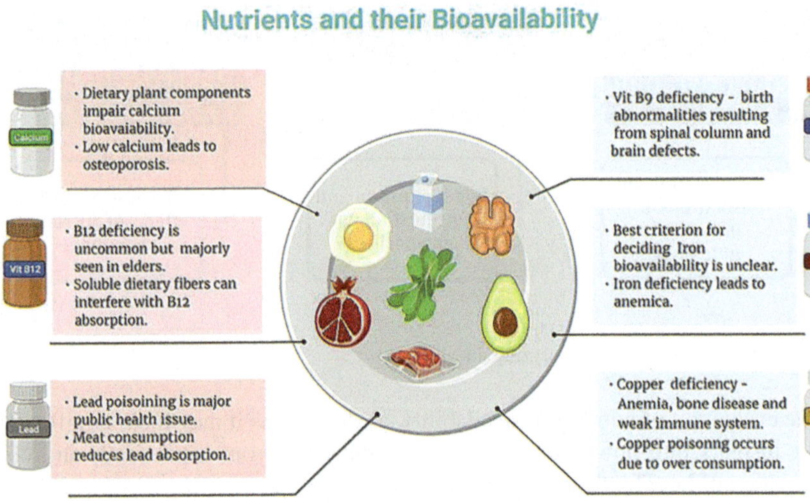

Fig. 14.5 Nutrients and their bioavailability

14.7 Role of Nanotechnological Applications in Enhancing Nutrient Bioavailability in the Human Body

The absorption rate or extent to which the body absorbs bioactive compounds such as functional foods or pharmaceutical products e.g. vitamins, minerals, fatty acids, flavonoids etc. depends on many factors involved in the metabolic cycles [13, 14]. The reduction in the bioavailability of any particular compound is due to several barriers arising inside the human body. These barriers include absorption rate, cellular barriers, and competitive biochemical present in the body. Delivering targeted nutrient to programmed location by passing the barriers such as escaping from the blood-brain involves careful consideration. Recent developments in nanotechnology have shown interest in developing nano-based products for efficient nutrient delivery [15–17]. Nanotechnology-based methods can be employed for enhancing the bioavailability of nutrients and functional foods. The use of food grade nanomaterial can enhance the efficacy and stability of bioactive compounds [18–21]. Table 14.2 depicts nanomaterial used for effective deliver of bioactive molecules and carriers used along with these nanomaterials [19].

14.7.1 Lipid Nano Carriers

Nanoencapsulation is one of the advanced and most reliable methods for delivering functional foods. High pressure homogenization and micro fluidization techniques can be used for the production of lipid-based nanoparticles [22]. Several dietary grade nutraceuticals are hydrophobic in nature; hence they have to be encapsulated

Table 14.2 Nanomaterial and carriers used for delivery of nutraceuticals

Material	Carrier
Nano-fibres	Liposomes
Nano-sheets	Solid lipid nanoparticles
Nano-whiskers	Cubosomes
Nano-tubes	Micro-emulsions

in the lipid based nanocarriers to incorporate them in the beverages. These dietary nutraceuticals include flavours, preservative, carotenoids, and vitamins. It is observed that small intestine absorbs nutraceuticals rapidly when they are combined with digestible lipids, reason being that digestible lipids enhances the quantity of micelles for solubilisation and transporting. Hence lipid-based nanoparticles has attracted interest of many research teams to optimize the production and characterization parameters [23].

14.7.2 Protein Nanoparticles

The advancement in the field of proteomics has helped us to utilize proteins as therapeutics. Protein nanoparticles offer many advantages, such as biocompatibility and biodegradability [24]. There are several techniques available for producing protein nanoparticles using single biopolymer. These techniques include:

(a) Desolvation,
(b) Precipitation,
(c) Thermal gelation,
(d) Self-assembly.

14.7.3 Polysaccharide Nanoparticles

In parallel to the production of lipid and protein nanoparticles, researchers focussed on polysaccharide nanoparticles. Polysaccharide nanoparticles are being produced using starch for targeted delivery of functional foods and nutraceuticals. Under optimized conditions, nanoparticles of various size, shape and compositions can be produced using naturally available starch. Nanoparticles produced using starch acts as shield to cover the pharmaceutical compounds and helps in the regulated release in the aqueous medium of gastro intestinal environment. Nanogels prepared using starch has three dimensional networks which holds large quantity of water. These self-assembled starch based nanogel formation is based on their interaction with other molecules covalently or by electrostatic binding. Polysaccharide nanogels typically consist of a three-dimensional network of cross-linked polysaccharide molecules that trap relatively high amounts of water [25].

14.7.4 Natural Nano Carriers

Nature has its own way of drug delivery, either it may be in the case of macro scale or in nano scale. There are many naturally available substances which can be used as carrier of our targeted nutraceuticals, functional foods, and pharmaceuticals. These naturally available carriers have gained much attention due to lower cost of production, easy availability, less toxic when compared to synthetic compounds. Some of the common natural nano carrier systems are casein micelles, nanocrystals and cyclodextrins [26].

14.8 Conclusion

Pharmaceutical and food industry has awakened the knowledge of nanomaterials to use them as carriers for nutraceuticals. Delivery method of the bioactive compound to the target organ and its bioavailability affects its efficacy. Also, physico-chemical property of any functional food or pharmaceutical compound play role in its bioavailability. Not all nutraceuticals bear the same properties; hence it is necessary to use an effective carrier system. This chapter emphasizes on the importance and classification of nutrients. Traditional, non-traditional and commercial available nutrients and their bioavailability using engineered nanomaterial technology like nano-encapsulation, desolvation, natural nano-carrier that can help overcome delivery issues. Nanotechnology has been successfully applied to enhance the food quality, nutrient bioavailability, and shelf life stability. Thus, the number of nano-food products and nano-pharmaceuticals available in the market is rapidly increasing due to its demand.

References

1. Bruinsma J, Hrabovszky J, Alexandratos N, Petri P. Crop production technology and input requirements in the agriculture of developing countries. Eur Rev Agric Econ. 1983;10(3):197–222.
2. Kah M, Beulke S, Tiede K, Hofmann T. Nanopesticides: state of knowledge, environmental fate and exposure modelling. Crit Rev Environ Sci Technol. 2013;43:1823–67. https://doi.org/10.1080/10643389.2012.671750.
3. Bodies B, Newsletter E, Information C. Bioavailability: how the nutrients in food become available to our bodies. 2021. HealthGuidance.org. HealthGuidance.org. https://www.healthguidance.org/entry/6265/1/bioavailability-how-the-nutrients-in-food-become-available-to-our-bodies.html.

4. Bleiweiss-Sande R, Chui K, Evans E, Goldberg J, Amin S, Sacheck J. Robustness of food processing classification systems. Nutrients. 2019;11(6):1344.
5. Comerford K, Papanikolaou Y, Jones J, Rodriguez J, Slavin J, Angadi S, Drewnowski A. Toward an evidence-based definition and classification of carbohydrate food quality: an expert panel report. Nutrients. 2021;13(8):2667.
6. Villena J, Kitazawa H. Probiotic microorganisms: a closer look. Microorganisms. 2017;5(2):17.
7. Legas A. The benefits of nano nutrients—NanoHydr8. NanoHydr8. 2021. https://nanohydr8.com/2021/03/03/the-benefits-of-nano-nutrients/.
8. Gupta C, Prakash D. Nutraceuticals for geriatrics. J Tradit Complement Med. 2015;5(1):5–14.
9. Harmsen J. Measuring bioavailability: from a scientific approach to standard methods. J Environ Qual. 2007;36(5):1420–8.
10. Huang S, Wang Z, Ma M. Measuring the bioavailable/toxic concentration of copper in natural water by using anodic stripping voltammetry and Vibrio-qinghaiensis sp.Nov.-Q67 bioassay. Chem Spec Bioavailab. 2003;15(2):37–45.
11. Du J, Pang J, You J. Bioavailability-based chronic toxicity measurements of permethrin to Chironomus dilutus. Environ Toxicol Chem. 2013;32(6):1403–11.
12. Eisenhauer B, Natoli S, Liew G, Flood V. Lutein and zeaxanthin—food sources, bioavailability and dietary variety in age-related macular degeneration protection. Nutrients. 2017;9(2):120.
13. McClements DJ, Jafari SM. Improving emulsion formation, stability and performance using mixed emulsifiers: a review. Adv Colloid Interface Sci. 2018;1(251):55–79.
14. Zou L, Zheng B, Zhang R, Zhang Z, Liu W, Liu C, Xiao H, McClements DJ. Enhancing the bioaccessibility of hydrophobic bioactive agents using mixed colloidal dispersions: curcumin-loaded zein nanoparticles plus digestible lipid nanoparticles. Food Res Int. 2016;1(81):74–82.
15. Huang Q, Yu H, Ru Q. Bioavailability and delivery of nutraceuticals using nanotechnology. J Food Sci. 2010;75(1):R50–7.
16. Joye IJ, Davidov-Pardo G, McClements DJ. Nanotechnology for increased micronutrient bioavailability. Trends Food Sci Technol. 2014;40(2):168–82.
17. Esfanjani AF, Jafari SM. Biopolymer nano-particles and natural nano-carriers for nano-encapsulation of phenolic compounds. Colloids Surf B Biointerfaces. 2016;1(146):532–43.
18. Quintanilla-Carvajal MX, Camacho-Díaz BH, Meraz-Torres LS, Chanona-Pérez JJ, Alamilla-Beltrán L, Jimenéz-Aparicio A, Gutiérrez-López GF. Nanoencapsulation: a new trend in food engineering processing. Food Eng Rev. 2010;2(1):39–50.
19. Cushen M, Kerry J, Morris M, Cruz-Romero M, Cummins E. Nanotechnologies in the food industry—recent developments, risks and regulation. Trends Food Sci Technol. 2012;24(1):30–46.
20. Katouzian I, Jafari SM. Nano-encapsulation as a promising approach for targeted delivery and controlled release of vitamins. Trends Food Sci Technol. 2016 Jul;1(53):34–48.
21. Romero GB, Brysch W, Keck CM, Müller RH. Nanocapsule formation by nanocrystals. In: Nanoencapsulation technologies for the food and nutraceutical industries. San Diego, CA: Academic; 2017. p. 165–86.
22. Fathi M, Mozafari MR, Mohebbi M. Nanoencapsulation of food ingredients using lipid based delivery systems. Trends Food Sci Technol. 2012;23(1):13–27.
23. McClements DJ. Encapsulation, protection, and release of hydrophilic active components: potential and limitations of colloidal delivery systems. Adv Colloid Interface Sci. 2015;1(219):27–53.
24. Hong S, Choi DW, Kim HN, Park CG, Lee W, Park HH. Protein-based nanoparticles as drug delivery systems. Pharmaceutics. 2020;12(7):604. https://doi.org/10.3390/pharmaceutics12070604.
25. Simi CK, Abraham TE. Encapsulation of crosslinked subtilisin microcrystals in hydrogel beads for controlled release applications. Eur J Pharm Sci. 2007;32(1):17–23.
26. Tavares GM, Croguennec T, Carvalho AF, Bouhallab S. Milk proteins as encapsulation devices and delivery vehicles: applications and trends. Trends Food Sci Technol. 2014;37(1):5–20.

Chapter 15
Prospects and Toxicological Concerns of Nanotechnology Application in the Food Industry

Abeer Mohamed Ali El Sayed, Chukwuebuka Egbuna, Kingsley C. Patrick-Iwuanyanwu, Chukwuemelie Zedech Uche, Johra Khan, and Eugene N. Onyeike

15.1 Introduction

In the food industry, nanotechnology can be applied to enhance food quality, shelf life, safety, and nutritional benefits [1]. Some nanomaterials used in the food industry are not intended to find their way into the final food product, e.g., those utilized in sensors, packaging, and antimicrobial treatments intended for sterilizing food manufacturing plants. Other nanomaterials are precisely constructed to be integrated into food products, such as nanoparticles used as delivery systems or to

---

A. M. A. El Sayed (✉)
Department of Pharmacognosy, Faculty of Pharmacy, Cairo University, Cairo, Egypt
e-mail: abeer.ali@pharma.cu.edu.eg

C. Egbuna
Nutritional Biochemistry and Toxicology Unit, Africa, Centre of Excellence, Centre for Public Health and Toxicological Research (PUTOR), University of Port Harcourt, Port Harcourt, Nigeria

Department of Biochemistry, Faculty of Natural Sciences, Chukwuemeka Odumegwu Ojukwu University, Uli, Nigeria

K. C. Patrick-Iwuanyanwu · E. N. Onyeike
Nutritional Biochemistry and Toxicology Unit, Africa, Centre of Excellence, Centre for Public Health and Toxicological Research (PUTOR), University of Port Harcourt, Port Harcourt, Nigeria

C. Z. Uche
Department of Medical Biochemistry and Molecular Biology, Faculty of Basic Medical Sciences, University of Nigeria, Enugu Campus, Nsukka, Nigeria

J. Khan
Department of Medical Laboratory Sciences, College of Applied Medical Sciences, Majmaah University, Majmaah, Saudi Arabia

Health and Basic Sciences Research Center, Majmaah University, Majmaah, Saudi Arabia

C. Egbuna et al. (eds.), *Application of Nanotechnology in Food Science, Processing and Packaging*, https://doi.org/10.1007/978-3-030-98820-3_15

modify optical, rheological properties. Herein, we focus on the properties and potential safety of ingested nanomaterials since they are most likely to cause health concerns. Nanoscale materials are naturally present in many commonly consumed foods, such as the emulsion micelles in milk or certain organelles found in plant or animal cells [2]. Artificial nanomaterials can be divided into four categories–Carbon-based, metal-based, dendrimers, and composites. Intentionally added to foods (such as nanoparticle-based delivery systems), or they may inadvertently find their way into foods (such as nanoparticles in packaging materials that leach into the food matrix) [3]. Silver is the most common nanomaterial used in products, followed by carbon-based nanomaterials and metal oxides such as TiO_2. Different types of nanoscale materials that may be found in foods and their potential origins are highlighted in Table 15.1.

Few Information concerning the safety of used nanomaterials in food and nutrition industries is available. The British Royal Society report notes that we may face a nanotoxicity crisis in the future. It advises that avoiding nanotechnology in products until there is a comprehensive understanding of the environmental and health risks of exposure to nanoparticles [4, 5].

The main concern regarding human exposure to nanoparticles is that there are different entry routes such as digestion, inhalation, or skin absorption. After absorption, nanoparticles may enter the bloodstream and settle in different tissues such as the brain or trigger immune responses [6]. Despite all these debates, nanotechnology has already been applied in food packaging, agriculture technologies, and food processing, as well as the nature of food, so the public is seeking safety assurances from governments and food producers [7].

Table 15.1 Different types of nanoscale materials that may be found in foods, and their potential origins

Nanoscale material	Origin	Features
Casein micelles	Natural	Protein–mineral clusters
Cell organelles	Natural	Ribosomes, vacuoles, lysosome etc.
Oil bodies	Natural	Phospholipid/protein-coated triglyceride droplets
Lipid nanoparticles	Artificial	Solid particles or liquid droplets coated by emulsifiers
Protein nanoparticles	Artificial	Clusters of protein molecules held together by physical or covalent interactions
Carbohydrate nanoparticles	Artificial	Small solid fragments extracted from starch, cellulose, or chitosan. Clusters of polysaccharide molecules are held together by physical or covalent interactions
Iron oxide	Artificial	Nanoparticles used to fortify foods with iron.
Titanium dioxide	Artificial	Nanoparticles used as whitening agents
Silicon dioxide	Artificial	Nanoparticles used to control powder flowability
Silver	Artificial	Nanoparticles used as antimicrobials in foods, coatings, and packaging

Recently, investing in food industry products was devoted to nanotechnology [8], in agriculture and food processing. Advocates emphasize that this can improve the quality, nutritional value, safety, and quantity of food to meet the needs of a growing population [9]. Herein we describe some of nanotechnology's possible effects on humans and the environment. The use of nanotechnology in food irrespective of its wide benefits confers the possible adverse environmental, social, and health risks as these particles are believed to enter the ecosystem through the delivery of pesticides in agriculture or through application in processed food such as the packaging sector, thus raising the toxicity concerns about their usage [10]. The enhanced risk of artificial nanoparticles is due to the higher reactivity of these nanoparticles and increased bioavailability of smaller particles to our bodies leading to long-term pathological effects.

Nanomaterials can be introduced to food through:

1. Direct incorporation of nanoparticles in novel food as nanoemulsions, nanocapsules, and nano antimicrobial films.
2. By use of nanomaterials in food manufacturing, processing, preservation, and trackings such as the use of nanolaminates, nanosensors, and CNTs.

The level of human exposure to nanoparticles greatly depends on the specific area where it is used in the food industry and the concentration of usage with exposure risk being higher in the fields where nanomaterials are added directly to food products as carriers of novel food ingredients. Some of the toxic effects of nanoparticles used in food are presented in Table 15.2. The migration of nanoparticles from food

Table 15.2 Some toxic effects of nanoparticles used in food

Nanoparticle	Toxicity	Purpose in food
TiO_2	Little impact as assessed by bacterial respiration, fatty acid profiles, and phylogenetic composition Oxidative stress, DNA damage Suppressed IDO activity and IFN-c production	As food additives (E171-1 and E171-6a)
Nanoclay	Released nanoclays did not show toxicity	Food packaging
ZnO	Cytotoxicity on human pulmonary adenocarcinoma cell line LTEP-a-2 Delay in human neutrophil apoptosis	Food packaging
Ag	Oxidative stress, cytotoxicity endothelial cell injury and dysfunction	Food packaging and coating
NiO	Inflammation and genotoxic effect in lung epithelial cells	Biosensors
FeO	Decrease the cell viability	Enzyme immobilization, protein purification, and food analysis
Silica	Increase ROS, LDH, malondialdehyde oxidative stress, and mitochondrial damage	Packaging, additive (E551)
CuO	Decrease in cell viability, increase in LDH, and lipid peroxidation	Antimicrobial agent in packaging
Al_2O_3	DNA damage	Packaging

packaging materials and the behavior of nanoparticles upon entering the body are still being evaluated at an extensive level [11].

Food, by its nature, is a pool that presents enormous possibilities for biochemical interactions, and the incorporation of a highly reactive species of nanoparticles into food may trigger different reactions. The interaction of nanoparticles with such functional ingredients and other constituents is unclear and needs to be explored. Besides a lot of advantages of nanotechnology to the food industry, safety issues associated with the nanomaterial cannot be neglected. Safety concerns associated with nanomaterial emphasizing the possibility of nanoparticles migrating from the packaging material into the food and their impact on consumer's health are discussed by many researchers [12, 13]. The physicochemical properties in nanostates are completely different from that are in macrostate. Moreover, the small size of these nanomaterials may increase the risk for bioaccumulation within body organs and tissues [14]. For instance, silica nanoparticles which are used as anti-caking agents can be cytotoxic in human lung cells when subjected to exposure [15]. There are a lot of factors that affect dissolution including surface morphology of the particles, concentration, surface energy, aggregation, and adsorption. Since every nanomaterial has its individual property, therefore, toxicity will likely be established on a case-by-case basis [16]. Further, regulatory authorities must develop some standards for commercial products to ensure product quality, health and safety, and environmental regulations. The transparency of safety issues and environmental impact should be the priority while dealing with the development of nanotechnology in food systems and therefore compulsory testing of nano foods is required before they are released to the market.

15.1.1 Is Nano Safe in Foods?

Credits to nanotechnology, plenty of new products, and nanomaterials for food can be developed. Nano-iron, for example, could be added to foods to fight anemia and nano-packaging methods can be developed to improve the shelf life of products. In principle, nanoparticles in packaging may leach into food products and therefore be ingested as part of the human diet.

Are there specific health risks from nanoproducts? Out of three human studies, only one showed a passage of inhaled nanoparticles into the bloodstream. Materials which by themselves are not very harmful could be toxic if they are inhaled in the form of nanoparticles. The effects of inhaled nanoparticles in the body may include lung inflammation and heart problems.

What are the possible dangers of nanotechnology?

- Nanoparticles may damage the lungs.
- Nanoparticles can get into the body through the skin, lungs, and digestive system.

15.2 Factors Affecting the Gastrointestinal Fate and Toxicity of Food-Grade Nanoparticles

A major factor that has been frequently ignored in the studies of the biological fate of ingested food nanoparticles is their interactions with various components within complex food matrices and GIT. These interactions may occur within the food itself, or during the passage of the food nanoparticles through the GIT. The interaction of a food or GIT component with nanoparticles may alter their physicochemical properties in the GI tract and therefore their biological fate and function. Indeed, the results of many previous studies have been highly limited because they used unrealistic test systems that ignored food matrix and GIT effects [3, 17].

15.2.1 Food Matrix Effects

Prior to ingestion, nanoparticles are typically dispersed within food matrices that vary considerably in their compositions, structures, and properties. The physicochemical and structural properties of nanoparticles may therefore be changed considerably when they are dispersed in food products, which would play an important role in determining their subsequent GIT fate and toxicity. For example, the interfacial composition and properties of food-grade nanoparticles changes appreciably when they are added to foods or when they enter the GIT because of the adsorption of surface-active molecules from the surrounding environment [18]. Moreover, it has been reported that certain flavonoids in foods can be tightly bound to the surface of inorganic nanoparticles [19]. The interaction between these food components and nanoparticles may significantly alter the biological fate of these nanoparticles. Although knowledge of food matrix effects is critical for understanding the gastrointestinal fate of food nanoparticles, this important factor is currently ignored in most studies. Consequently, this should be an important focus for future research in this area.

15.2.2 GIT Effects

After ingestion, nanoparticles travel through the complicated environment of the GIT before they are absorbed or exhibit their toxic effects [20] (Fig. 15.1). If the nanoparticles are not absorbed in the upper GIT, then they will reach the colon (pH 6–7) where they will encounter colonic bacteria and undigested food components. If the nanoparticles are originally trapped within a food when they are ingested, then they may be released into the GIT fluids as the food matrix is disrupted and digested [21]. The GIT region where they are released will therefore depend on the composition and structure of the food.

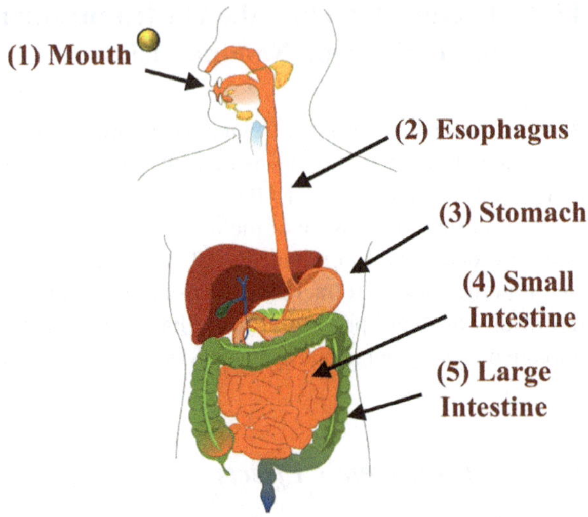

Fig. 15.1 Nanoparticles travel through the complicated environment of the GIT before they are absorbed or exhibit their toxic effects

Some of the most important properties of GIT fluids that may alter nanoparticle characteristics are emphasized here**:**

(a) **PH and ionic strength**: The pH and ionic composition of the gastrointestinal fluids depends on the nature of the food consumed and the specific GIT region (mouth, stomach, small intestine, or colon). These parameters determine the surface potential and electrostatic interactions of nanoparticles, which influences their aggregation state and interactions with other components.
(b) **Surface-active components:** Gastrointestinal fluids contain surface-active components that may arise from the ingested food or the GIT, such as surfactants, proteins, bile salts, phospholipids, and FFAs. These surface-active components may adsorb to nanoparticle surfaces and alter their interfacial properties and subsequently their biological fate [22].
(c) **Enzyme activity:** Gastrointestinal fluids contain digestive and metabolic enzymes that may change the properties of certain types of nanoparticles. For example, nanoparticles containing starches, proteins, or lipids may be digested by amylases, proteases, or lipases. Consequently, the properties of the nanoparticles reaching specific regions within the GIT may be very different from those of the ingested nanoparticles.
(d) **Biopolymers**: Gastrointestinal fluids contain biopolymers that may also alter the properties of nanoparticles. These biopolymers may arise from the ingested food or be secreted by the GIT, e.g., proteins, polysaccharides, and glycoproteins. Biopolymers may adsorb to nanoparticle surfaces and change their interfacial properties, or they may alter their aggregation state by promoting or opposing flocculation [23].
(e) **Mechanical forces**: Ingested nanoparticles are contained within gastrointestinal fluids that are subjected to various kinds of mechanical forces as they pass through the GIT which may alter the properties of the nanoparticles. Mechanical

forces may alter the aggregation state of nanoparticles by breaking down weakly flocculated systems.

As a result of these factors, the properties of nanoparticles are changed appreciably as they pass through the GIT, which will alter their GIT fate and potential toxicity. For example, there may be changes in the composition, dimensions, surface properties, physical state, and aggregation state of nanoparticles, which should be considered when establishing their potential toxicity. The interfacial properties of inorganic (magnetite) nanoparticles co-ingested with bread were altered in a way that promoted their uptake by intestinal epithelium cells [24]. The presence of a digested food matrix enhanced the absorption of silver nanoparticles by intestinal epithelium cells [25]. These findings demonstrated that the characteristics of the nanoparticles inside the GIT may be appreciably different from those of the original nanoparticles, which is often ignored in biological fate and toxicity assessments of food nanoparticles potentially leading to unrealistic and misleading results.

15.3 Mechanisms of Action of Nanoparticle Toxicity

This section highlights some of the most important mechanisms of nanoparticle toxicity. Ingested nanoparticles may cause toxicity due to numerous physicochemical and physiological mechanisms depending on their compositions, structures, and properties.

The direct contact of nanomaterials used as food additives/functional/nutritional ingredients may pose threats to human health. The production of reactive oxidative species (ROS) acts as one of the main toxicological mechanisms causing cellular damage and death [26]. Overproduction of ROS can lead to autophagy [27], neuron damage [28], and severe damage to DNA [29], and potentially mutagenesis, carcinogenesis, and aging-related diseases in humans. Allergic reactions and damage from metal ion release from nanomaterials are also possible adverse outcomes upon exposure to food nano-products [30]. Additionally, the accumulation of nanomaterials in the edible parts (seeds) of plants [31] and the human body [32] may cause severer problems at a higher concentration and long-term interactions.

15.3.1 Interference with GIT Normal Function

The small size of nanoparticles means they have a high specific surface area, which offers a large area for adsorption of any surface-active components in the GIT. Consequently, high levels of nanoparticles could reduce the rate or extent of starch, lipid, or protein digestion within the GIT. For example, digestive or metabolic enzymes could adsorb to nanoparticle surfaces thereby altering their normal GIT function. Many globular proteins are denatured after adsorption to particle surfaces due to the change in their thermodynamic environment, which could lead to a reduction in the catalytic activity of some enzymes.

The concentration of inorganic nanoparticles in the small intestine is likely to be a fraction of a percent, and so this effect is only likely to be important for relatively large lipid droplets at relatively low concentrations. In addition, the effect of a nanoparticle is likely to be difficult to be predicted for several reasons: first, the inorganic nanoparticles may aggregate in the GIT; second, the lipase molecules may adsorb more strongly to the lipid droplet surfaces than to the inorganic nanoparticle's surfaces; third, there may be other surface-active substances in the GIT that compete with the lipase for the surfaces of the inorganic nanoparticles.

There has been little research in this area, and so it is difficult to assess any potentially harmful effects associated with this mechanism. At the worst, one might expect that there would be a reduction in the rate of lipid, protein, or starch digestion, but that these components would eventually be fully digested due to the bodies' ability to secrete additional enzymes and other digestive components when needed. Due to the relatively low levels of inorganic nanoparticles normally ingested, the authors do not anticipate that this mechanism will be a major health concern.

Some types of inorganic nanoparticles may also be able to physically disrupt important structures within the GIT, such as the tight junctions or microvilli, thereby altering normal nutrient absorption and the protective function of the epithelium cells [33]. The presence of nanoparticles in the GIT may also stimulate an immune response, which could have adverse effects on human health, and so this possibility should be tested for food-grade nanoparticles [34].

15.3.2 Accumulation Within Specific Tissues

Certain types of ingested nanoparticles are absorbed within the GIT and accumulate in numerous tissues [35]. Apparently, these nanoparticles travel across the mucus layer and are then absorbed by active or passive transport mechanisms. After they have been absorbed into the cells, the nanoparticles may be metabolized, transferred out of the cells, or accumulate within the cells. The accumulation of nanoparticles within specific tissues may lead to long-term problems if they exhibit toxic effects above a certain accumulation threshold. This mechanism of action is likely to be most important for inorganic nanoparticles that are biopersistent.

15.3.3 Cytotoxicity and Cellular Malfunction

Nanoparticles may produce toxicity in cells through a variety of different mechanisms, depending on their composition and structure [33]. One of the most important factors contributing to the toxicity of inorganic nanoparticles is their ability to generate ROS, such as singlet oxygen, superoxide, hydrogen peroxide, and hydroxyl radicals [36]. These ROS may then cause damage to cell membranes, organelles, and the nucleus by interacting with lipids, proteins, or nucleic acids [37, 38]. As a result, many biochemical functions required to maintain cell viability, such as ATP production, DNA replication, and gene expression, may be adversely affected [39]. Several studies have reported the ability of inorganic nanoparticles to increase the generation of ROS in cells and to produce cytotoxicity, including silicon dioxide nanoparticles, [40] ZnO nanoparticles, [41] and silver nanoparticles [35]. Some inorganic nanoparticles produce toxicity by generating ions (such as Ag + from silver nanoparticles or Zn 2+ from zinc oxide nanoparticles) that interact with the normal functioning cellular components (such as proteins, nucleic acids, or lipids) required to maintain biochemical processes. These mechanisms of action are most likely to be important for inorganic nanoparticles that are absorbed by the intestinal cells since most organic nanoparticles are digested before being absorbed. However, it is still unclear about the extent to which inorganic nanoparticles would produce cytotoxicity when they are consumed as part of a complex diet under normal conditions.

15.3.4 Altered Location of Bioactive Release

The encapsulation of bioactive agents within nanoparticles may alter the location of their release and absorption within the GIT. For example, a bioactive agent that is normally released in the mouth, stomach, or small intestine could be released within the colon. As a result, the physiological response and biological impact of the bioactive agent may be altered by nanoencapsulation, which could have potentially adverse health effects. For example, the encapsulation of digestible lipids within nanolaminated dietary fiber coatings may inhibit the rate and extent of lipid digestion in the upper GIT, [42] so that high levels of undigested lipids reach the colon. These lipids may then be fermented by the colonic bacteria, which could cause gastrointestinal problems. Alternatively, an antimicrobial agent may be encapsulated within a nanoparticle that is not digested within the upper GIT, so that it reaches the colon, where it could alter the nature of the colonic microflora, which could again have adverse health effects. These effects are likely to be highly system-specific, depending on the nature of the encapsulated bioactive and nanoparticle used, and would therefore need to be established on a case-by-case basis.

15.3.5 Enhancement of Oral Bioavailability

One of the most widely studied applications of nanotechnology in the food industry is for the encapsulation and delivery of hydrophobic bioactive agents, such as certain nutrients and nutraceuticals [43]. Numerous *in vitro* and *in vivo* studies have shown that delivering these bioactive agents within nanoparticles can greatly increase their bioavailability. For illustration, nanoemulsions have been shown to increase the bioavailability of carotenoids, curcumin, coenzyme Q10, ω-3 fatty acids, and fat-soluble vitamins [43, 44]. There are a few different physicochemical mechanisms that may be responsible for this improvement.

In particular, the nanoparticles may increase the bioaccessibility, chemical stability, and/or absorption of the encapsulated bioactive agents [45]. In general, nanoparticles tend to be digested or dissolved more rapidly in the GIT and/or release any encapsulated components more rapidly because of their small size and high surface area. A change in the exposure level of bioactive agents within the blood could have potentially adverse health effects. The biological effects of many bioactive agents depend on their exposure levels in the blood and specific tissues. If the exposure level is too low, then the bioactive agent will have a little biological impact. If the exposure level is too high, then it may be toxic. Thus, the concentration should be within a certain intermediate level to have the most beneficial biological effects. This effect is likely to be highly system-dependent. It will depend on the toxicity profile of the bioactive agent. Some bioactive agents can be consumed at relatively high levels and have little toxicity, and therefore the ability of nanoparticles to boost their bioavailability should not have any adverse consequences. On the other hand, boosting the bioavailability of some bioactive agents could cause health problems. Vitamin E (a mixture of tocopherols and tocotrienols) is essential for maintaining human health and performance. However, the consumption of high doses of vitamin E may increase the risk of various chronic diseases [46]. Much of the studies establishing the upper limits for the adverse health effects of bioactive agents have not considered the nature of the delivery systems used. Consequently, the level where toxic effects are observed could be appreciably lower in cases where nanoparticle delivery systems greatly increase the bioavailability of the bioactive agents being tested.

Nanoparticles may increase the bioavailability of bioactive agents through two different approaches: delivery systems or excipient systems [40]. In both cases, the delivery or excipient system is specifically designed to increase the bioavailability of the bioactive agents by increasing the bioaccessibility or absorption, or by modulating any transformations (such as chemical or biochemical reactions) of the bioactive agents in the GIT.

15.3.6 Interference with Gut Microbiota

Nanoparticles that reach the colon may interact with colonic bacteria and alter their viability, thereby changing the relative proportions of different bacterial species present [33]. The type of bacteria populating the human colon is known to play a major role in human health and wellbeing [47]. Consequently, any change in the gut microbiota due to the presence of food-grade nanoparticles could have adverse health effects. This is an important area that requires further research to determine the impact of specific nanoparticle characteristics on the gut microbiota and the resulting health implications.

15.4 Toxicity Measurement of Nanoparticles Used in the Food Industry

Nanomaterials have unique properties such as high surface area, which make them more chemically active than bulk material so they could participate in most biological reactions that may have a harmful effect on human health or the environment. Nanostructures in nutrition or related industries must not create any direct or indirect damage to human health. Some features of nanoparticles are more important in unintentional side effects observed.

15.4.1 Size

Size is an important characteristic of the irreplaceable properties of nanoparticles. Size determines the surface area of nanoparticles. The effect of surface area on the respiratory response has been shown [48]. It has been reported that the size of particles is an important factor in observed dermal-cell *in vitro* cytotoxicity [49]. Absorbed nanoparticles in different absorption routes could trigger an immune system response [50]. The small size of these particles allows them to pass through different biological barriers and settle in tissues like the central nervous system [51]. The size of the nanoparticles in different routes of exposure should be considered in assessing the safety of nanomaterials that are to be used in food and food-related industries.

15.4.2 Chemical Composition

During the production of nanoparticles, many reagents are used that could be toxic. Some may remain in the final product and result in exposure to toxins that are unrelated to the nanomaterials themselves. For instance, some observed toxic effects of carbon nanotubes and semiconductor nanoparticles are related to residual reagents during synthesis. The remaining reagents and impurities may hinder our understanding of the possible side effects of carbon nanotubes. Iron ions and impurities can accelerate oxidative stress in cells [52]. Crystallinity is another important aspect of chemical composition. Titanium oxide has three different levels of crystallinity that each has different cytotoxic effects [53].

15.4.3 Surface Structure

There are many factors in the surfaces of nanostructures that could affect their cytotoxicity. Hydrophobicity, charge, roughness, and, most importantly, surface chemistry are factors that could change the toxicological effects of absorbed nanoparticles in the human body [54]. The coating of nanoparticles with hydrophilic polymer-like polyethylene glycol decreases the toxic effects of bare particles [55]. Evidence indicates that positively charged nanoparticles are more toxic than negative or neutral nanoparticles [56]. Different types of coatings or functionalization groups on the surface of nanoparticles are referred to as surface chemistry. Surface chemistry is one of the most important factors affecting the interaction of nanoparticles and biological systems [57].

15.4.4 Solubility

Solubility is also important in the toxicity of nanoparticles. For instance, soluble (hydrophilic) titanium oxide nanoparticles are more toxic than insoluble titanium oxide nanoparticles [58]. Some soluble nickel compounds are recognized as carcinogenic agents [59]. A detailed report on the solubility of the oxide nanoparticle's toxicity has been published [60]. Thus, understanding the toxicity and biological activity of nanoparticles requires an understanding of these factors and many others that must be considered in applying nanotechnology in food and related industries. In other words, all factors regarding the toxicity and environmental activity of nanoparticles should be investigated. Nanoparticle uptake routes and pathways are also important and must be considered in nanosafety investigations [61].

15.5 Harmful Effects of Nanoparticles on Humans

The use of nanotechnology in food irrespective of its wide benefits confers the possible adverse environmental, social, and health risks as these particles are believed to enter the ecosystem through the delivery of pesticides in agriculture or through application in processed food such as the packaging sector, thus raising the toxicity concerns about their usage [10].

The level of human exposure to nanoparticles greatly depends on the specific area where it is used in the food industry and the concentration of usage with exposure risk being higher in the fields where nanomaterials are added directly to food products as carriers of novel food ingredients. The migration of nanoparticles from food packaging materials and the behavior of nanoparticles upon entering the body are still being evaluated at an extensive level [11]. Nanoparticles can cause oxidative stress to human body cells and can traverse from lungs to blood, cell nuclei, and central nervous system leading to the inflammation of the gastrointestinal tract, Parkinson's syndrome, Alzheimer's disease, as well as the impairment of the DNA. Adverse effects on the kidney, liver and other vital organs have been reported due to long-term exposure to nanoparticles [62].

15.6 Prospects in Nanotoxicology Research

As one of the main characteristics of a nanoparticle is the enhancement of its reactivity, it is quite possible that, when a nontoxic nanoparticle is incorporated in food, it may get converted to a harmful form or vice versa.

Food has different roles in the body and the composition of the food is important with respect to that role. A food may contain a functional ingredient that is specific to that food; for instance, beef contains vitamin B12. During the processing of such food, the main aim is to reduce the loss of such functional ingredients. Food, by its nature, is a pool that presents enormous possibilities for biochemical interactions, and the incorporation of a highly reactive species of nanoparticles into food may trigger different reactions. The interaction of nanoparticles with such functional ingredients and other constituents is another area of research that needs to be explored.

15.7 Conclusion

The transparency of safety issues and environmental impact should be the priority while dealing with the development of nanotechnology in food systems and therefore compulsory testing of nano foods is required before they are released to the market. The main role of nanotoxicology is to provide clear guidelines and roadmaps for reducing risks in the optimal use of nanomaterials. Exposures routes in industrial workers and consumers of food products that contain nanomaterials must be studied carefully. With a precise understanding of the properties of nanomaterials such as size, dose, surface chemistry, and structures, we will have useful and safe food products.

References

1. Sozer N, Kokini JL. Nanotechnology and its applications in the food sector. Trends Biotechnol. 2008;27:82–9.
2. Livney YD. Milk proteins as vehicles for bioactives. Curr Opin Colloid Interface Sci. 2010;15:73–83.

3. Bellmann S, Carlander D, Fasano A, Momcilovic D, Scimeca JA, Waldman JW, et al. Mammalian gastrointestinal tract parameters modulating the integrity, surface properties, and absorption of food-relevant nanomaterials. Nanomed Nanobiotechnol. 2015;7:609–22.
4. Fellows PJ. Food processing technology. 4th ed. Cambridge: Woodhead Publishing; 2017.
5. Dowling A, Clift R, Grobert N, Hutton D, Oliver R, O'Neill O, et al. Nanoscience and nanotechnologies: opportunities and uncertainties. London: The Royal Society & The Royal Academy of Engineering Report; 2004. p. 61–4.
6. Egbuna C, Parmar VK, Jeevanandam J, Ezzat SM, Patrick-Iwuanyanwu KC, Adetunji CO, Khan J, et al. Toxicity of nanoparticles in biomedical application: nanotoxicology. J Toxicol. 2021;2021:9954443. https://doi.org/10.1155/2021/9954443.
7. Kuzma J, VerHage P. Nanotechnology in agriculture and food production: anticipated applications: project on emerging nanotechnologies. 2006. http://www.nanotechproject.org/process/assets/files/2706/94_pen4_agfood.pdf. Accessed 6 May 2014.
8. Bowman DM, Hodge GA. Nanotechnology and public interest dialogue: some international observations. Bull Sci Technol Soc. 2007;27(2):118–32.
9. Scott N, Chen H, Rutzke CJ. Nanoscale science and engineering for agriculture and food systems: a report submitted to cooperative state research, education and extension service, the United States Department of Agriculture: National Planning Workshop; November 18–19 2002; Washington, DC: USDA; 2003.
10. Kalpana SR, Anshul S, Rao NH. Nanotechnology in food processing sector—an assessment of emerging trends. J Food Sci Technol. 2013;50:831–41.
11. Magnuson BA, Jonaitis TS, Card JW. A brief review of the occurrence, use, and safety of food-related nanomaterials. J Food Sci. 2011;76:R126–33.
12. Editorial SV. Environmental impacts of engineered nanoparticles. Environ Toxicol Chem. 2010;29:2389–90.
13. Jain A, Shivendu R, Nandita D, Chidambaram R. Nanomaterials in food and agriculture: an overview on their safety concerns and regulatory issues. Crit Rev Food Sci Nutr. 2018;58(2):297–317.
14. Savolainen K, Pylkkänen L, Norppa H, Falck G, Lindberg H, Tuomi T, et al. Nanotechnologies, engineered nanomaterials and occupational health and safety—a review. Saf Sci. 2010;6:1–7.
15. Athinarayanan J, Periasamy VS, Alsaif MA, Al-Warthan AA. Alshatwi a a presence of nanosilica (E551) in commercial food products: TNF-mediated oxidative stress and altered cell cycle progression in human lung fibroblast cells. Cell Biol Toxicol. 2014;30:89–100.
16. Mahler GJ, Esch MB, Tako E, Southard TL, Archer SD, Glahn RP, et al. Oral exposure to polystyrene nanoparticles affects iron absorption. Nat Nanotechnol. 2012;7:264–71.
17. McClements DJ, DeLoid G, Pyrgiotakis G, Shatkin JA, Xiao H, Demokritou P. The role of the food matrix and gastrointestinal tract in the assessment of biological properties of ingested engineered nanomaterials (iENMs): state of the science and knowledge gaps. NanoImpact. 2016;3–4:47–57.
18. Monopoli MP, Åberg C, Salvati A, Dawson KA. Biomolecular coronas provide the biological identity of nanosized materials. Nat Nanotechnol. 2012;7:779–86.
19. Cao X, Ma C, Gao Z, Zheng J, He L, McClements DJ, Xiao H. Characterization of the interactions between titanium dioxide nanoparticles and polymethoxyflavones using surface-enhanced Raman spectroscopy. J Agric Food Chem. 2016;64(49):9436–41.
20. Yada RY, Buck N, Canady R, DeMerlis C, Duncan T, Janer G, Juneja L, Lin M, McClements DJ, Noonan G, Oxley J, Sabliov C, Tsytsikova L, Vázquez-Campos S, Yourick J, Zhong Q, Thurmond S. Engineered nanoscale food ingredients: evaluation of current knowledge on material characteristics relevant to uptake from the gastrointestinal tract. Comp Rev Food Sci Food Saf. 2014;2014(13):730–44.
21. Silletti E, Vingerhoeds MH, Norde W, Van Aken GA. The role of electrostatics in saliva-induced emulsion flocculation. Food Hydrocoll. 2007;21:596–606.
22. Giri K, Shameer K, Zimmermann MT, Saha S, Chakraborty PK, Sharma A, Arvizo RR, Madden BJ, Mccormick DJ, Kocher J-PA, Bhattacharya R, Mukherjee P. Understanding

protein–nanoparticle interaction: a new gateway to disease therapeutics. Bioconjug Chem. 2014;25:1078–90.

23. Das P, Saulnier E, Carlucci C, Allen-Vercoe E, Shah V, Walker VK. Interaction between a broad-spectrum antibiotic and silver nanoparticles in a human gut ecosystem. J Nanomed Nanotechnol. 2016;7:1–7.
24. Di Silvio D, Rigby N, Bajka B, Mackie A, Bombelli FB. Effect of protein corona magnetite nanoparticles derived from bread in vitro digestion on CaCo-2 cells morphology and uptake. Int J Biochem Cell Biol. 2016;75:212–22.
25. Lichtenstein D, Ebmeyer J, Knappe P, Juling S, Böhmert L, Selve S, Niemann B, Braeuning A, Thünemann AF, Lampen A. Impact of food components during in vitro digestion of silver nanoparticles on cellular uptake and cytotoxicity in intestinal cells. Biol Chem. 2015;396(11):1255–64.
26. He X, Aker WG, Leszczynski J, Hwang H-M. Using a holistic approach to assess the impact of engineered nanomaterials inducing toxicity in aquatic systems. J Food Drug Anal. 2014;22:128–46.
27. Khan MI, Mohammad A, Patil G, Naqvi SAH, Chauhan LKS, Ahmad I. Induction of ROS, mitochondrial damage and autophagy in lung epithelial cancer cells by iron oxide nanoparticles. Biomaterials. 2012;33:1477–88.
28. Long TC, Tajuba J, Sama P, Saleh N, Swartz C, Parker J, et al. Nanosize titanium dioxide stimulates reactive oxygen species in brain microglia and damages neurons in vitro. Environ Health Perspect. 2007;115:1631.
29. He X, Aker WG, Hwang H-M. An *in vivo* study on the photo-enhanced toxicities of S-doped TiO_2 nanoparticles to zebrafish embryos (*Danio rerio*) in terms of malformation, mortality, rheotaxis dysfunction, and DNA damage. Nanotoxicology. 2014;8:185–95.
30. He X, Hwang H-M. Nanotechnology in food science: functionality, applicability, and safety assessment. J Food Drug Anal. 2016;24:671–81.
31. Tripathi KM, Bhati A, Singh A, Sonker AK, Sarkar S, Sonkar SK. Sustainable changes in the contents of metallic micronutrients in first generation gram seeds imposed by carbon nano-onions: life cycle seed to seed study. ACS Sustain Chem Eng. 2017;5:2906–16.
32. Singh N, Manshian B, Jenkins GJ, Griffiths SM, Williams PM, Maffeis TG, et al. Nano genotoxicology: the DNA damaging potential of engineered nanomaterials. Biomaterials. 2009;30:3891–914.
33. Frohlich EE, Frohlich E. Cytotoxicity of nanoparticles contained in food on intestinal cells and the gut microbiota. Int J Mol Sci. 2016;17:1–22.
34. Orfi E, Szebeni J. The immune system of the gut and potential adverse effects of oral nanocarriers on its function. Adv Drug Deliv Rev. 2016;106:402–9.
35. Gaillet S, Rouanet JM. Silver nanoparticles: their potential toxic effects after oral exposure and underlying mechanisms—a review. Food Chem Toxicol. 2015;77:58–63.
36. Wu HH, Yin JJ, Wamer WG, Zeng MY, Lo YM. Reactive oxygen species-related activities of nano-iron metal and nano-iron oxides. J Food Drug Anal. 2014;22:86–94.
37. Sharma VK, Siskova KM, Zboril R, Gardea-Torresdey JL. Organic-coated silver nanoparticles in biological and environmental conditions: fate, stability and toxicity. Adv Colloid Interface Sci. 2014;204:15–34.
38. Athinarayanan J, Periasamy VS, Alsaif MA, Al-Warthan AA, Alshatwi AA. Presence of nanosilica (E551) in commercial food products: TNF-mediated oxidative stress and altered cell cycle progression in human lung fibroblast cells. Cell Biol Toxicol. 2014;30:89–100.
39. Choi J, Kim H, Kim P, Jo E, Kim HM, Lee MY, Jin SM, Park K. Toxicity of zinc oxide nanoparticles in rats treated by two different routes: single intravenous injection and single oral administration. J Toxicol Environ Health A. 2015;78(4):226–43.
40. Salvia-Trujillo L, Martin-Belloso O, McClements DJ. Excipient nanoemulsions for improving oral bioavailability of bioactives. Nanomaterials. 2016;6:1–16.
41. McClements DJ. Design of nano-laminated coatings to control bioavailability of lipophilic food components. J Food Sci. 2010;75:R30–42.

42. Joye IJ, Davidov-Pardo G, McClements DJ. Nanotechnology for increased micronutrient bioavailability. Trends Food Sci Technol. 2014;40:168–82.
43. Walker R, Decker EA, McClements DJ. Development of food-grade nanoemulsions and emulsions for delivery of omega-3 fatty acids: opportunities and obstacles in the food industry. Food Funct. 2015;6:42–55.
44. Katouzian I, Jafari SM. Nano-encapsulation as a promising approach for targeted delivery and controlled release of vitamins. Trends Food Sci Technol. 2016;53:34–48.
45. McClements DJ. Reduced-fat foods: the complex science of developing diet-based strategies for tackling overweight and obesity. Adv Nutr. 2015;6:338S–52S.
46. Miller ER 3rd, Pastor-Barriuso R, Dalal D, Riemersma RA, Appel LJ, Guallar E. Meta-analysis: high-dosage vitamin E supplementation may increase all-cause mortality. Ann Intern Med. 2005;142(1):37–46.
47. Human Microbiome Project Consortium. Structure, function and diversity of the healthy human microbiome. Nature. 2012;486:207–21.
48. Oberdörster G. Pulmonary effects of inhaled ultrafine particles. Int Arch Occup Environ Health. 2000;74(1):1–8. https://doi.org/10.1007/s004200000185.
49. Amini SM, Kharrazi S, Hadizadeh M, Fateh M, Saber R. Effect of gold nanoparticles on photodynamic efficiency of 5-aminolevolenic acid photosensitiser in epidermal carcinoma cell line: an in vitro study. Institution of Engineering and Technology [Internet]. http://digital--library.theiet.org/content/journals/10.1049/IET-NBT.2013.0021
50. Kyung OY, Grabinski CM, Schrand AM, Murdock RC, Wang W, Gu B, et al. Toxicity of amorphous silica nanoparticles in mouse keratinocytes. J Nanopart Res. 2009;11(1):15–24. https://doi.org/10.1007/s11051-008-9417-9.
51. Borm PJ, Kreyling W. Toxicological hazards of inhaled nanoparticles—potential implications for drug delivery. J Nanosci Nanotechnol. 2004;4(5):521–31.
52. Murray A, Kisin E, Leonard S, Young S, Kommineni C, Kagan V, et al. Oxidative stress and inflammatory response in dermal toxicity of single-walled carbon nanotubes. Toxicology. 2009;257(3):161–71.
53. Braydich-Stolle LK, Schaeublin NM, Murdock RC, Jiang J, Biswas P, Schlager JJ, et al. Crystal structure mediates mode of cell death in TiO2 nanotoxicity. J Nanopart Res. 2009;11(6):1361–74.
54. Kirchner C, Liedl T, Kudera S, Pellegrino T, Muñoz Javier A, Gaub HE, et al. Cytotoxicity of colloidal CdSe and CdSe/ZnS nanoparticles. Nano Lett. 2005;5(2):331–8.
55. Gupta AK, Wells S. Surface-modified superparamagnetic nanoparticles for drug delivery: preparation, characterization, and cytotoxicity studies. NanoBiosci IEEE Trans. 2004;3(1):66–73.
56. El Badawy AM, Silva RG, Morris B, Scheckel KG, Suidan MT, Tolaymat TM. Surface charge-dependent toxicity of silver nanoparticles. Environ Sci Technol. 2010;45(1):283–7.
57. Ryman-Rasmussen JP, Riviere JE, Monteiro-Riviere NA. Surface coatings determine cytotoxicity and irritation potential of quantum dot nanoparticles in epidermal keratinocytes. JID. 2006;127(1):143–53.
58. Oberdörster G. Pulmonary effects of inhaled ultrafine particles. Int Arch Occup Einviron Health. 2001;74:1–8.
59. Salnikow K, Kasprzak KS. Ascorbate depletion: a critical step in nickel carcinogenesis? Environ Health Perspect. 2005;113(5):577.
60. Brunner TJ, Wick P, Manser P, Spohn P, Grass RN, Limbach LK, et al. In vitro cytotoxicity of oxide nanoparticles: comparison to asbestos, silica, and the effect of particle solubility. Environ Sci Technol. 2006;40(14):4374–81.
61. Li J, Li Q, Xu J, Li J, Cai X, Liu R, et al. Comparative study on the acute pulmonary toxicity induced by 3 and 20 nm TiO_2 primary particles in mice. Environ Toxicol Pharmacol. 2007;24(3):239–44.
62. Momin J, Jayakumar C, Prajapati J. Potential of nanotechnology in functional foods. Emirates J Food Agric. 2013;25:10.

Index

C. Egbuna et al. (eds.), *Application of Nanotechnology in Food Science, Processing and Packaging*, https://doi.org/10.1007/978-3-030-98820-3

O

P

The manufacturer's authorised representative in the EU is Springer
Nature Customer Service Centre GmbH, Europaplatz 3, 69115 Heidelberg,
Germany. If you have any concerns regarding our products, please
contact ProductSafety@springernature.com

Printed and bound by CPI Group (UK) Ltd, Croydon, CR0 4YY
30/09/2026
02225329-0008